AF356654

The Application of Quantum Mechanics in Reactivity of Molecules

The Application of Quantum Mechanics in Reactivity of Molecules

Editor

Sérgio F. Sousa

MDPI • Basel • Beijing • Wuhan • Barcelona • Belgrade • Manchester • Tokyo • Cluj • Tianjin

Editor
Sérgio F. Sousa
Faculdade de Medicina da
Universidade do Porto
Portugal

Editorial Office
MDPI
St. Alban-Anlage 66
4052 Basel, Switzerland

This is a reprint of articles from the Special Issue published online in the open access journal *Applied Sciences* (ISSN 2076-3417) (available at: https://www.mdpi.com/journal/applsci/special_issues/Appl_Quantum_Mechanics).

For citation purposes, cite each article independently as indicated on the article page online and as indicated below:

LastName, A.A.; LastName, B.B.; LastName, C.C. Article Title. *Journal Name* **Year**, *Volume Number*, Page Range.

ISBN 978-3-0365-0658-6 (Hbk)
ISBN 978-3-0365-0659-3 (PDF)

Contents

About the Editor . . . . **vii**

Sérgio F. Sousa
Special Issue on "The Application of Quantum Mechanics in Reactivity of Molecules"
Reprinted from: *Appl. Sci.* **2021**, *11*, 1132, doi:10.3390/app11031132 . . . . **1**

Surajit Nandi, Bernardo Ballotta, Sergio Rampino and Vincenzo Barone
A General User-Friendly Tool for Kinetic Calculations of Multi-Step Reactions within the Virtual
Multifrequency Spectrometer Project
Reprinted from: *Appl. Sci.* **2020**, *10*, 1872, doi:10.3390/app10051872 . . . . **3**

Ol'ha O. Brovarets' and Dmytro M. Hovorun
A Never-Ending Conformational Story of the Quercetin Molecule: Quantum-Mechanical
Investigation of the O3'H and O4'H Hydroxyl Groups Rotations
Reprinted from: *Appl. Sci.* **2020**, *10*, 1147, doi:10.3390/app10031147 . . . . **17**

Samah Moubarak, N. Elghobashi-Meinhardt, Daria Tombolelli and Maria Andrea Mroginski
Probing the Structure of [NiFeSe] Hydrogenase with QM/MM Computations
Reprinted from: *Appl. Sci.* **2020**, *10*, 781, doi:10.3390/app10030781 . . . . **39**

Cecilia Muraro, Marco Dalla Tiezza, Chiara Pavan, Giovanni Ribaudo, Giuseppe Zagotto and Laura Orian
Major Depressive Disorder and Oxidative Stress: In Silico Investigation of Fluoxetine Activity
against ROS
Reprinted from: *Appl. Sci.* **2019**, *9*, 3631, doi:10.3390/app9173631 . . . . **51**

Roman F. Nalewajski
Phase Equalization, Charge Transfer, Information Flows and Electron Communications in
Donor–Acceptor Systems
Reprinted from: *Appl. Sci.* **2020**, *10*, 3615, doi:10.3390/app10103615 . . . . **67**

**Redouane Khaoulaf, Puja Adhikari, Mohamed Harcharras, Khalid Brouzi,
Hamid Ez-Zahraouy and Wai-Yim Ching**
Atomic-Scale Understanding of Structure and Properties of Complex Pyrophosphate Crystals
by First-Principles Calculations
Reprinted from: *Appl. Sci.* **2019**, *9*, 840, doi:10.3390/app9050840 . . . . **99**

Bhaskar Chilukuri, Ursula Mazur and K. W. Hipps
Structure, Properties, and Reactivity of Porphyrins on Surfaces and Nanostructures with
Periodic DFT Calculations
Reprinted from: *Appl. Sci.* **2020**, *10*, 740, doi:10.3390/app10030740 . . . . **115**

Roman F. Nalewajski
Understanding Electronic Structure and Chemical Reactivity:
Quantum-Information Perspective
Reprinted from: *Appl. Sci.* **2019**, *9*, 1262, doi:10.3390/app9061262 . . . . **141**

About the Editor

Sérgio F. Sousa (PhD) is a computational chemist specialized in computational enzymatic catalysis (QM/MM), docking, virtual screening and molecular dynamics simulations. He currently leads the BioSIM research group at the Faculty of Medicine, University of Porto.

Editorial

Special Issue on "The Application of Quantum Mechanics in Reactivity of Molecules"

Sérgio F. Sousa

UCIBIO/REQUIMTE, BioSIM—Departamento de Biomedicina, Faculdade de Medicina da Universidade do Porto, Alameda Prof. Hernâni Monteiro, 4200-319 Porto, Portugal; sergiofsousa@med.up.pt

Citation: Sousa, S.F. Special Issue on "The Application of Quantum Mechanics in Reactivity of Molecules". *Appl. Sci.* **2021**, *11*, 1132. https://doi.org/10.3390/app11031132

Received: 4 January 2021
Accepted: 19 January 2021
Published: 26 January 2021

Publisher's Note: MDPI stays neutral with regard to jurisdictional claims in published maps and institutional affiliations.

Over the last few decades, the increase in computational resources, coupled with the popularity of competitive quantum mechanics alternatives (particularly DFT (Density Functional Theory methods)), has promoted the widespread penetration of quantum mechanics applications into a variety of fields targeting the reactivity of molecules. This Special Issue attempts to illustrate the conceptual diversity of the applications of quantum mechanics in the study of the electronic structure of molecules and their reactivity. It is composed of eight selected articles, of which two are review articles.

The article by Nandi et al. [1] describes the implementation of a computer program for the chemical kinetics of multi-step reactions and its integration with the graphical interface of the Virtual Multifrequency Spectrometer tool. This program is based on the ab initio modeling of the molecular species involved and adopts the transition-state theory for each elementary step of the reaction. A master-equation approach accounting for the complete reaction scheme is adopted. Some features of the software are illustrated through specific examples.

Brovarets et al. [2] addressed the conformational diversity of the quercetin molecule, an effective pharmaceutical compound of plant origin. In particular, the authors employed DFT(B3LYP) and MP2 to investigate the conformational mobility of quercetin, focusing on the rotation of the hydroxyl groups in the 3$'$ and 4$'$ positions. New pathways associated with the transformations of the conformers of the quercetin molecule into each other and associated with the torsional mobility of the O3$'$H and O4$'$H hydroxyl groups are described, highlighting the dynamical nature of this molecule.

Moubarak and co-workers [3] investigated the geometry and vibrational behavior of selenocysteine [NiFeSe] hydrogenase isolated from *Desulfovibrio vulgaris Hildenborough* using a hybrid quantum mechanical (QM)/molecular mechanical (MM) approach. The authors employed DFT (BP86 functional) to describe the QM region and CHARMM36 for the treatment of the remainder of the enzyme (MM region). The results provide an explanation for the experimental vibrational spectra, suggesting a mixture of conformers and Fe^{2+} and Fe^{3+} oxidation states.

The study by Muraro et al. [4] examined, through quantum mechanics, the antioxidant and scavenging activity of fluoxetine, a well-known and widely prescribed antidepressant drug. In particular, the authors employed the semi-empirical quantum mechanical method GFN2-xTB for conformational analysis, while the characterization of the intermediates was performed using DFT (M06-2X density functional) and SMD to account for the solvation effects. The results suggest that the antioxidant capacity of fluoxetine is due to its efficiency in increasing the concentration of free serotonin, and not due its direct ROS scavenging activity.

The article by Nalewajski [5] discussed phase equalization, charge transfer, information flows and electron communications in donor–acceptor systems, exploring the mutual relationship between the phase component of the electronic wavefunction and its current descriptor.

Khaoaulaf et al. [6] studied the electronic structure and mechanical and optical properties of five pyrophosphate crystals with very complex structures. For that, the authors employed first-principles density functional theory calculations with different density functionals and used the results to understand and rationalize the structure and properties of those complexes, providing important clues for the understanding of pyrophosphates.

Chilukuri et al. [7] present a detailed review on the use of periodic density functional theory (PDFT) calculations in the study of the structure, electronic properties and reactivity of porphyrins on ordered two-dimensional surfaces and in the formation of nanostructures. In particular, the authors focused on examples of the application of PDFT calculations for bridging the gaps in the experimental studies on porphyrin nanostructures and self-assembly on 2D surfaces, also illustrating the diversity in terms of the density functionals used.

Finally, Nalewajski [8] reviewed different applications of quantum mechanics and information theory to problems of chemical reactivity. Particular emphasis was placed on the equivalence of variational principles for the constrained minima of the system's electronic energy and its kinetic energy component.

Together, these eight contributions constitute a rather diverse collection on the applications of quantum mechanics in the reactivity of molecules, presenting very distinct examples of applications and of perspectives, highlighting the growth and multiplicity of the field.

Funding: This research was funded by Fundação para a Ciência e a Tecnologia (grant UIDB/04378/2020).

Institutional Review Board Statement: Not applicable.

Informed Consent Statement: Not applicable.

Conflicts of Interest: No conflict of interest.

References

1. Nandi, S.; Ballotta, B.; Rampino, S.; Barone, V. A general user-friendly tool for kinetic calculations of multi-step reactions within the virtual multifrequency spectrometer project. *Appl. Sci.* **2020**, *10*, 1872. [CrossRef]
2. Brovarets', O.O.; Hovorun, D.M. A never-ending conformational story of the quercetin molecule: Quantum-mechanical investigation of the O3'H and O4'H hydroxyl groups rotations. *Appl. Sci.* **2020**, *10*, 1147. [CrossRef]
3. Moubarak, S.; Elghobashi-Meinhardt, N.; Tombolelli, D.; Mroginski, M.A. Probing the structure of [NiFeSe] hydrogenase with QM/MM computations. *Appl. Sci.* **2020**, *10*, 781. [CrossRef]
4. Muraro, C.; Tiezza, M.D.; Pavan, C.; Ribaudo, G.; Zagotto, G.; Orian, L. Major depressive disorder and oxidative stress: In silico investigation of fluoxetine activity against ROS. *Appl. Sci.* **2019**, *9*, 3631. [CrossRef]
5. Nalewajski, R.F. Phase equalization, charge transfer, information flows and electron communications in donor-acceptor systems. *Appl. Sci.* **2020**, *10*, 3615. [CrossRef]
6. Khaoulaf, R.; Adhikari, P.; Harcharras, M.; Brouzi, K.; Ez-Zahraouy, H.; Ching, W.-Y. Atomic-scale understanding of structure and properties of complex pyrophosphate crystals by first-principles calculations. *Appl. Sci.* **2019**, *9*, 840. [CrossRef]
7. Chilukuri, B.; Mazur, U.; Hipps, K.W. Structure, properties, and reactivity of porphyrins on surfaces and nanostructures with periodic DFT calculations. *Appl. Sci.* **2020**, *10*, 740. [CrossRef]
8. Nalewajski, R.F. Understanding electronic structure and chemical reactivity: Quantum-information perspective. *Appl. Sci.* **2019**, *9*, 1262. [CrossRef]

 *applied sciences*

Article

A General User-Friendly Tool for Kinetic Calculations of Multi-Step Reactions within the Virtual Multifrequency Spectrometer Project

Surajit Nandi, Bernardo Ballotta, Sergio Rampino * and Vincenzo Barone

SMART Laboratory, Scuola Normale Superiore, Piazza dei Cavalieri 7, 56126 Pisa, Italy; surajit.nandi@sns.it (S.N.); bernardo.ballotta@sns.it (B.B.); vincenzo.barone@sns.it (V.B.)
* Correspondence: sergio.rampino@sns.it

Received: 5 February 2020; Accepted: 5 March 2020; Published: 9 March 2020

Abstract: We discuss the implementation of a computer program for accurate calculation of the kinetics of chemical reactions integrated in the user-friendly, multi-purpose Virtual Multifrequency Spectrometer tool. The program is based on the ab initio modeling of the involved molecular species, the adoption of transition-state theory for each elementary step of the reaction, and the use of a master-equation approach accounting for the complete reaction scheme. Some features of the software are illustrated through examples including the interconversion reaction of hydroxyacetone and 2-hydroxypropanal and the production of HCN and HNC from vinyl cyanide.

Keywords: chemical kinetics; reaction rate; RRKM theory; master equation

1. Introduction

Calculation of chemical kinetics in the gas phase by accurate theoretical models is extremely important in research areas like atmospheric chemistry, combustion chemistry, and astrochemistry. As a matter of fact, the accurate prediction of reaction rates and the evolution of the involved species in a given set of physical conditions is a key feature for understanding the presence of a molecular or ionic species in that environment. Sometimes, the reactions involved are too fast to be tracked by laboratory experiment or the associated physical conditions are simply not reproducible, hence the understanding of those reactions relies on accurate theoretical treatments capable of predicting the evolution of the species involved in a reaction network.

A rigorous treatment of the time evolution of a chemical reaction should be based on quantum-dynamics calculations modeling the motion of the involved nuclei on ab initio potential-energy surfaces [1–3]. However, exact quantum-dynamics methods are only applicable to very small systems made up of three or four atoms (see, for instance, Refs. [4–6]). For the remaining systems—the vast majority—one can either opt for approximate methods, such as the Multi-Configuration Time-Dependent Hartree (MCTDH) [7] or the Ring Polymer Molecular Dynamics [8] (which can however extend this limit only marginally), or for a (quasi-)classical treatment of the nuclear motion [9–11].

On the other hand, a less expensive yet reliable route to chemical kinetics is the adoption of statistical models, such as the popular transition-state theory (TST) in one of its variants, which successfully exploits information on the energetics of a limited set of important points of the potential energy surface to predict the kinetics of chemical reactions. The usual procedure in this framework involves the calculation of transition states and intermediates of a given reaction and a description of the motions at molecular level of these species. Then, classical or semiclassical transition state theory (TST) is applied to calculate the reaction rates of each of the elementary steps making up the whole reaction (the Rice–Ramsperger–Kassel–Marcus (RRKM) [12–14] theory, shortly summarized in the following, is usually adopted for unimolecular reactions in the gas-phase). Finally, the time evolution of the relative abundances of each of the reactant, intermediate, and product species is calculated using methods based on either master-equation or stochastic approaches.

In this paper, we discuss the implementation of the computer program StarRate for kinetics calculations of multi-step chemical reactions, and its integration in the graphical interface of the user-friendly, multipurpose framework Virtual Multifrequency Spectrometer (VMS) [15]. The Virtual Multifrequency Spectrometer (VMS) is a tool that aims at integrating a wide range of computational and experimental spectroscopic techniques with the final goal of disclosing the static and dynamic physical-chemical properties of molecular systems, and is composed of two parts: VMS-Comp, which provides access to the latest developments in the field of computational spectroscopy [16,17], and VMS-Draw, which provides a powerful graphical user interface (GUI) for an intuitive interpretation of theoretical outcomes and a direct comparison to experiment [18]. We discuss the integration of StarRate within the VMS tool and illustrate some features of the developed software through two important reactions: the single-step interconversion of hydroxyacetone and 2-hydroxypropanal, and the more challenging multi-step dissociation of vinyl cyanide. It is worth mentioning here that the reported calculations were performed for the purpose of illustrating the developed computational software, and that providing new accurate results on the above reactions for comparison with experiment is beyond the scope of this work.

The article is organized as follows. Section 2 is devoted to computational details of the developed software. In Sections 3 and 4, we address the kinetics of the above mentioned reactions. In Section 5, conclusions are drawn and perspectives are outlined.

2. Computational Details: StarRate and the VMS Tool

StarRate is an object-based, modern Fortran program for modeling the kinetics of multistep reactions. At its current stage of development, StarRate targets multi-step unimolecular reactions (which can however dissociate, irreversibly, to multiple products). (The implementation of procedures for the treatment of bimolecular entrance channels is currently in progress). From a technical point of view, the program is written in the so-called 'F language' [19,20], a carefully crafted subset of Fortran 95, and is conceived in an object-based programming paradigm. As described in deeper detail in Ref. [21], StarRate is structured in three main modules, namely `molecules`, `steps` and `reactions`, which reflect the entities associated with a multi-step chemical reaction. All of these modules contain a defined data-type and some related procedures to access and operate on it. The main program, StarRate, controls the sequences of the calling of the procedures contained in each of the three main modules.

Another important module of StarRate is `in_out`, which handles the input and output operations of the program. Input data are accessed by StarRate through an XML interface based on the same versatile hierarchical data structure that is adopted by VMS. (The current version of VMS reads data in the JSON format, so that a straightforward conversion from XML to JSON is in order. This can be easily done, for instance, using xml2json (https://github.com/hay/xml2json), or online converters such as https://www.convertjson.com/xml-to-json.htm). At the beginning, the user has to prepare a very simple input file encoding a reaction scheme (see the starrate.inp box in Figure 1) and gather all the files, one for each molecular species, containing data deriving from electronic-structure calculations. These can be either in an internal standard format (similar to that adopted in the EStokTP [22] package) or directly output files of quantum-chemistry packages, as exemplified in Figure 1 for the case of the Gaussian package. Currently StarRate supports output files from this quantum-chemistry package (.log extension), though support for other popular electronic-structure programs is presently being pursued (see also Refs. [23,24] on the issue of interoperability and common data formats in quantum chemistry). Then, a Python script is run which extracts data from the output files generated by the quantum-chemistry calculations and, driven by the reaction scheme, collects the information in the proper sections of the XML file.

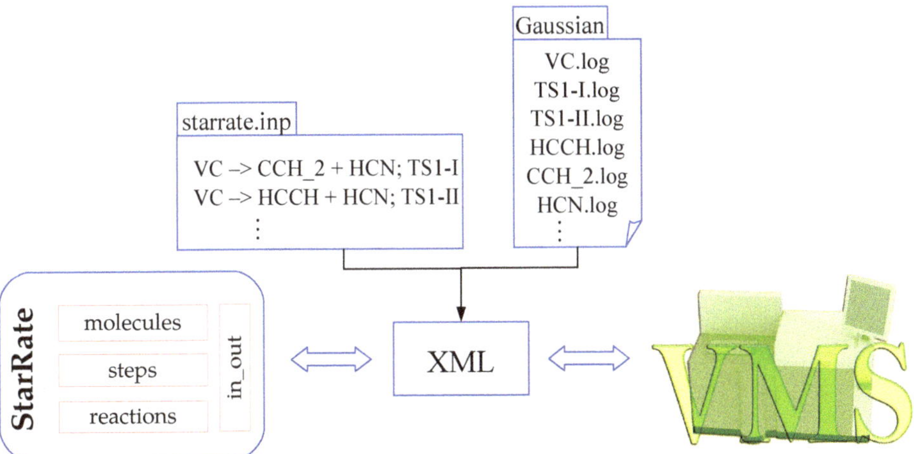

Figure 1. Diagram showing the interoperability between electronic-structure calculations, StarRate, and Virtual Multifrequency Spectrometer (VMS) through a dedicated hierarchical data structure XML interface (see Figure 2).

The structure of the XML interface is schematized in Figure 2. The whole XML document develops under a root element named `escdata`. The `escdata` has one child element for each molecule named `section_run`. Each of these elements contains three nodes: `program_info`, `section_system`, and `system_single_configuration_calculation`. All the information regarding a molecule is handled by these three sibling nodes. The `program_info` node contains two subnodes which keep track of quantum chemical software name and .log file location. The `section_system` node contains basic information which does not require quantum chemical information (viz., molecular charge, spin multiplicity, atom label, atomic numbers, rotational constants). The last sibling, `system_single_configuration_calculation`, contains information which requires quantum chemical calculations (viz., vibrational constants, SCF energy, density of state data). Finally, the last `section_run` collects information on physical conditions and on the reaction under study. For illustrative purposes, the actual .xml file for the reaction studied in Section 3 is given as Supplementary Material.

Once the XML has been generated, the StarRate program comes into play. The module `in_out` reads the XML file (a well-built external Fortran library, `FoX_dom` [25], is used for XML parsing) and saves the data for each molecule and step as structured arrays of the `molecules` and `steps` modules, respectively. Some information, such as vibrational frequencies, rotational constants, and electronic energy, is collected from the electronic-structure calculations; some other information such as densities of states (see later on) and single-step microcanonical rate coefficients are either also read as input data or computed internally to StarRate. Lastly, the `reactions` module solves the kinetics for the overall reactions using a chemical master equation method. At the end of the calculations, VMS is used to access, visualize and analyze the data produced thanks to the shared XML interface (see Figure 1).

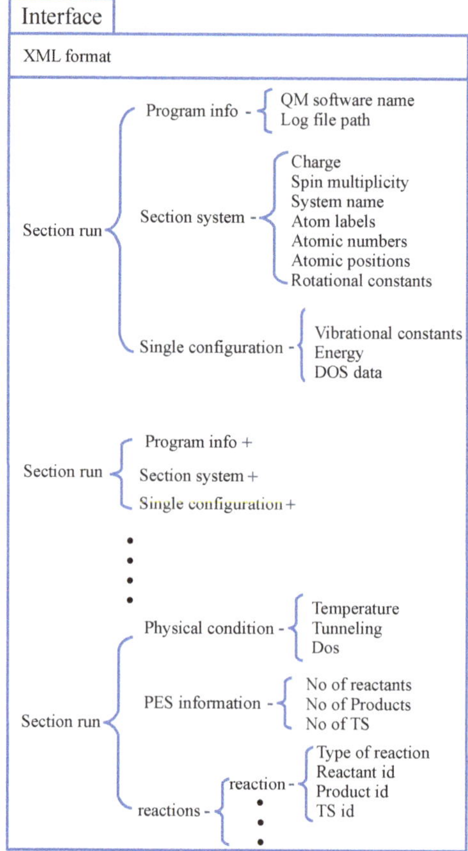

Figure 2. Hierarchical data structure of the XML interface.

3. Elementary Steps: The Interconversion Reaction of Hydroxyacetone and 2-Hydroxypropanal

The interconversion reaction between hydroxyacetone and 2-hydroxypropanal is an important reaction in the context of atmospheric chemistry because the hydroxyacetone represent the simplest form of photochemically oxidised volatile organic compounds [26]. In a recent study, Sun et al. [27] have considered the interconversion mechanisms on several hydroxycarbonyl compounds, and much attention has been focused on the interconversion reaction between hydroxyacetone and 2-hydroxypropanal. This isomerization reaction can occur through three different mechanisms, 2 high-barrier multistep processes and, as shown in Figure 3, a direct mechanism via double hydride shift involving a low-barrier concerted transition state.

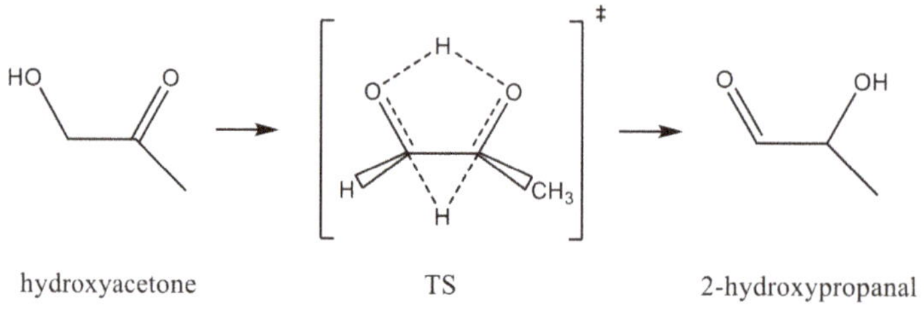

Figure 3. Direct isomerization reaction mechanism between hydroxyacetone and 2-hydroxypropanal.

In their work, Sun et al. also supposed that hydroxycarbonyl compounds can adsorb solar radiation, as carbonyl compounds, from 320 to 220 nm and then undergo an internal conversion to the vibrationally excited ground state with an energy more than sufficient to overcome the isomerization barrier, and computed RRKM microcanonical rate coefficients in order to understand how much the isomerization reaction is favored with respect to collisional deactivation and fragmentation processes at a given excitation energy.

Within the RRKM theory [12–14], the microcanonical rate coefficient for the reaction of Figure 3 is given by the equation [28]

$$k(E) = \frac{N^{\ddagger}(E)}{h\rho(E)} \tag{1}$$

where

$$N^{\ddagger}(E) = \int_0^E \rho^{\ddagger}(E') \, dE' \tag{2}$$

In Equations (1) and (2), h is Planck's constants, $N^{\ddagger}(E)$ is the sum of states of the transition state (TS) (computed by excluding the normal mode with imaginary frequency under the assumption that the motion along the reaction coordinates is separable from that of the other modes), and $\rho(E)$ and $\rho^{\ddagger}(E)$ are the density of states (DOS, i.e., the number of rovibrational states per energy interval) of the reactant molecule and transition state, respectively. As apparent, a central quantity in this framework is the molecular rovibrational density of states of the involved molecular species. This can be easily worked out by convoluting its rotational and vibrational counterparts [29]. In the present version of the program, a classical expression is used for the rotational DOS (see [21] for details), while the vibrational DOS is evaluated at uncoupled anharmonic level by adoption of the Stein–Rabinovitch [30] extension of the Beyer–Swinehart algorithm [31].

An improved version of Equation (1) accounts for the tunneling correction by using a modified version of the sum of states $N^{\ddagger}(E)$ of the TS. A common and efficient way of including tunneling is by means of the asymmetric Eckart barrier [32]. Within this model, the sum of states of the transition state is redefined by

$$N^{\ddagger}_{\text{tunn}}(E) = \int_{-V_0}^{E-V_0} \rho^{\ddagger}(E - E') P_{\text{tunn}}(E') \, dE' \tag{3}$$

where $N^{\ddagger}_{\text{tunn}}(E)$ is a tunneling-corrected version of the sum of state of the TS and V_0 is the classical energy barrier for the forward reaction. The quantity $P_{\text{tunn}}(E')$ is the tunneling coefficient at the energy E', and is given by the expression

$$P_{\text{tunn}}(E') = \frac{\sinh(a)\sinh(b)}{\sinh^2((a+b)/2) + \cosh^2(c)} \tag{4}$$

where, a, b, and c are parameters defined by:

$$a = \frac{4\pi\sqrt{E' + V_0}}{h\nu_i(V_0^{-\frac{1}{2}} + V_1^{-\frac{1}{2}})}, \quad b = \frac{4\pi\sqrt{E' + V_1}}{h\nu_i(V_0^{-\frac{1}{2}} + V_1^{-\frac{1}{2}})}, \quad c = 2\pi\sqrt{\frac{V_0 V_1}{(h\nu_i)^2} - \frac{1}{16}}. \tag{5}$$

Here, V_1 is the classical energy barrier for the reverse reaction, and ν_i is the magnitude of the imaginary frequency of the saddle point (in Equation (5), $h = 1$ if the energies are expressed, as in this work, in cm^{-1}).

For illustrative purposes, we computed the microcanonical rate coefficient for the direct and inverse reaction of Figure 3, both with and without tunneling correction. To this purpose, the three molecular species were modeled by density-functional theory with the B2PLYP-D3/jun-cc-pVTZ model chemistry. According to our calculations, the forward reaction is exoergic by 1719 cm^{-1} with a barrier of 16448 cm^{-1} (the barrier for the backward reaction is 14729 cm^{-1}). The resulting microcanonical rate coefficients $k(E)$ are plotted in Figure 4 (on a logarithmic scale) for the forward (blue line) and backward (red line) reaction as a function of the energy relative to the hydroxyacetone zero-point energy. In the same figure, the dashed curves are the tunneling-corrected ones. As apparent, the tunneling correction enhances the reaction rate both in the forward and backward direction, more visibly nearby the threshold region, thus lowering the actual value of the reaction threshold in both directions.

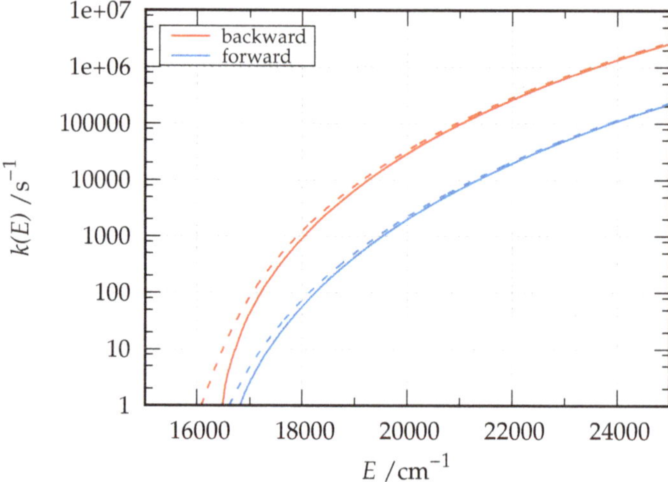

Figure 4. Microcanonical rate coefficients for the backward (red color) and forward (blue color) isomerization reaction of hydroxyacetone and 2-hydroxypropanal (Figure 3) as a function of the energy relative to the reactant zero-point energy. Dashed lines are the tunneling-corrected versions of the rate coefficients.

The thermal rate coefficient can easily be computed from the microcanonical rate coefficients using the following equation:

$$k(T) = \frac{1}{Q(T)} \int_0^\infty k(E)\rho(E)e^{-E/k_{\mathrm{B}}T}\,\mathrm{d}E\,, \tag{6}$$

with $Q(T)$ being the partition function of the reactant species. The computed thermal rate coefficients for the forward reaction (isomerization reaction of hydroxyacetone to 2-hydroxypropanal) in the temperature range 151–501 K are given as Arrhenius plot ($\log_{10} k(T)$ versus $1/T$) in Figure 5.

Results are reported both neglecting (blue triangles) and including (blue circles) tunneling. These data can be fitted to the popular Arrhenius equation:

$$k(T) = Ae^{-\frac{E_a}{RT}} \tag{7}$$

(with A being the pre-exponential factor, E_{a} the activation energy, and R the gas constant) or to the more refined Arrhenius–Kooij formula [33] (also known as modified Arrhenius equation [34]) allowing for a temperature dependence of the pre-exponential factor:

$$k(T) = \alpha(T/300)^\beta e^{-\gamma/T}\,, \tag{8}$$

which essentially implies a linear variation of the activation energy with the temperature, $E_{\mathrm{a}}/R = \gamma + \beta T$. The Arrhenius best-fitting curve for both the tunneling-corrected and no-tunneling data are shown in Figure 5 as dashed black line and solid black line, respectively. The Arrhenius–Kooij best-fitting curve for the tunneling-corrected data is also reported as a red dashed line in the same figure, while the best-fitting parameters together with the associated coefficient of determination $\mathcal{R}^2$ are given in Table 1.

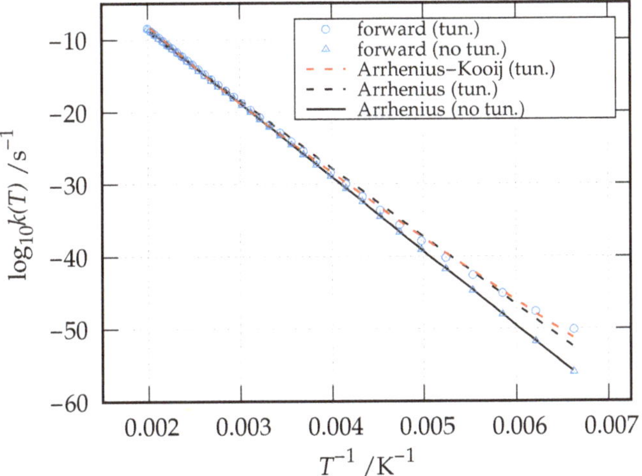

Figure 5. Arrhenius plot of the computed thermal rate coefficient (both neglecting and including tunneling) for the isomerization reaction of hydroxyacetone to 2-hydroxypropanal (Figure 3) and of the Arrhenius and Arrhenius–Kooij best-fitting curves.

Table 1. Results of the fit of the Arrhenius and Arrhenius–Kooij equations to the computed thermal rate coefficients for the isomerization reaction of hydroxyacetone to 2-hydroxypropanal in the temperature range 151–501 K.

Arrhenius (Equation (7)), No Tunneling		
A (s^{-1})	E_a/R (K)	$\mathcal{R}^2$
1.01×10^{12}	23636.0	1.0000
Arrhenius (Equation (7)), with Tunneling		
A (s^{-1})	E_a/R (K)	$\mathcal{R}^2$
9.12×10^{9}	21752.9	0.9973
Arrhenius–Kooij (Equation (8)), with Tunneling		
α (s^{-1})	β	γ (K)	$\mathcal{R}^2$
1.20×10^{1}	21.4	16000.4	0.9992

As evident from Figure 5 and Table 1, the Arrhenius equation perfectly fits the thermal rate coefficients calculated by neglecting tunneling, yielding a $\mathcal{R}^2 = 1.0000$ and an activation energy of 16428 cm^{-1} that compares well with the computed reaction barrier. On the contrary, the tunneling-corrected thermal rate coefficients show a deviation from linearity with decreasing temperatures. As a result, the Arrhenius expression yields a worse best-fitting curve ($\mathcal{R}^2 = 0.9973$) and a lower activation energy of 15120 cm^{-1}, while a better fitting is obtained through the Arrhenius–Kooij equation ($\mathcal{R}^2 = 9.9992$), which gives an activation energy of 14839 cm^{-1} at $T = 250$ K and of 13352 cm^{-1} at $T = 150$ K.

4. Multi-Step Reactions: The Dissociation of Vinyl Cyanide

The dissociation of vinyl cyanide (VC, C_3H_3N), is particularly interesting because it involves multiple reaction channels and different sets of products (HCN, HNC, HCCH, and :CCH$_2$) and hence it serves as a very good test case for master-equation based kinetic models. The potential-energy surface for this reaction has been investigated in a recent work by Homayoon et al. [35] through ab initio CCSD and CCSD(T) calculations with the 6-311+G(2d,2p) and 6-311++G(3df,3pd) basis sets. In the same work, a reaction scheme involving ten unimolecular steps, three of which reversible, was proposed. The ten reaction steps are summarized in Table 2, while the associated reaction diagram is given in Figure 6.

Table 2. Reaction steps involved in the dissociation mechanism of vinyl cyanide considered in this work. The associated activation energies (relative to the zero-point energy of the reactant of each step) are also given.

	Reaction Step	E_a/cm^{-1}
1	$VC \xrightarrow{k_{1f}} {:}CCH_2 + HCN$	35185
2	$VC \xrightarrow{k_{2f}} HCCH + HCN$	41271
3	$VC \underset{k_{3b}}{\overset{k_{3f}}{\rightleftharpoons}} Int1\text{-}III$	20076
4	$Int1\text{-}III \xrightarrow{k_{4f}} HCCH + HCN$	32947
5	$VC \underset{k_{5b}}{\overset{k_{5f}}{\rightleftharpoons}} Int1\text{-}IV$	37494
6	$Int1\text{-}IV \xrightarrow{k_{6f}} HCCH + HNC$	12521
7	$VC \underset{k_{7b}}{\overset{k_{7f}}{\rightleftharpoons}} Int1\text{-}V$	36724
8	$Int1\text{-}V \xrightarrow{k_{8f}} {:}CCH_2 + HNC$	19727
9	$Int1\text{-}III \xrightarrow{k_{9f}} {:}CCH_2 + HNC$	31128
10	$Int1\text{-}III \xrightarrow{k_{10f}} HCCH + HNC$	32912

As shown in Figure 6 and Table 2, VC can directly dissociate to product sets $:CCH_2 + HCN$ and $HCCH + HCN$ through steps 1 and 2 (the only direct dissociation paths), or lead to formation of reaction intermediates Int1-III (the most stable one), Int1-IV, and Int1-V, further evolving to products. On the other hand, product HNC can only be formed via indirect dissociation paths involving the above-mentioned intermediates. A screenshot of VMS showing the structures of all the molecular species involved in this reaction is given in Figure 7.

Within a master-equation approach (see for instance [36]), to determine the time evolution of the relative abundance of the involved species, initially a matrix, $\mathbf{K}$, is set up by opportunely combining the microcanonical rate coefficients at a specified energy. In particular, the diagonal elements K_{ii} contain the loss rate of species i, while the off-diagonal elements K_{ij} contain the rate of formation of species i from species j. The rate of change in the concentration of each species is given by the vector differential equation:

$$\frac{d\mathbf{c}}{dt} = \mathbf{Kc} \tag{9}$$

where **c** is the vector of the concentrations of the species at time t. This is a linear differential equation and can be solved by diagonalization of **K**. In terms of the eigenvector matrix **Z** and eigenvalue vector Λ, the solution of Equation (9) reads:

$$\mathbf{c}(t) = \mathbf{Z}e^{\Lambda t}\mathbf{Z}^{-1}\mathbf{c}(0) \tag{10}$$

where $\mathbf{c}(0)$ is the concentration vector at $t = 0$. In this model, a fundamental hypothesis is that collisional relaxation occurs on time scales much shorter than those that characterize phenomenological kinetics [37]. It is worth mentioning here that a more general version of the master equation would involve diagonalizing a much larger matrix explicitly including collisional relaxation [36]. However, if the above-mentioned hypothesis holds, the resulting eigenvalues would appear in two separated sets: one made up by so-called internal energy relaxation eigenvalues (IEREs) and one made up by so-called chemically significant eigenvalues (CSEs). These last eigenvalues that relate to the phenomenological kinetics of interest in interstellar space and atmospheric studies would be identical to those obtained by solving Equation (9).

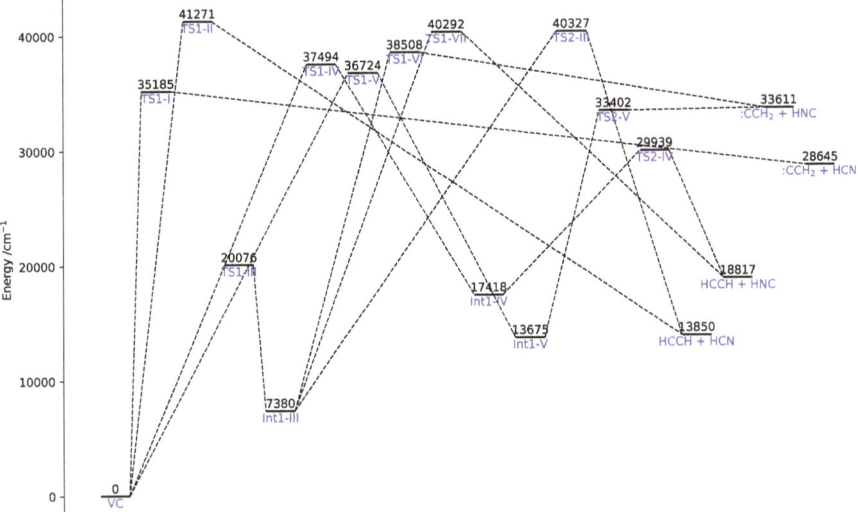

Figure 6. Reaction diagram for the dissociation of vinyl cyanide yielding HCN, HNC, :CCH₂, and HCCH. All energies are relative to the reactant zero-point energy.

By using the methodology described, we computed the evolution of species with respect to time through the StarRate program using the structural parameters of the species given in Ref. [35], and computing the microcanonical rate coefficients through Equation (1). For the reader's convenience, we give the full form of the matrix **K** for this reaction (please note that expressions in square brackets, though spanning several rows, relate to single matrix elements and are shown as such to give a compact picture of the matrix):

$$
\mathbf{K} = \begin{pmatrix}
 & \text{VC} & \text{HCCH} & \text{:CCH}_2 & \text{HCN} & \text{Int1-III} & \text{Int1-IV} & \text{HNC} & \text{Int1-V} & \\
-\begin{bmatrix} k_{1f} + k_{2f} \\ + k_{3f} + k_{5f} \\ + k_{7f} \end{bmatrix} & 0 & 0 & 0 & k_{3b} & k_{5b} & 0 & k_{7b} & \text{VC} \\
k_{2f} & 0 & 0 & 0 & \begin{bmatrix} k_{4f} + k_{10f} \end{bmatrix} & k_{6f} & 0 & 0 & \text{HCCH} \\
k_{1f} & 0 & 0 & 0 & k_{9f} & 0 & 0 & k_{8f} & \text{:CCH}_2 \\
\begin{bmatrix} k_{1f} + k_{2f} \end{bmatrix} & 0 & 0 & 0 & k_{4f} & 0 & 0 & 0 & \text{HCN} \\
k_{3f} & 0 & 0 & 0 & -\begin{bmatrix} k_{3b} + k_{4f} \\ + k_{9f} + k_{10f} \end{bmatrix} & 0 & 0 & 0 & \text{Int1-III} \\
k_{5f} & 0 & 0 & 0 & 0 & -\begin{bmatrix} k_{5b} + k_{6f} \end{bmatrix} & 0 & 0 & \text{Int1-IV} \\
0 & 0 & 0 & 0 & \begin{bmatrix} k_{9f} + k_{10f} \end{bmatrix} & k_{6f} & 0 & k_{8f} & \text{HNC} \\
k_{7f} & 0 & 0 & 0 & 0 & 0 & 0 & -\begin{bmatrix} k_{7b} + k_{8f} \end{bmatrix} & \text{Int1-V}
\end{pmatrix}
$$

In our calculations, the initial concentration of VC was taken as 1.0 and the concentration of other species was set to 0.0. The relative abundances of the involved species as a function of time are plotted for two energies, namely $E = 51764 \ \text{cm}^{-1}$ and $E = 62000 \ \text{cm}^{-1}$, in Figures 8 and 9, respectively.

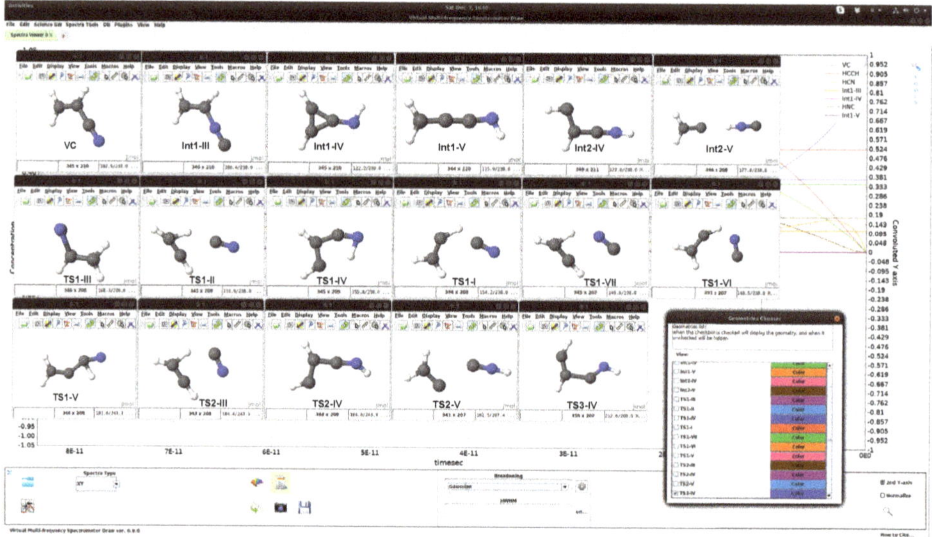

Figure 7. Structures of the reactant molecule (top left corner), intermediates (remaining frames in the top row), and transition states (second and third rows) visualized through the VMS software.

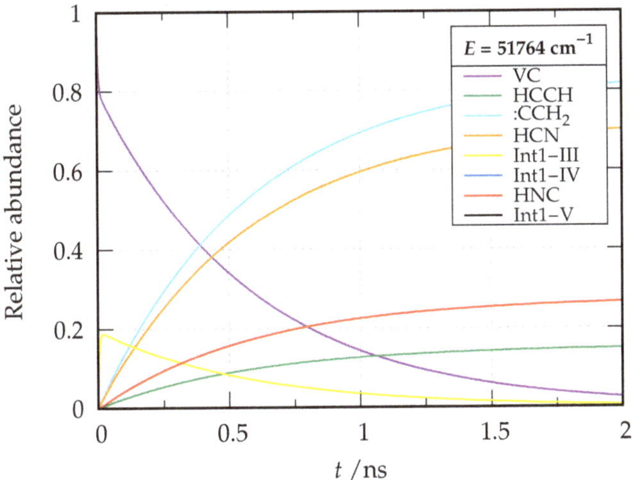

Figure 8. Relative abundance of the species involved in the dissociation of vinyl cyanide as a function of time at energy $E = 51764$ cm^{-1} relative to the reactant zero-point energy.

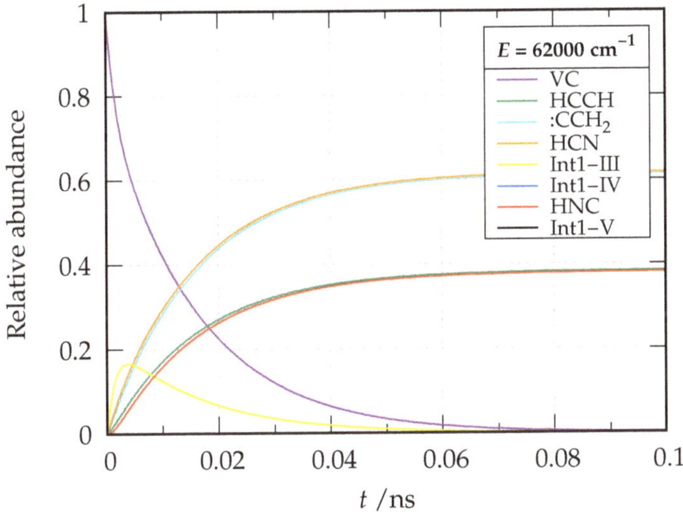

Figure 9. Relative abundance of the species involved in the dissociation of vinyl cyanide as a function of time at energy $E = 62000$ cm^{-1} relative to the reactant zero-point energy.

By inspection of the plots, a first remark is that there is a sudden spike of the concentration for Int1-III in a very small time range for both energies. This is because Step 3 involves a low activation energy (20076 cm^{-1}) compared to Int1-IV (37494 cm^{-1}) and Int1-V (36724 cm^{-1}), and the intermediate Int1-III is relatively stable compared to the other two (the stability of Int1-III is also reflected by the long sigmoidal tail of the plot). At the considered energies, these last two species are virtually never present and as soon as formed evolve into products. The reaction paths involving these two intermediates (the only ones leading to formation of HNC) become increasingly important at $E = 62000$ cm^{-1}; in fact, while the branching ratio HCN/HNC tends to a value of about 2.5 at $E = 51764$ cm^{-1}, a branching ratio of 1.9 is obtained at $E = 62000$ cm^{-1}. The predicted branching ratio at $E = 51764$ cm^{-1} nicely compares with the experimental estimate of 3.3 of Ref. [38] and the theoretical one of 1.9 of Ref. [35] (where, however, only vibrational densities and sum of states were taken into account in the calculation of the microcanonical rate coefficients), substantially improving over former theoretical calculations yielding a branching ratio of over 120 [39].

An interesting feature offered by VMS is that of visualizing matrices in a 'heat-map' fashion through a color palette reflecting the actual value of the elements. A heat-map of **K** for the rate coefficient at $E = 62000$ cm^{-1} is given in Figure 10. Looking at the first column and recalling the meaning of the K_{ij} elements, one can see that VC is directly converted to all the remaining species except HNC. The highest conversion rate (darkest gray) is towards Int1-III, while the rate of formation of the other two intermediates from VC is slower due to higher activation energies. Direct conversion to product sets containing HCN is also fast. On the contrary, as already mentioned, HNC is not formed directly from VC (blank square in position 7,1). Looking at the whole matrix, the darkest squares are those of the matrix elements connecting Int1-IV and Int1-V to products, due to the associated lowest activation energy. This is in line with the fact that, as also shown by Figure 8, these intermediates evolve rapidly to products.

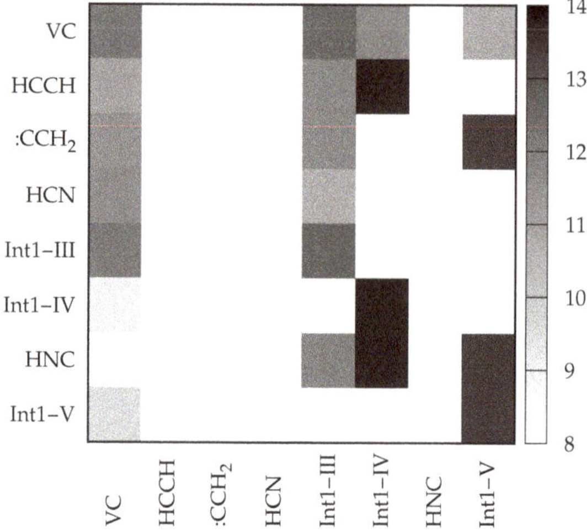

Figure 10. Color map of the transition **K** matrix at energy $E = 62000$ cm^{-1} relative to the reactant zero-point energy as visualized in the VMS software. For each element of the matrix, the value of $|\log_{10} K_{ij}|$ in s^{-1} is plotted according to the gray scale on the right-hand-side of the plot area.

5. Conclusions

In this paper, the implementation of a computer program for chemical kinetics of multi-step reactions and its integration with the graphical interface of the Virtual Multifrequency Spectrometer has been discussed. The developed computational machinery is built around an input/output interface using a hierarchical data structure based on the XML language and shared in common with VMS. Details are given on the implementation of the calculation of microcanonical rate coefficients for the single steps of a unimolecular reaction, and on the modeling of the time evolution of the relative abundance of the involved species. The main features of the program have been illustrated through two example reactions, namely the atmospherically relevant interconversion between hydroxyacetone and 2-hydroxypropanal, and the production of HCN and HNC by dissociation of vinyl cyanide. Work is ongoing in our laboratory to account for bimolecular entrance channels, enhance the potentialities of the program, and integrate it in virtual-reality environments [40,41].

Supplementary Materials: The following are available online at http://www.mdpi.com/2076-3417/10/5/1872/s1, Sample XML file for the interconversion reaction of hydroxyacetone and 2-hydroxypropanal.

Author Contributions: conceptualization, S.R. and V.B.; methodology, S.N. and S.R.; software, S.N. and S.R.; validation, S.N. and B.B.; formal analysis, S.N. and B.B.; investigation, B.B.; writing—original draft preparation, S.N., B.B. and S.R.; writing—review and editing, S.R. and V.B.; supervision, S.R. and V.B.; funding acquisition, S.R. and V.B. All authors have read and agreed to the published version of the manuscript.

Funding: The research leading to these results has received funding from Scuola Normale Superiore through project "DIVE: Development of Immersive approaches for the analysis of chemical bonding through Virtual-reality Environments" (SNS18_B_RAMPINO) and program "Finanziamento a supporto della ricerca di base" (SNS_RB_RAMPINO).

Acknowledgments: The authors are grateful to Daniele Licari for fruitful discussions and to the technical staff of the SMART Laboratory for managing the computational facilities at the Scuola Normale Superiore.

Conflicts of Interest: The authors declare no conflict of interest.

References

1. Althorpe, S.C.; Clary, D.C. Quantum Scattering Calculations on chemical reactions. *Ann. Rev. Phys. Chem.* **2003**, *54*, 493–529. [CrossRef] [PubMed]

2. Rampino, S.; Faginas Lago, N.; Laganà, A.; Huarte-Larrañaga, F. An extension of the grid empowered molecular simulator to quantum reactive scattering. *J. Comput. Chem.* **2012**, *33*, 708–714. [CrossRef] [PubMed]

3. Rampino, S. Configuration-Space Sampling in Potential Energy Surface Fitting: A Space-Reduced Bond-Order Grid Approach. *J. Phys. Chem. A* **2016**, *120*, 4683–4692. [CrossRef] [PubMed]

4. Rampino, S.; Skouteris, D.; Laganà, A. Microscopic branching processes: The $O + O_2$ reaction and its relaxed potential representations. *Int. J. Quantum Chem.* **2010**, *110*, 358–367. [CrossRef]

5. Rampino, S.; Skouteris, D.; Laganà, A. The $O + O_2$ reaction: Quantum detailed probabilities and thermal rate coefficients. *Theor. Chem. Acc. Theory Comput. Model.* **2009**, *123*, 249–256. [CrossRef]

6. Laganà, A.; Faginas Lago, N.; Rampino, S.; Huarte Larrañaga, F.; García, E. Thermal rate coefficients in collinear versus bent transition state reactions: The $N + N_2$ case study. *Phys. Scr.* **2008**, *78*, 058116. [CrossRef]

7. Beck, M.H.; Jäckle, A.; Worth, G.A.; Meyer, H.D. The multiconfiguration time-dependent Hartree (MCTDH) method: A highly efficient algorithm for propagating wavepackets. *Phys. Rep.* **2000**, *324*, 1–105. [CrossRef]

8. Craig, I.R.; Manolopoulos, D.E. Chemical reaction rates from ring polymer molecular dynamics. *J. Chem. Phys.* **2005**, *122*, 084106. [CrossRef]

9. Rampino, S.; Pastore, M.; Garcia, E.; Pacifici, L.; Laganà, A. On the temperature dependence of the rate coefficient of formation of C_2^+ from $C + CH^+$. *Mon. Not. R. Astron. Soc.* **2016**, *460*, 2368–2375. [CrossRef]

10. Pacifici, L.; Pastore, M.; Garcia, E.; Laganà, A.; Rampino, S. A dynamics investigation of the $C + CH^+ \rightarrow C_2^+ + H$ reaction on an *ab initio* bond-order like potential. *J. Phys. Chem. A* **2016**, *120*, 5125–5135. [CrossRef]

11. Rampino, S.; Suleimanov, Y.V. Thermal Rate Coefficients for the Astrochemical Process $C + CH^+ \rightarrow C2^+ + H$ by Ring Polymer Molecular Dynamics. *J. Phys. Chem. A* **2016**, *120*, 9887–9893. [CrossRef] [PubMed]

12. Rice, O.K.; Ramsperger, H.C. Theories of Unimolecular Gas Reactions at Low Pressures. *J. Am. Chem. Soc.* **1927**, *49*, 1617–1629. [CrossRef]

13. Kassel, L.S. Studies in Homogeneous Gas Reactions. I. *J. Phys. Chem.* **1927**, *32*, 225–242. [CrossRef]

14. Marcus, R.A. Unimolecular Dissociations and Free Radical Recombination Reactions. *J. Chem. Phys.* **1952**, *20*, 359–364. [CrossRef]

15. Barone, V. The virtual multifrequency spectrometer: A new paradigm for spectroscopy. *WIREs Comput. Mol. Sci.* **2016**, *6*, 86–110. [CrossRef] [PubMed]

16. Bloino, J.; Biczysko, M.; Barone, V. General Perturbative Approach for Spectroscopy, Thermodynamics, and Kinetics: Methodological Background and Benchmark Studies. *J. Chem. Theory Comput.* **2012**, *8*, 1015–1036. [CrossRef] [PubMed]

17. Bloino, J.; Barone, V. A second-order perturbation theory route to vibrational averages and transition properties of molecules: General formulation and application to infrared and vibrational circular dichroism spectroscopies. *J. Chem. Phys.* **2012**, *136*, 124108. [CrossRef]

18. Licari, D.; Baiardi, A.; Biczysko, M.; Egidi, F.; Latouche, C.; Barone, V. Implementation of a graphical user interface for the virtual multifrequency spectrometer: The VMS-Draw tool. *J. Comput. Chem.* **2015**, *36*, 321–334. [CrossRef]

19. F Syntax Rules. Available online: http://www.fortran.com/F/F_bnf.html (accessed on 5 February 2020).
20. Metcalf, M.; Reid, J. *The F Programming Language*; Oxford University Press, Inc.: New York, NY, USA, 1996.
21. Nandi, S.; Calderini, D.; Bloino, J.; Rampino, S.; Barone, V. A Modern-Fortran Program for Chemical Kinetics on Top of Anharmonic Vibrational Calculations. *Lect. Notes Comput. Sci.* **2019**, *11624*, 401–412.
22. Cavallotti, C.; Pelucchi, M.; Georgievskii, Y.; Klippenstein, S.J. EStokTP: Electronic Structure to Temperature- and Pressure-Dependent Rate Constants–A Code for Automatically Predicting the Thermal Kinetics of Reactions. *J. Chem. Theory Comput.* **2019**, *15*, 1122–1145. [CrossRef]
23. Rossi, E.; Evangelisti, S.; Laganà, A.; Monari, A.; Rampino, S.; Verdicchio, M.; Baldridge, K.K.; Bendazzoli, G.L.; Borini, S.; Cimiraglia, R.; et al. Code interoperability and standard data formats in quantum chemistry and quantum dynamics: The Q5/D5Cost data model. *J. Comput. Chem.* **2014**, *35*, 611–621. [CrossRef] [PubMed]
24. Rampino, S.; Monari, A.; Rossi, E.; Evangelisti, S.; Laganà, A. A priori modeling of chemical reactions on computational grid platforms: Workflows and data models. *Chem. Phys.* **2012**, *398*, 192–198. [CrossRef]
25. FoX in Fortran Wiki. Available online: http://fortranwiki.org/fortran/show/FoX (accessed on 5 February 2020).
26. Butkovskaya, N.I.; Pouvesle, N.; Kukui, A.; Mu, Y.; Le Bras, G. Mechanism of the OH-Initiated Oxidation of Hydroxyacetone over the Temperature Range 236–298 K. *J. Phys. Chem. A* **2006**, *110*, 6833–6843. [CrossRef]
27. Sun, J.; Thus, S.; da Silva, G. The gas phase aldose-ketone isomerization mechanism: Direct interconversion of the model hydroxycarbonyls 2-hydroxypropanal and hydroxyacetone. *Int. J. Quantum Chem.* **2017**, *117*, e25434. [CrossRef]
28. Brouard, M. *Reaction Dynamics*; Oxford Chemistry Primers, OUP Oxford: Oxford, UK, 1998.
29. Green, N.J.B. Chapter 1—Introduction. In *Unimolecular Kinetics*; Green, N.J., Ed.; Elsevier: Amsterdam, The Netherlands, 2003; Volume 39, pp. 1–53.
30. Stein, S.; Rabinovitch, B.S. Accurate evaluation of internal energy level sums and densities including anharmonic oscillators and hindered rotors. *J. Chem. Phys.* **1973**, *58*, 2438–2445. [CrossRef]
31. Beyer, T.; Swinehart, D.F. Algorithm 448: Number of Multiply-restricted Partitions. *Commun. ACM* **1973**, *16*, 379. [CrossRef]
32. Miller, W.H. Tunneling corrections to unimolecular rate constants, with application to formaldehyde. *J. Am. Chem. Soc.* **1979**, *101*, 6810–6814. [CrossRef]
33. Kooij, D.M. Über die Zersetzung des Gasförmigen Phosphorwasserstoffs. *Zeitschrift für Physikalische Chemie Abteilung B* **1893**, *12*, 155–161.
34. Laidler, K. A glossary of terms used in chemical kinetics, including reaction dynamics (IUPAC Recommendations 1996). *Pure Appl. Chem.* **1996**, *68*, 149–192. [CrossRef]
35. Homayoon, Z.; Vázquez, S.A.; Rodríguez-Fernández, R.; Martínez-Núñez, E. Ab Initio and RRKM Study of the HCN/HNC Elimination Channels from Vinyl Cyanide. *J. Phys. Chem. A* **2011**, *115*, 979–985. [CrossRef]
36. Robertson, S.H.; Pilling, M.J.; Jitariu, L.C.; Hillier, I.H. Master equation methods for multiple well systems: application to the 1-,2-pentyl system. *Phys. Chem. Chem. Phys.* **2007**, *9*, 4085–4097. [CrossRef] [PubMed]
37. Barker, J.R.; Weston, R.E. Collisional Energy Transfer Probability Densities $P(E, J; E', J')$ for Monatomics Colliding with Large Molecules. *J. Phys. Chem. A* **2010**, *114*, 10619–10633. [CrossRef] [PubMed]
38. Wilhelm, M.J.; Nikow, M.; Letendre, L.; Dai, H.L. Photodissociation of vinyl cyanide at 193 nm: Nascent product distributions of the molecular elimination channels. *J. Chem. Phys.* **2009**, *130*, 044307. [CrossRef] [PubMed]
39. Derecskei-Kovacs, A.; North, S.W. The unimolecular dissociation of vinylcyanide: A theoretical investigation of a complex multichannel reaction. *J. Chem. Phys.* **1999**, *110*, 2862–2871. [CrossRef]
40. Salvadori, A.; Fusè, M.; Mancini, G.; Rampino, S.; Barone, V. Diving into chemical bonding: An immersive analysis of the electron charge rearrangement through virtual reality. *J. Comput. Chem.* **2018**, *39*, 2607–2617. [CrossRef]
41. Martino, M.; Salvadori, A.; Lazzari, F.; Paoloni, L.; Nandi, S.; Mancini, G.; Barone, V.; Rampino, S. Chemical Promenades: Exploring Potential-energy Surfaces with Immersive Virtual Reality. *J. Comput. Chem.* **2020**. [CrossRef]

Article

A Never-Ending Conformational Story of the Quercetin Molecule: Quantum-Mechanical Investigation of the O3′H and O4′H Hydroxyl Groups Rotations

Ol'ha O. Brovarets' [1,*] **and Dmytro M. Hovorun** [1,2]

[1] Department of Molecular and Quantum Biophysics, Institute of Molecular Biology and Genetics, National Academy of Sciences of Ukraine, 150 Akademika Zabolotnoho Street, 03680 Kyiv, Ukraine; dhovorun@imbg.org.ua

[2] Department of Molecular Biotechnology and Bioinformatics, Institute of High Technologies, Taras Shevchenko National University of Kyiv, 2-h Akademika Hlushkova Avenue, 03022 Kyiv, Ukraine

* Correspondence: o.o.brovarets@imbg.org.ua

Received: 8 January 2020; Accepted: 2 February 2020; Published: 8 February 2020

Abstract: The quercetin molecule is known to be an effective pharmaceutical compound of a plant origin. Its chemical structure represents two aromatic A and B rings linked through the C ring containing oxygen and five OH hydroxyl groups attached to the 3, 3′, 4′, 5, and 7 positions. In this study, a novel conformational mobility of the quercetin molecule was explored due to the turnings of the O3′H and O4′H hydroxyl groups, belonging to the B ring, around the exocyclic C-O bonds. It was established that the presence of only three degrees of freedom of the conformational mobility of the O3′H and O4′H hydroxyl groups is connected with their concerted behavior, which is controlled by the non-planar (in the case of the interconverting planar conformers) or locally non-planar (in other cases) TSs$^{O3′H/O4′H}$ transition states, in which O3′H and O4′H hydroxyl groups are oriented by the hydrogen atoms towards each other. We also explored the number of the physico-chemical and electron-topological characteristics of all intramolecular-specific contacts—hydrogen bonds and attractive van der Waals contacts at the conformers and also at the transition states. Long-terms perspectives for the investigations of the structural bases of the biological activity of this legendary molecule have been shortly described.

Keywords: Quercetin molecule; conformational mobility; hydroxyl group; transition state; concerted rotation of the hydroxyl groups; quantum-chemical calculations; quantum technology

---

1. Introduction

The quercetin molecule (3, 3′, 4′, 5, 7—pentahydroxyflavone, $C_{15}H_{10}O_7$) is an important flavonoid compound, which is found in many foods and plants, in particular in honey [1], and is known to act as a natural drug molecule with a wide range of treatment properties—antioxidant, anti-toxic, etc.—and is also involved in drug delivery from the site of administration to the therapeutic target [2–9]. The structure of the quercetin contains two aromatic A and B rings linked through the C ring containing oxygen and five OH hydroxyl groups attached to the 3, 3′, 4′, 5 and 7 positions (see Scheme 1) [10–15].

In a previous study [16], by using the quantum-mechanical (QM) calculations at the MP2/6-311++G(2df,pd)//B3LYP/6-311++G(d,p) level of QM theory together with Bader's "Quantum Theory of Atoms in Molecules", for the first time, all possible conformers were established, corresponding to local minima on the potential energy hypersurface of the isolated quercetin molecule.

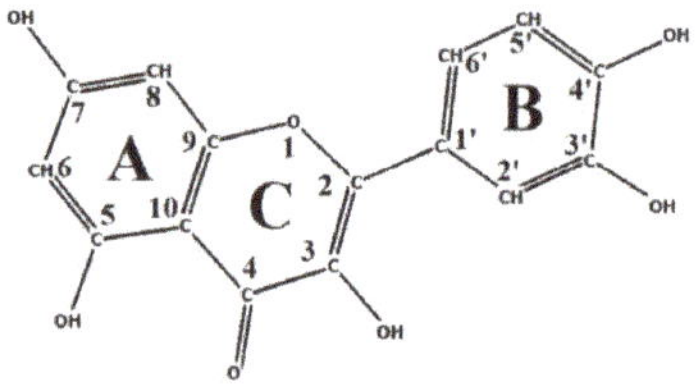

Scheme 1. Chemical structure of the quercetin molecule and standard numeration of its atoms.

Altogether, 48 stable conformers were established, which have been divided into four different conformational subfamilies by their structural properties: subfamily I—conformers **1–12**; subfamily II—conformers **13-18**, **20**, **23**, **24**, **26**, **29**, and **30**; subfamily III—conformers **19**, **21**, **22**, **25**, **27**, **28**, and **31–36**; subfamily IV—conformers **37–48** [16]. It was shown that these 48 stable conformers (24 planar structures (C_s point symmetry) and 24 non-planar structures (C_1 point symmetry)) represent a comprehensive set of the theoretically possible structures.

Conformers of quercetin are polar structures with a dipole moment, which varies within the range from 0.35 to 9.87 Debye for different conformers with different direction for each.

Their relative Gibbs free energies are arranged within the range from 0.00 to 25.30 kcal·mol^{-1} under normal conditions in vacuum.

One half of these structures (24 conformers) possesses planar structure (C_s point symmetry), whereas the other half (24 conformers) does not have symmetry at all (C_1 point symmetry) (C3C2C1'C2' = 40.9–44.3 degree; C2C3O3H = 9.4–16.3 degree).

We also defined their physico-chemical characteristics, in particular, structural, energetic, and polar, which are necessary for understanding of the biological mechanisms of action of this molecule. Intramolecular specific contacts have also been explored in detail.

Bader's "Quantum Theory of Atoms in Molecules" analysis shows that conformers of the quercetin molecule differ from each other by the intramolecular specific contacts (two or three), stabilizing all possible conformers of the molecule—H-bonds (both classical OH ... O and so-called unusual CH ... O and OH ... C) and attractive van der Waals contacts O ... O. Energies of these cooperative intramolecular specific contacts have been estimated [16].

Also, it was theoretically modeled the conformational interconversions [17–21] in the 24 pairs of the conformers of the quercetin molecule through the rotation of its almost non-deformable (A+C) and B rings around the C2-C1' bond through the quasi-orthogonal transition state (TS) with low values of the imaginary frequencies (28-33/29-36 cm^{-1}) and Gibbs free energies of activation in the range of 2.17 to 5.68/1.86 to 4.90 kcal·mol^{-1} in the continuum with dielectric permittivity $\varepsilon = 1/\varepsilon = 4$ under normal conditions. Also, we studied the changes of the number of physico-chemical characteristics of all intramolecular specific contacts—hydrogen bonds and attractive van der Waals contacts during these conformational rearrangements.

This study is a logical development of the previous investigations [16–18] and is devoted to the novel interconversions between the conformers of the quercetin molecule due to the rotations of the O3'H and O4'H hydroxyl groups around the exocyclic C-O bonds outside.

As a result, it was found that different conformers of the quercetin molecule are tightly interconnected with each other through the set of the TSs. Moreover, these conformational transformations are assisted by the intramolecular H-bonds and van der Waals contacts.

In reality, it is not as easy a task as it seems, as the question of "Why two neighboring hydroxyl groups in the aromatic ring have only three, but not four conformational degrees of freedom?" remains without answer [22,23].

The main idea of this investigation is summarized in the following statements.

We suggest that the conformational mobility of the C ring of the quercetin molecule, which contains two neighboring O3'H and O4'H hydroxyl groups, that is, conversion of one stable

O3'HO4'H/O4'HO3'H configuration into the other O4'HO3'H/O3'HO4'H and *vice versa*, is realized by two significantly different pathways from the topological and energetical point of view.

First, (this pathway is more or less evident) it occurs by the restricted rotations of the hydroxyls by the angle of 180 degrees through the corresponding TSs and through the high-energetical dynamically stable O3'HO4'H/O4'HO3'H configuration. The other pathway is quite unusual—it is realized through one conformational transition, which has concerted character and is controlled by the non-planar TSs$^{O3'HHO4'/O4'HHO3'}$ with the high values of the imaginary frequency.

A previously suggested idea has been completely confirmed by careful QM investigation—we have identified for the first time the aforementioned pathways of the conformational variability of the quercetin molecule and documented their structural properties, including symmetrical, polar, energetical, and kinetic characteristics, which are quite important for the understanding of the structural grounds of the biological activity of the quercetin molecule.

2. Computational Methods

Calculations of the geometrical structures of the TSs of the conformational interconversions and their vibrational spectra, corresponding to the local minima on the potential (electronic) energy hyper surface, have been performed at the DFT B3LYP/6-311++G(d,p) level of QM theory [24–26] by Gaussian'09 program package [27], which was successfully approved in our previous studies for the calculations of the heterocyclic compounds [28,29]. A scaling factor of 0.9668 has been used to correct the harmonic frequencies for the investigated structures [30]. Intrinsic reaction coordinate (IRC) calculations in the forward and reverse directions from each TS, which have been confirmed by the presence of one and only one imaginary frequency in the vibrational spectra, have been performed using Hessian-based predictor–corrector integration algorithm [31].

All calculations were performed for the quercetin molecule as their intrinsic property, that is adequate for modeling of the processes occurring in real systems [16,17,32].

Electronic and Gibbs free energies under normal conditions have been calculated by single point calculations at the MP2/6-311++G(2df,pd) level of theory [33–35].

The time $\tau_{99.9\%}$ necessary to reach 99.9% of the equilibrium concentration of the reactant and product in the system of the reversible first-order forward (k_f) and reverse (k_r) reactions can be estimated by the formula [36]

$$\tau_{99.9\%} = \frac{ln10^3}{k_f + k_r} \tag{1}$$

The lifetime, τ, of the conformers has been calculated using the formula $1/k_r$, where the values of the forward k_f and reverse k_r rate constants for the tautomerization reactions were obtained as [36]

$$k_{f,r} = \Gamma \cdot \frac{k_B T}{h} e^{-\frac{\Delta\Delta G_{f,r}}{RT}} \tag{2}$$

where the quantum tunneling effect has been accounted by Wigner's tunneling correction [37], successfully used for the double proton reactions in DNA base pairs [28]:

$$\Gamma = 1 + \frac{1}{24}\left(\frac{h\nu_i}{k_B T}\right)^2 \tag{3}$$

where k_B—Boltzmann's constant, h—Planck's constant, $\Delta\Delta G_{f,r}$—Gibbs free energy of activation for the conformational transition in the forward (f) and reverse (r) directions, and ν_i—magnitude of the imaginary frequency associated with the vibrational mode at the TS.

The topology of the electron density was analyzed using the program package AIM'2000 [38] with all default options and wave functions obtained at the level of theory used for geometry optimization. The presence of the (3,−1) bond critical point (BCP), bond path between hydrogen donor and acceptor, and positive value of the Laplacian at this BCP ($\Delta\rho > 0$) were considered altogether as criteria for the formation of the H-bond and attractive van der Waals contact [39].

In this work, standard numeration of atoms has been used [16,17]. Numeration of the conformers, which are highlighted in bold in the text, have been used as in the work [16].

3. Obtained Results and Their Discussion

In this study, we logically continued to investigate the conformational mobility [16–18] of the quercetin molecule and extend this approach to the rotations of the hydroxyl groups in the 3′ and 4′ positions, which are carefully presented in Tables 1–3 and Figures 1–3. The most obvious methods of the conformational interconversions between the 48 conformers [16] of the quercetin molecule were considered and investigated in detail through the rotations of the O3′H and O4′H hydroxyl groups around the exocyclic C-O covalent bonds. In this case, the TSs have been formed gradually, starting from the 48 conformers of the quercetin molecule [16] by the single or concerted rotations of the O3′H and O4′H hydroxyl groups—designated as $TS^{O3'H}$, $TS^{O4'H}$, and $TS^{O3'H/O4'H}$, respectively.

Therefore, detailed analysis of the obtained results enabled us to obtain the following observations and their discussion. As individual, the concerted rotational transitions of the O3′H and O4′H hydroxyl groups proceed through the mirror-symmetrical pathways, which are controlled by the mirror-symmetrical TSs. Totally, we have revealed 48 TSs—16 TSs in each case (Figures 1–3; Tables 1–3).

Individual conformational transitions are controlled by the non-planar $TS^{O3'H}$ and $TS^{O4'H}$ (C_1 point symmetry) with non-orthogonal structure (see HO3′C3′C2′ (78.7–83.3 degree) and HO4′C4′C5′ (80.2–82.1 degree) dihedral angles in Tables 1 and 2 and Figures 1 and 2). Their non-orthogonal structure, most probably, could be connected with the non-symmetrical surrounding of the free electronic pairs of the oxygen atoms of the hydroxyl groups. The TSs for the concerted conformational transformations—$TSs^{O3'H/O4'H}$—possess non-planar structure in the case of the planar conformers **1-12**, **19**, **21**, **22**, **25**, **27**, **28**, and **31–36** and local non-planar structure for the non-planar conformers, **13-18**, **20**, **23**, **24**, **26**, **29**, **30**, and **37–48**, which mutually interconvert (Figure 3, Table 3).

Gibbs free energies of activation for these processes form the following order; $\Delta\Delta G_{TS}^{O4'H}$ (3.33–7.05) $< \Delta\Delta G_{TS}^{O3'H}$ (4.23–7.08) $< \Delta\Delta G_{TS}^{O3'H/O4'H}$ (4.41–7.56 kcal·mol^{-1} under normal conditions) (Tables 1–3). The imaginary frequencies are in the following ranges: 366.7–391.5 ($TS^{O3'H}$), 328.8–363.5 ($TS^{O4'H}$), and 454.5–483.1 ($TS^{O3'H/O4'H}$). Without exception, 48 conformers of the quercetin molecule have been established to be the dynamically stable structures, based on the investigated conformational transitions. During their lifetime ($\tau = (1.05–2.53)\cdot10^{-11}$ s) (Tables 1–3), the lowest frequency intramolecular vibrations can occur [16].

It is a characteristic feature that investigated conformational transitions are dipole-active, as they cause the changing of the dipole moment by the absolute value, so by the spatial orientation, and practically do not disturb the structure of the quercetin molecule and physico-chemical characteristics of its specific intramolecular interactions. Even the energy of the intramolecular C2′/C6′H . . . O3 and O3H . . . C2′/C6′H H-bonds between the B and C rings (Figures 1–3) change at these conformational transitions by no more than on ~4.7%. Interestingly, concerted conformational transitions, which are controlled by the $TSs^{O3'H/O4'H}$, proceed without intermediates on the hyperspace of the Gibbs free energy.

Moreover, we did not register any specific intramolecular interactions in the B ring of the quercetin molecule at the conformational motions of the O3′H and O4′H hydroxyl groups. All investigated conformational transitions are quite rapid processes, for which $1.04\cdot10^{-10} > \tau_{99.9\%} > 7.30\cdot10^{-11}$ s.

Therefore, provided investigation gives total assurance that the availability of the three conformational degrees of freedom for the O3′H and O4′H hydroxyl groups is connected with their concerted, coordinated behavior (Figure 3, Table 3).

Let us to make one important notion before going to the conclusions. It is known, that biological activity of the molecules, is caused by at least two interdependent reasons—their intramolecular structural variability and specific interaction with the targets of the different origin.

Table 1. Energetic, polar, structural, and kinetic characteristics of the conformational transitions in the isolated quercetin molecule via the mirror-symmetrical rotations of the O3′H hydroxyl group around the C3′-O3′ bond through the transition states (TSs) with a non-perpendicularly-oriented O3′H group, obtained at the MP2/6-311++G(2df,pd)//B3LYP/6-311++G(d,p) level of QM theory under normal conditions (see Figure 1).

TS of the Conformational Transition	μ_{TS} [a]	ν_i [b]	ΔG [c]	ΔE [d]	$\Delta\Delta G_{TS}$ [e]	$\Delta\Delta E_{TS}$ [f]	$\Delta\Delta G$ [g]	$\Delta\Delta E$ [h]	k_f [i]	k_r [j]	$\tau_{99.9\%}$ [k]	τ [l]	HO3′C3′C2′ [m]
$TS^{O3'H}_{2\leftrightarrow9}$	1.78	340.2	3.98	3.80	6.90	6.75	2.91	2.95	$6.01 \cdot 10^7$	$5.01 \cdot 10^{10}$	$1.38 \cdot 10^{-10}$	$2.00 \cdot 10^{-11}$	± 80.1
$TS^{O3'H}_{4\leftrightarrow11}$	4.02	333.4	4.15	4.07	7.08	6.73	2.93	2.66	$4.36 \cdot 10^7$	$4.82 \cdot 10^{10}$	$1.43 \cdot 10^{-10}$	$2.07 \cdot 10^{-11}$	± 79.1
$TS^{O3'H}_{7\leftrightarrow10}$	4.10	361.1	3.29	3.21	6.25	6.53	2.97	3.32	$1.80 \cdot 10^8$	$4.62 \cdot 10^{10}$	$1.49 \cdot 10^{-10}$	$2.17 \cdot 10^{-11}$	∓ 82.8
$TS^{O3'H}_{8\leftrightarrow12}$	5.93	362.0	3.32	3.23	6.25	6.51	2.93	3.28	$1.80 \cdot 10^8$	$4.91 \cdot 10^{10}$	$1.40 \cdot 10^{-10}$	$2.04 \cdot 10^{-11}$	∓ 82.9
$TS^{O3'H}_{14\leftrightarrow24}$	5.08	355.2	3.70	3.75	6.50	6.92	2.80	3.17	$1.17 \cdot 10^8$	$6.09 \cdot 10^{10}$	$1.13 \cdot 10^{-10}$	$1.64 \cdot 10^{-11}$	± 81.3
$TS^{O3'H}_{15\leftrightarrow26}$	6.24	340.9	3.79	3.97	6.47	6.86	2.68	2.89	$1.23 \cdot 10^8$	$7.44 \cdot 10^{10}$	$9.27 \cdot 10^{-11}$	$1.34 \cdot 10^{-11}$	∓ 79.3
$TS^{O3'H}_{17\leftrightarrow29}$	7.60	355.6	3.81	3.91	6.62	7.05	2.81	3.15	$9.70 \cdot 10^7$	$6.03 \cdot 10^{10}$	$1.14 \cdot 10^{-10}$	$1.66 \cdot 10^{-11}$	± 80.6
$TS^{O3'H}_{18\leftrightarrow30}$	8.65	343.3	3.79	3.94	6.50	6.87	2.71	2.93	$1.17 \cdot 10^8$	$7.04 \cdot 10^{10}$	$9.79 \cdot 10^{-11}$	$1.42 \cdot 10^{-11}$	∓ 79.5
$TS^{O3'H}_{21\leftrightarrow34}$	3.08	328.8	4.01	4.14	6.54	6.70	2.53	2.56	$1.09 \cdot 10^8$	$9.51 \cdot 10^{10}$	$7.26 \cdot 10^{-11}$	$1.05 \cdot 10^{-11}$	± 78.7
$TS^{O3'H}_{27\leftrightarrow33}$	6.23	363.5	3.25	3.03	6.27	6.29	3.02	3.27	$1.75 \cdot 10^8$	$4.25 \cdot 10^{10}$	$1.62 \cdot 10^{-10}$	$2.35 \cdot 10^{-11}$	∓ 83.3
$TS^{O3'H}_{31\leftrightarrow36}$	3.35	336.2	3.85	3.86	6.39	6.63	2.54	2.77	$1.42 \cdot 10^8$	$9.44 \cdot 10^{10}$	$7.30 \cdot 10^{-11}$	$1.06 \cdot 10^{-11}$	± 79.8
$TS^{O3'H}_{32\leftrightarrow35}$	5.97	362.6	3.23	3.05	6.29	6.38	3.06	3.33	$1.69 \cdot 10^8$	$3.96 \cdot 10^{10}$	$1.74 \cdot 10^{-10}$	$2.53 \cdot 10^{-11}$	∓ 83.2
$TS^{O3'H}_{39\leftrightarrow45}$	6.15	354.7	3.83	3.92	6.60	7.01	2.77	3.09	$9.97 \cdot 10^7$	$6.43 \cdot 10^{10}$	$1.07 \cdot 10^{-10}$	$1.56 \cdot 10^{-11}$	± 80.5
$TS^{O3'H}_{40\leftrightarrow46}$	7.86	344.1	3.70	3.82	6.45	6.80	2.75	2.98	$1.27 \cdot 10^8$	$6.60 \cdot 10^{10}$	$1.04 \cdot 10^{-10}$	$1.52 \cdot 10^{-11}$	± 80.5
$TS^{O3'H}_{42\leftrightarrow48}$	6.37	341.6	3.70	3.85	6.41	6.79	2.71	2.94	$1.36 \cdot 10^8$	$7.05 \cdot 10^{10}$	$9.78 \cdot 10^{-11}$	$1.42 \cdot 10^{-11}$	∓ 79.6
$TS^{O3'H}_{44\leftrightarrow47}$	4.47	353.8	1.46	1.51	4.23	4.62	2.77	3.12	$5.47 \cdot 10^9$	$6.46 \cdot 10^{10}$	$9.85 \cdot 10^{-11}$	$1.55 \cdot 10^{-11}$	± 81.1

[a] The dipole moment of the TS, Debye. [b] The imaginary frequency at the TS of the conformational transition, cm^{-1}. [c] The Gibbs free energy of the initial relative to the terminal conformer of the quercetin molecule (T = 298.15 K), $kcal \cdot mol^{-1}$. [d] The electronic energy of the initial relative to the terminal conformer of the quercetin molecule, $kcal \cdot mol^{-1}$. [e] The Gibbs free energy barrier for the forward conformational transformation of the quercetin molecule, $kcal \cdot mol^{-1}$. [f] The electronic energy barrier for the forward conformational transformation of the quercetin molecule, $kcal \cdot mol^{-1}$. [g] The Gibbs free energy barrier for the reverse conformational transformation of the quercetin molecule, $kcal \cdot mol^{-1}$. [h] The electronic energy barrier for the reverse conformational transformation of the quercetin molecule, $kcal \cdot mol^{-1}$. [i] The rate constant for the forward conformational transformation, s^{-1}. [j] The rate constant for the reverse conformational transformation, s^{-1}. [k] The time necessary to reach 99.9% of the equilibrium concentration between the reactant and the product of the reaction of the conformational transformation, s. [l] The lifetime of the product of the conformational transition, s. [m] The dihedral angle, which describes at the TS the orientation of the O3′H hydroxyl group relatively the B ring of the quercetin molecule, degree; sings "±" correspond to enantiomers.

Table 2. Energetic, polar, structural, and kinetic characteristics of the conformational transitions in the isolated quercetin molecule via the mirror-symmetrical rotations of the O4'H hydroxyl group around the C4'-O4' bond through the transition states (TSs) with a non-perpendicularly-oriented O4'H group, obtained at the MP2/6-311++G(2df,pd)//B3LYP/6-311++G(d,p) level of QM theory under normal conditions (see Figure 2).

TS of the Conformational Transition	μ_{TS}	ν_i	ΔG	ΔE	$\Delta\Delta G_{TS}$	$\Delta\Delta E_{TS}$	$\Delta\Delta G$	$\Delta\Delta E$	k_f	k_r	$\tau_{99.9\%}$	τ	HO4'C4'C5'
$TS^{O4'H}_{1\leftrightarrow10}$	2.41	381.6	4.20	4.30	7.05	7.38	2.84	3.08	$4.75\cdot10^7$	$5.77\cdot10^{10}$	$1.20\cdot10^{-10}$	$1.73\cdot10^{-11}$	±81.1
$TS^{O4'H}_{3\leftrightarrow9}$	2.82	391.5	3.92	4.03	6.86	7.23	2.94	3.21	$6.60\cdot10^7$	$4.95\cdot10^{10}$	$1.39\cdot10^{-10}$	$2.02\cdot10^{-11}$	∓81.9
$TS^{O4'H}_{5\leftrightarrow12}$	3.63	379.7	4.24	4.37	6.93	7.35	2.69	2.98	$5.80\cdot10^7$	$7.41\cdot10^{10}$	9.32E-11	$1.35\cdot10^{-11}$	±80.8
$TS^{O4'H}_{6\leftrightarrow11}$	3.99	389.2	3.86	4.00	6.92	7.14	3.07	3.14	$5.90\cdot10^7$	$3.98\cdot10^{10}$	$1.73\cdot10^{-10}$	$2.51\cdot10^{-11}$	∓82.1
$TS^{O4'H}_{13\leftrightarrow24}$	4.78	375.8	4.07	4.13	6.81	7.22	2.74	3.09	$7.12\cdot10^7$	$6.89\cdot10^{10}$	$1.00\cdot10^{-10}$	$1.45\cdot10^{-11}$	∓80.4
$TS^{O3'H}_{16\leftrightarrow30}$	6.83	374.4	4.01	4.11	6.78	7.16	2.77	3.04	$7.42\cdot10^7$	$6.52\cdot10^{10}$	$1.06\cdot10^{-10}$	$1.53\cdot10^{-11}$	±80.7
$TS^{O4'H}_{19\leftrightarrow33}$	4.64	371.6	4.12	4.49	6.74	7.25	2.62	2.76	$7.98\cdot10^7$	$8.40\cdot10^{10}$	$8.22\cdot10^{-11}$	$1.19\cdot10^{-11}$	±80.2
$TS^{O4'H}_{20\leftrightarrow26}$	4.64	371.6	0.63	0.72	6.71	7.51	6.08	6.79	$8.36\cdot10^7$	$2.42\cdot10^8$	$2.12\cdot10^{-8}$	$4.13\cdot10^{-9}$	±80.2
$TS^{O4'H}_{22\leftrightarrow34}$	5.05	380.8	4.04	3.97	6.71	7.00	2.67	3.03	$8.43\cdot10^7$	$7.74\cdot10^{10}$	$8.91\cdot10^{-11}$	$1.29\cdot10^{-11}$	∓81.8
$TS^{O4'H}_{23\leftrightarrow29}$	7.10	375.3	0.55	0.60	3.33	3.73	2.78	3.13	$2.52\cdot10^{10}$	$6.41\cdot10^{10}$	$7.73\cdot10^{-11}$	$1.56\cdot10^{-11}$	∓80.7
$TS^{O4'H}_{25\leftrightarrow35}$	5.37	373.7	4.12	4.35	6.84	7.22	2.71	2.88	$6.77\cdot10^7$	$7.17\cdot10^{10}$	$9.62\cdot10^{-11}$	$1.39\cdot10^{-11}$	±80.5
$TS^{O4'H}_{28\leftrightarrow36}$	5.70	382.7	4.07	4.06	6.74	7.08	2.67	3.02	$7.99\cdot10^7$	$7.71\cdot10^{10}$	$8.95\cdot10^{-11}$	$1.30\cdot10^{-11}$	∓81.6
$TS^{O4'H}_{37\leftrightarrow45}$	6.56	369.1	4.03	4.05	6.75	7.11	2.72	3.06	$7.80\cdot10^7$	$7.09\cdot10^{10}$	$9.74\cdot10^{-11}$	$1.41\cdot10^{-11}$	∓80.4
$TS^{O4'H}_{38\leftrightarrow46}$	6.19	367.5	4.03	4.16	6.72	7.10	2.69	2.95	$8.14\cdot10^7$	$7.40\cdot10^{10}$	$9.32\cdot10^{-11}$	$1.35\cdot10^{-11}$	±80.2
$TS^{O4'H}_{41\leftrightarrow48}$	5.09	366.7	4.00	4.10	6.69	7.05	2.69	2.94	$8.64\cdot10^7$	$7.42\cdot10^{10}$	$9.30\cdot10^{-11}$	$1.35\cdot10^{-11}$	±80.4
$TS^{O4'H}_{43\leftrightarrow47}$	5.49	369.4	1.83	1.88	4.73	5.10	2.90	3.22	$2.37\cdot10^9$	$5.21\cdot10^{10}$	$1.27\cdot10^{-10}$	$1.92\cdot10^{-11}$	∓80.2

For designations see Table 1.

Table 3. Energetic, polar, structural, and kinetic characteristics of the conformational transitions in the isolated quercetin molecule via the mirror-symmetrical concerted rotations of the O3′H and O4′H hydroxyl groups around the C3′-O3′ and C4′-O4′ bonds through the non-planar or locally non-planar transition states (TSs), obtained at the MP2/6-311++G(2df,pd)//B3LYP/6-311++G(d,p) level of QM theory under normal conditions (see Figure 3).

TS of the Conformational Transition	μ_{TS}	ν_i	ΔG	ΔE	$\Delta\Delta G_{TS}$	$\Delta\Delta E_{TS}$	$\Delta\Delta G$	$\Delta\Delta E$	k_f	k_r	$\tau_{99.9\%}$	τ	HO3′C3′C4′/HO4′C4′C3′
$TS^{O3'H/O4'H}_{1\leftrightarrow7}$	2.97	457.1	0.92	1.09	7.50	7.99	6.59	6.90	$2.31\cdot10^7$	$1.09\cdot10^8$	$5.23\cdot10^{-8}$	$9.17\cdot10^{-9}$	$\pm12.5/\mp14.4$
$TS^{O3'H/O4'H}_{2\leftrightarrow3}$	3.73	470.5	0.07	-0.23	7.03	7.19	6.96	7.42	$5.16\cdot10^7$	$5.80\cdot10^7$	$6.30\cdot10^{-8}$	$1.72\cdot10^{-8}$	$\pm12.3/\mp14.1$
$TS^{O3'H/O4'H}_{4\leftrightarrow6}$	6.15	467.7	0.27	0.07	7.18	7.19	6.91	7.12	$4.04\cdot10^7$	$6.38\cdot10^7$	$6.63\cdot10^{-8}$	$1.57\cdot10^{-8}$	$\pm12.5/\mp14.3$
$TS^{O3'H/O4'H}_{5\leftrightarrow8}$	5.56	459.1	0.92	1.14	7.40	7.96	6.48	6.82	$2.74\cdot10^7$	$1.29\cdot10^8$	$4.41\cdot10^{-8}$	$7.73\cdot10^{-9}$	$\pm12.4/\mp14.3$
$TS^{O3'H/O4'H}_{13\leftrightarrow14}$	6.37	483.1	0.37	0.37	6.56	7.55	6.19	7.18	$1.16\cdot10^8$	$2.17\cdot10^8$	$2.07\cdot10^{-8}$	$4.61\cdot10^{-9}$	$\pm8.3/\mp9.3$
$TS^{O3'H/O4'H}_{15\leftrightarrow20}$	6.09	466.9	3.13	3.25	6.30	7.13	3.17	3.88	$1.79\cdot10^8$	$3.53\cdot10^{10}$	$1.95\cdot10^{-10}$	$2.83\cdot10^{-11}$	$\pm10.0/\mp12.3$
$TS^{O3'H/O4'H}_{16\leftrightarrow18}$	8.76	469.1	0.22	0.17	6.49	7.26	6.27	7.09	$1.29\cdot10^8$	$1.88\cdot10^8$	$2.18\cdot10^{-8}$	$5.31\cdot10^{-9}$	$\pm9.1/\mp11.3$
$TS^{O3'H/O4'H}_{17\leftrightarrow23}$	8.97	473.2	3.25	3.31	6.32	7.21	3.07	3.90	$1.73\cdot10^8$	$4.19\cdot10^{10}$	$1.64\cdot10^{-10}$	$2.39\cdot10^{-11}$	$\pm10.3/\mp11.2$
$TS^{O3'H/O4'H}_{19\leftrightarrow27}$	4.38	455.1	0.87	1.46	7.46	8.11	6.59	6.64	$2.50\cdot10^7$	$1.09\cdot10^8$	$5.17\cdot10^{-8}$	$9.20\cdot10^{-9}$	$\pm13.9/\mp15.6$
$TS^{O3'H/O4'H}_{21\leftrightarrow22}$	5.41	461.5	0.03	0.17	6.75	7.21	6.72	7.04	$8.30\cdot10^7$	$8.73\cdot10^7$	$4.06\cdot10^{-8}$	$1.15\cdot10^{-8}$	$\pm14.1/\mp15.6$
$TS^{O3'H/O4'H}_{25\leftrightarrow32}$	3.43	454.5	0.86	1.29	7.56	8.08	6.70	6.78	$2.11\cdot10^7$	$9.02\cdot10^7$	$6.21\cdot10^{-8}$	$1.11\cdot10^{-8}$	$\pm14.0/\mp15.8$
$TS^{O3'H/O4'H}_{28\leftrightarrow31}$	4.51	465.7	0.22	0.20	6.89	7.34	6.67	7.14	$6.56\cdot10^7$	$9.50\cdot10^7$	$4.30\cdot10^{-8}$	$1.05\cdot10^{-8}$	$\pm13.8/\mp15.3$
$TS^{O3'H/O4'H}_{37\leftrightarrow39}$	7.18	468.7	6.58	7.32	6.58	7.32	6.38	7.19	$1.11\cdot10^8$	$1.55\cdot10^8$	$2.60\cdot10^{-8}$	$6.46\cdot10^{-9}$	$\pm12.3/\mp13.0$
$TS^{O3'H/O4'H}_{38\leftrightarrow40}$	6.91	470.9	0.33	0.33	6.56	7.36	6.23	7.03	$1.14\cdot10^8$	$2.01\cdot10^8$	$2.19\cdot10^{-8}$	$4.98\cdot10^{-9}$	$\pm9.6/\mp11.7$
$TS^{O3'H/O4'H}_{41\leftrightarrow42}$	4.74	467.2	0.30	0.25	6.56	7.33	6.26	7.07	$1.15\cdot10^8$	$1.90\cdot10^8$	$2.27\cdot10^{-8}$	$5.26\cdot10^{-9}$	$\pm10.7/\mp12.9$
$TS^{O3'H/O4'H}_{43\leftrightarrow44}$	5.12	476.2	0.37	0.37	4.41	5.29	4.04	4.92	$4.37\cdot10^9$	$8.13\cdot10^9$	$5.53\cdot10^{-10}$	$1.23\cdot10^{-10}$	$\pm10.6/\mp11.3$

For designations see Table 1.

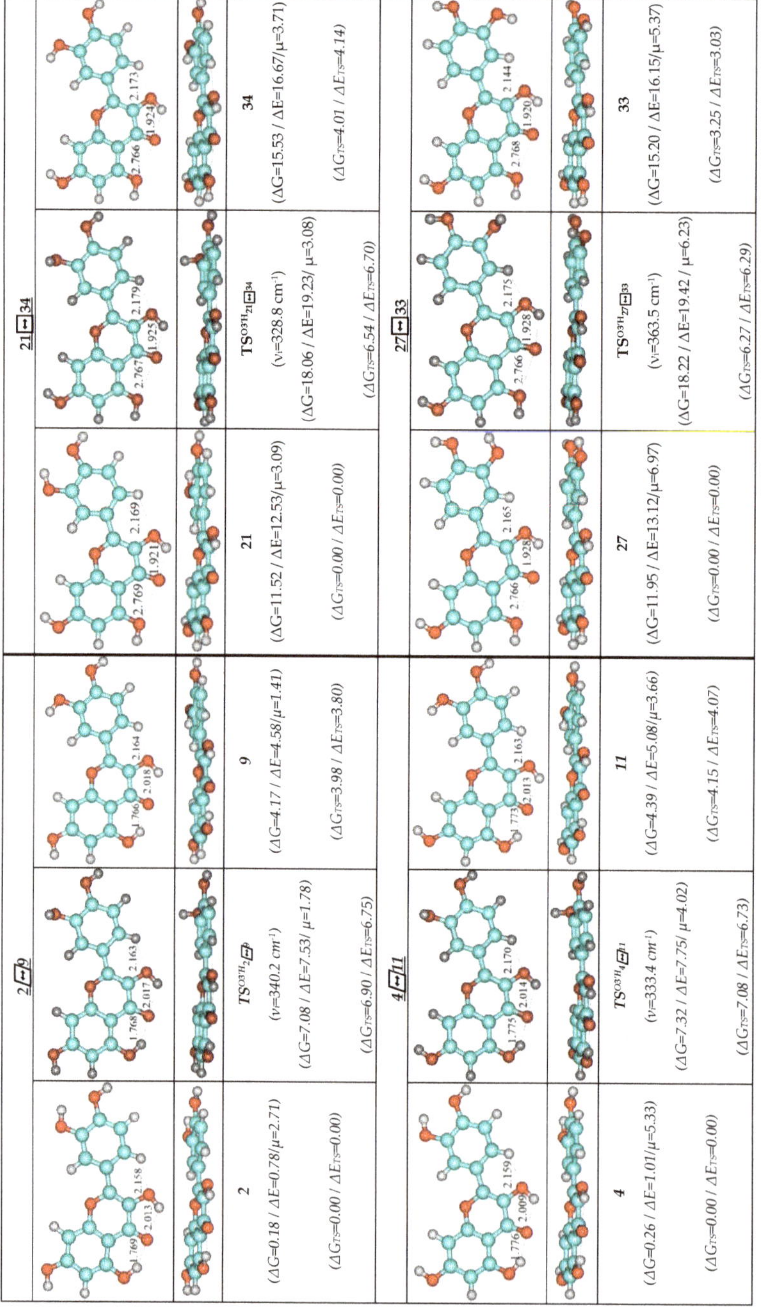

Figure 1. Cont.

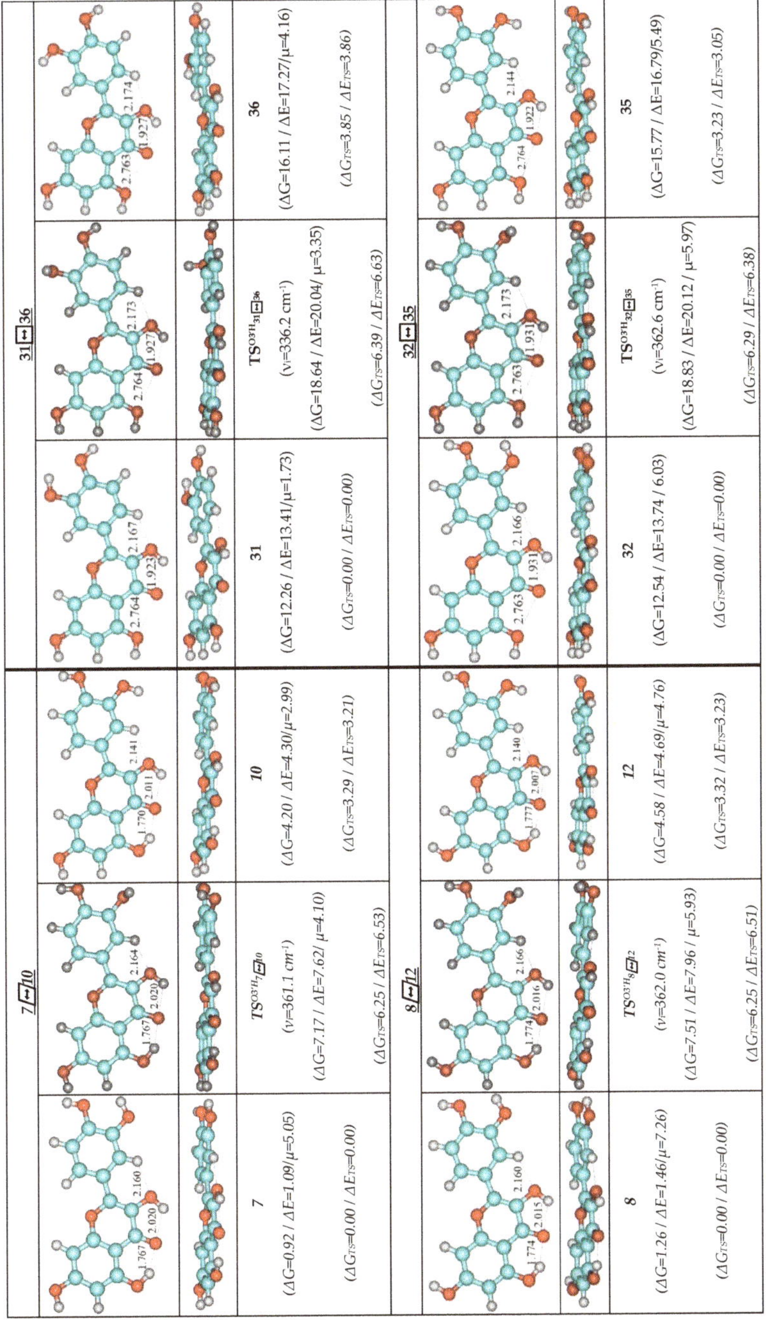

Figure 1. *Cont.*

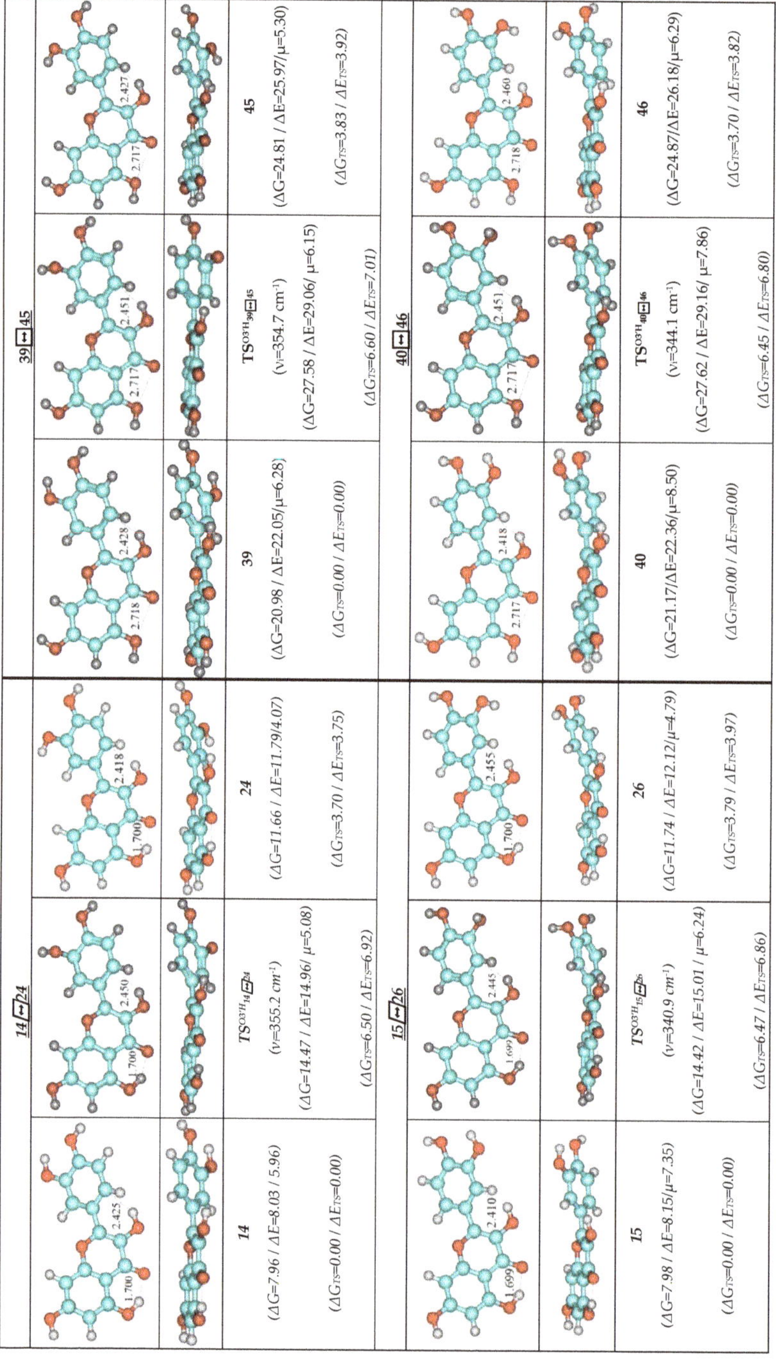

Figure 1. *Cont.*

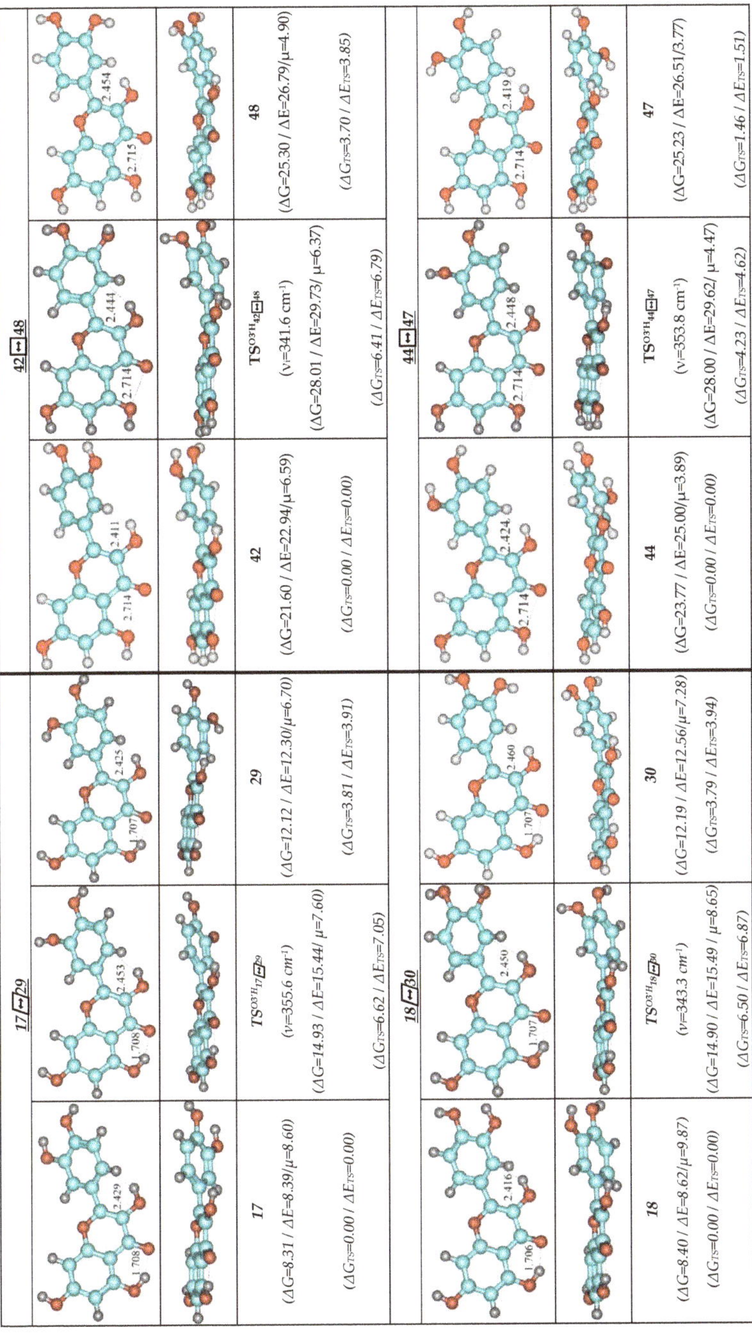

Figure 1. Geometrical structures of the quercetin molecule conformers and TSs with non-perpendicularly-oriented hydroxyl groups of their mutual interconversions via the mirror-symmetrical rotation of the O3'H hydroxyl group around the C3'-O3' bonds, obtained at the MP2/6-311++G(2df,pd)//B3LYP/6-311++G(d,p) level of QM theory under normal conditions. Relative Gibbs free ΔG and electronic ΔE energies (in kcal·mol^{-1}) (upper row represents energies relatively the conformer **1**, whereas the lower row presents the initial conformer for each transformation), dipole moments μ (Debye), and imaginary frequencies at TSs are provided at the MP2/6-311++G(2df,pd)//B3LYP/6-311++G(d,p) level of QM theory). Dotted lines indicate specific intramolecular contacts; their lengths are presented in Angstrom.

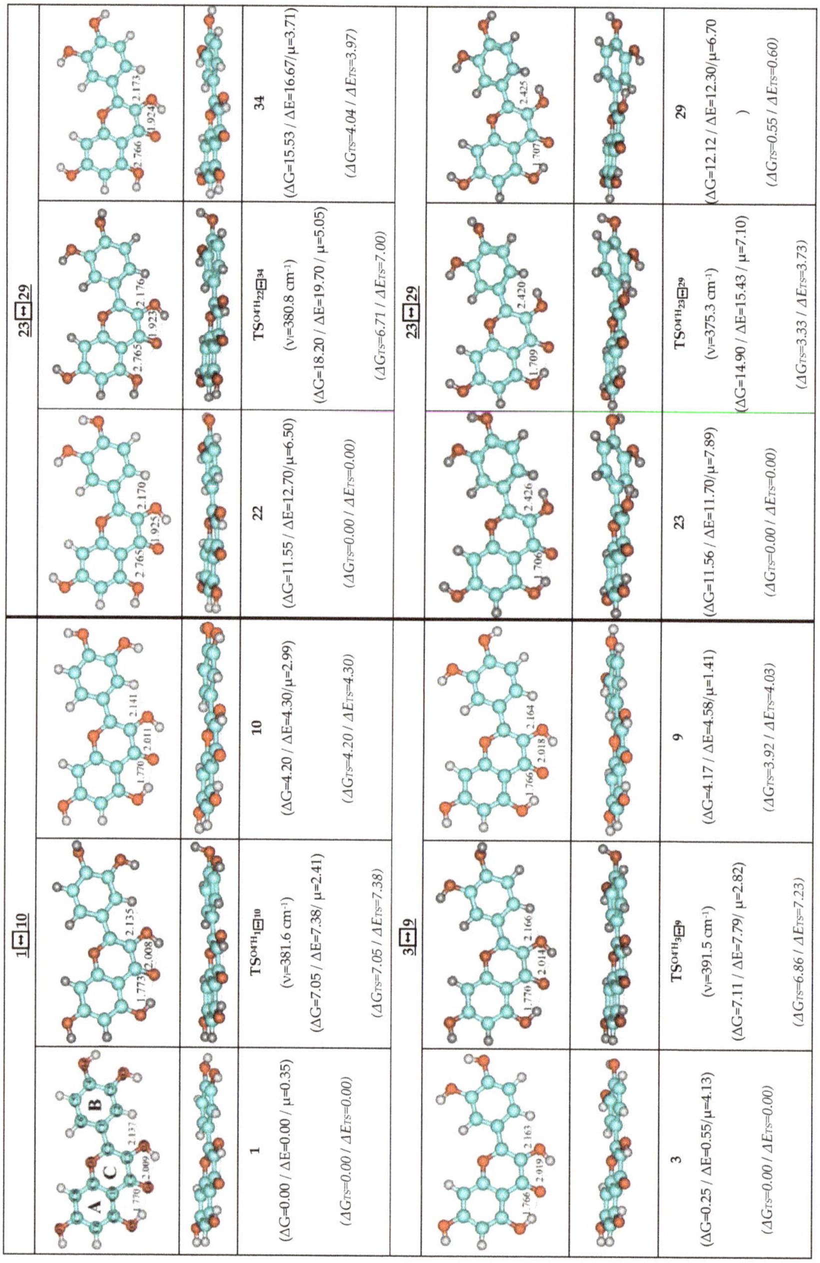

Figure 2. *Cont.*

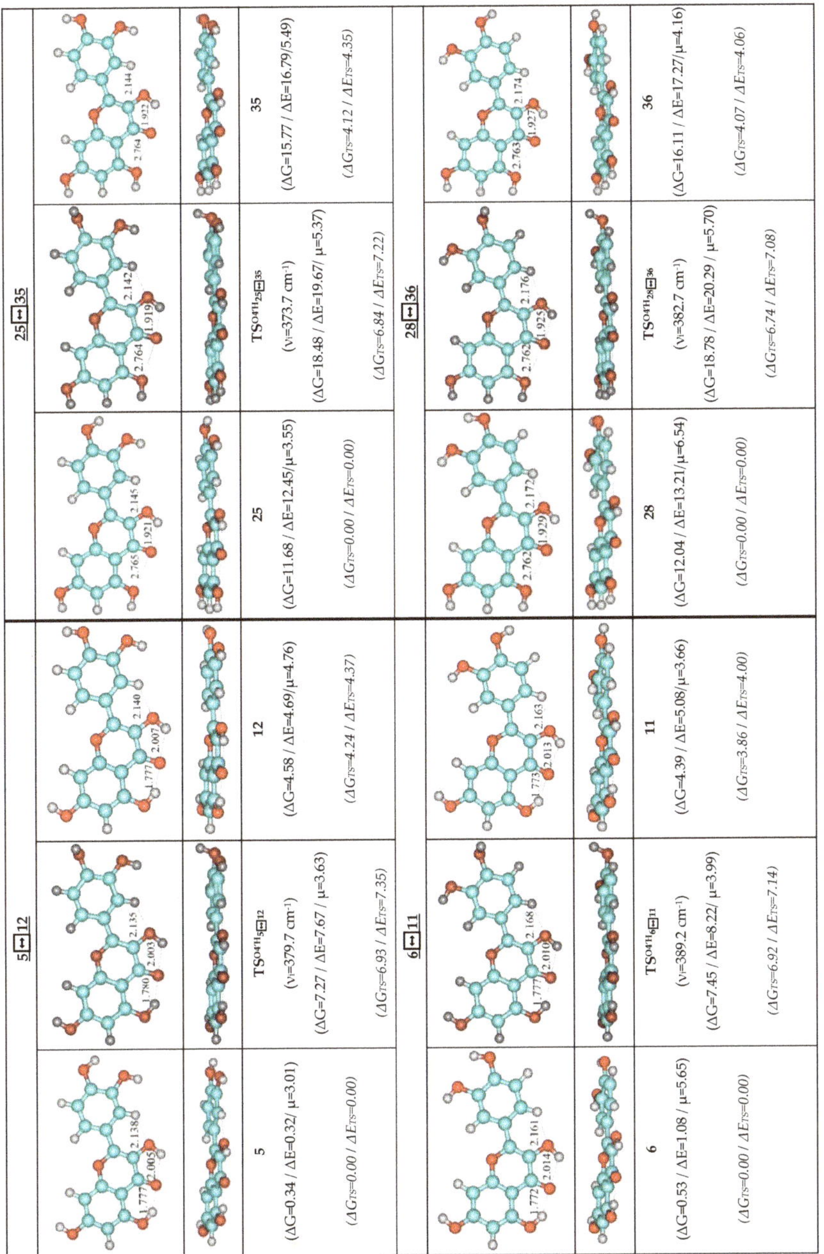

Figure 2. *Cont.*

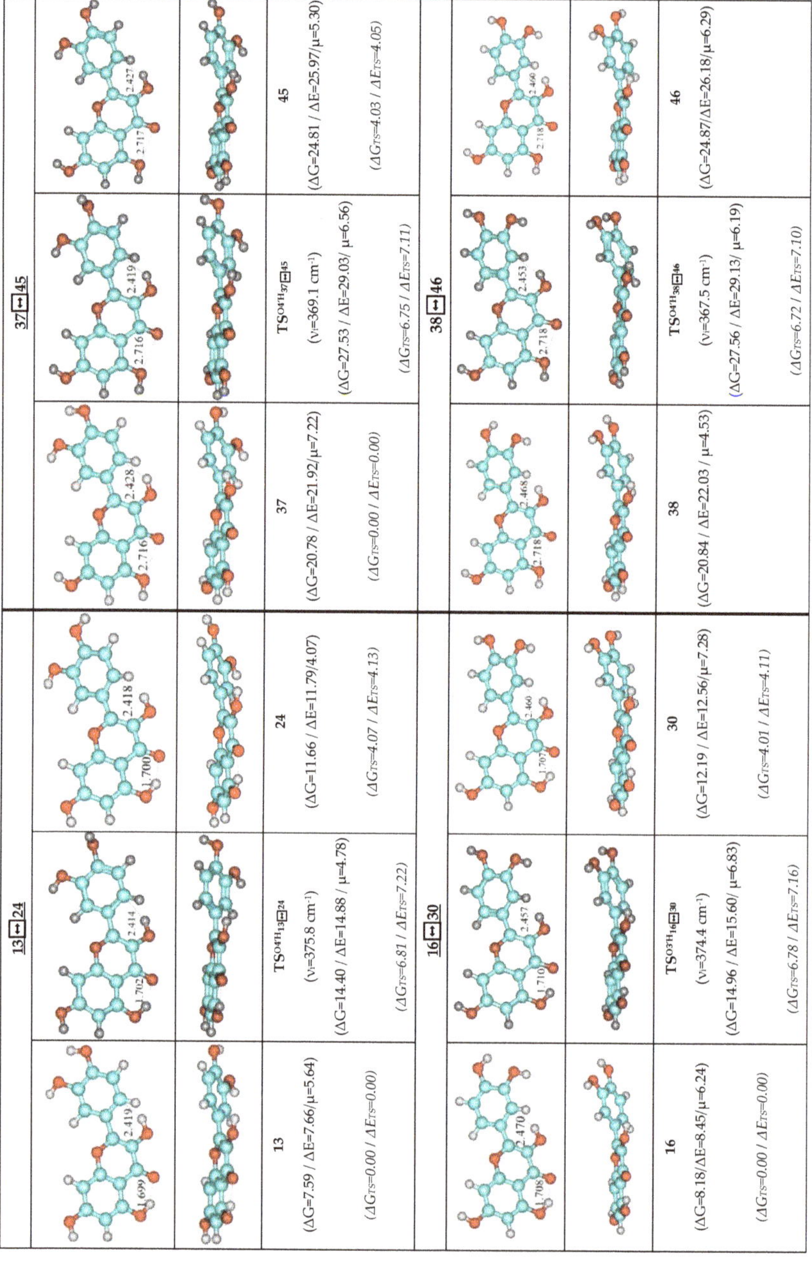

Figure 2. *Cont.*

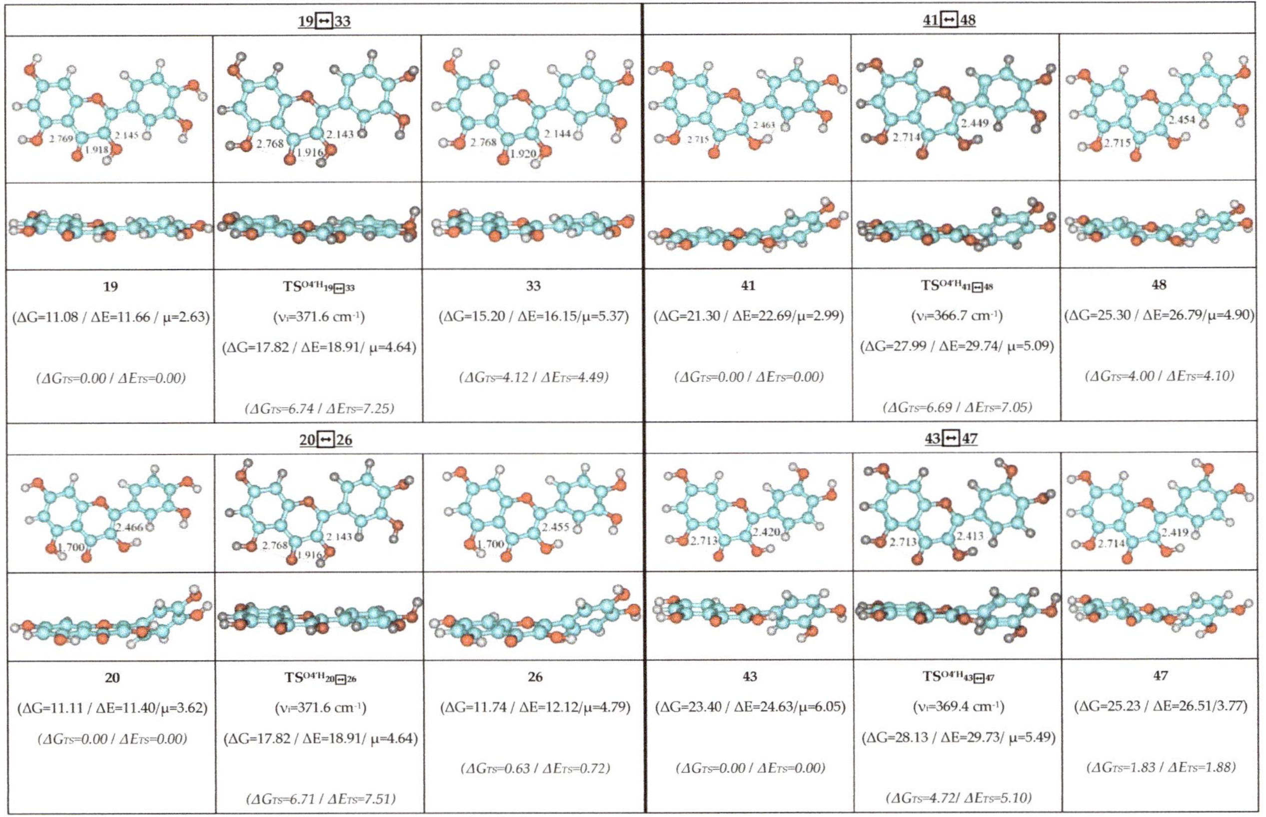

Figure 2. Geometrical structures of the quercetin molecule conformers and TSs with a non-perpendicularly-oriented hydroxyl groups of their mutual interconversions via the mirror-symmetrical rotation of the O4'H hydroxyl group around the C4'-O4' bond, obtained at the MP2/6-311++G(2df,pd)//B3LYP/6-311++G(d,p) level of QM theory under normal conditions. For designations see Figure 1.

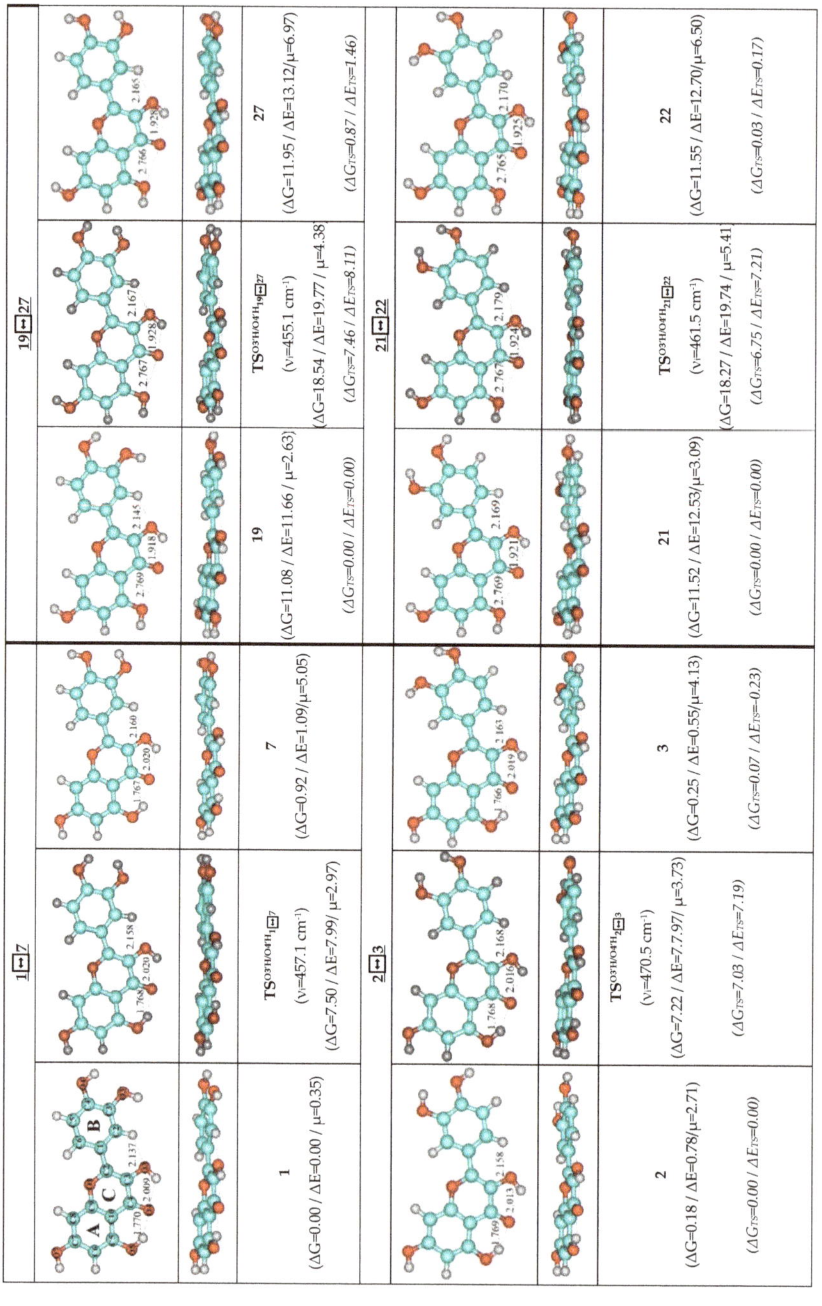

Figure 3. *Cont.*

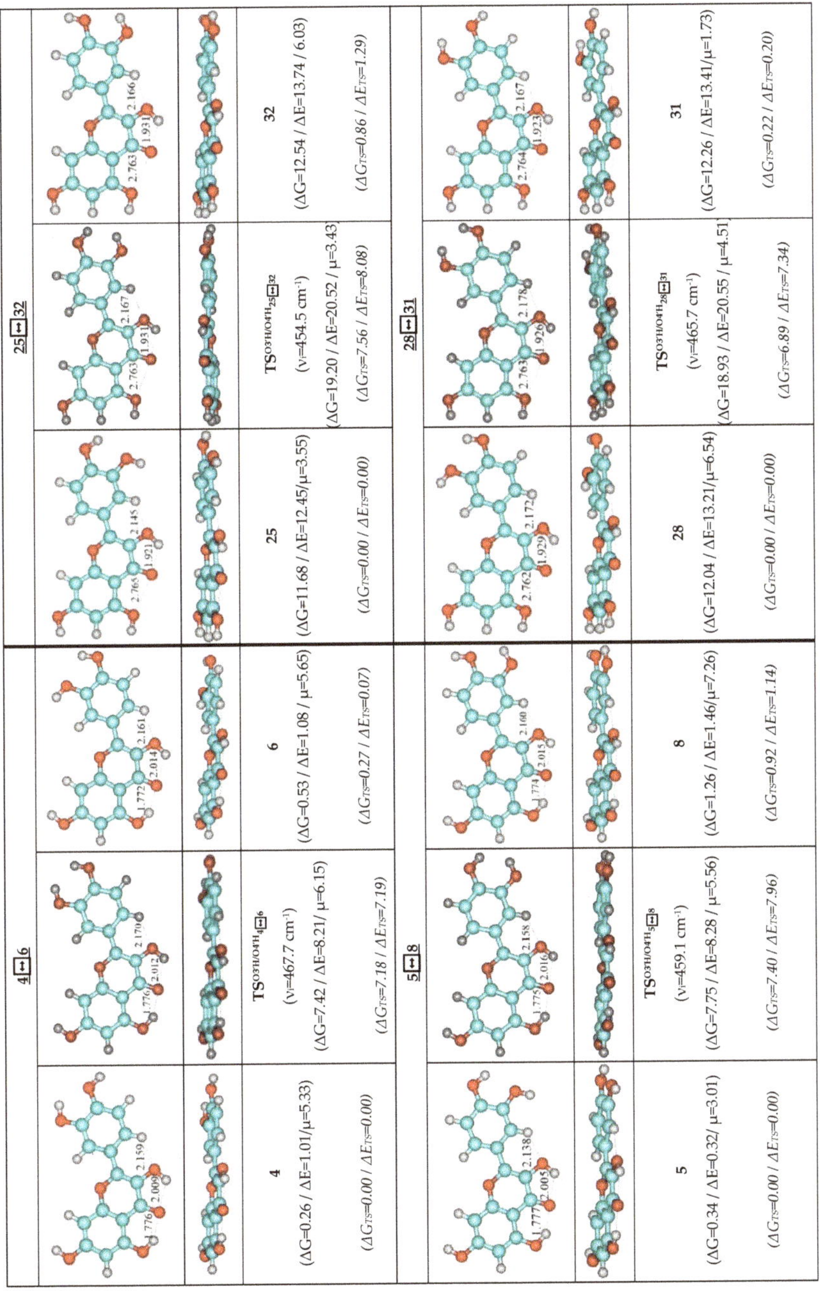

Figure 3. *Cont.*

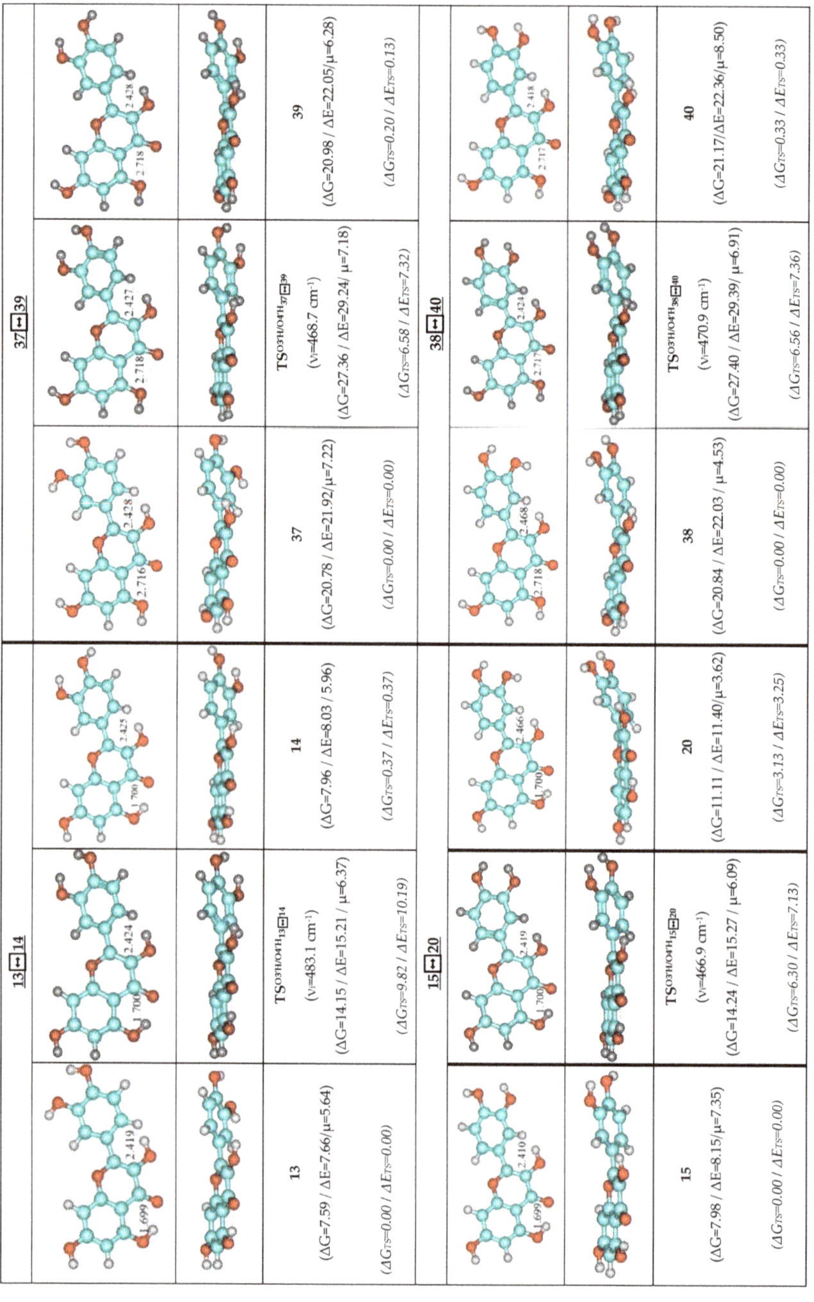

Figure 3. *Cont.*

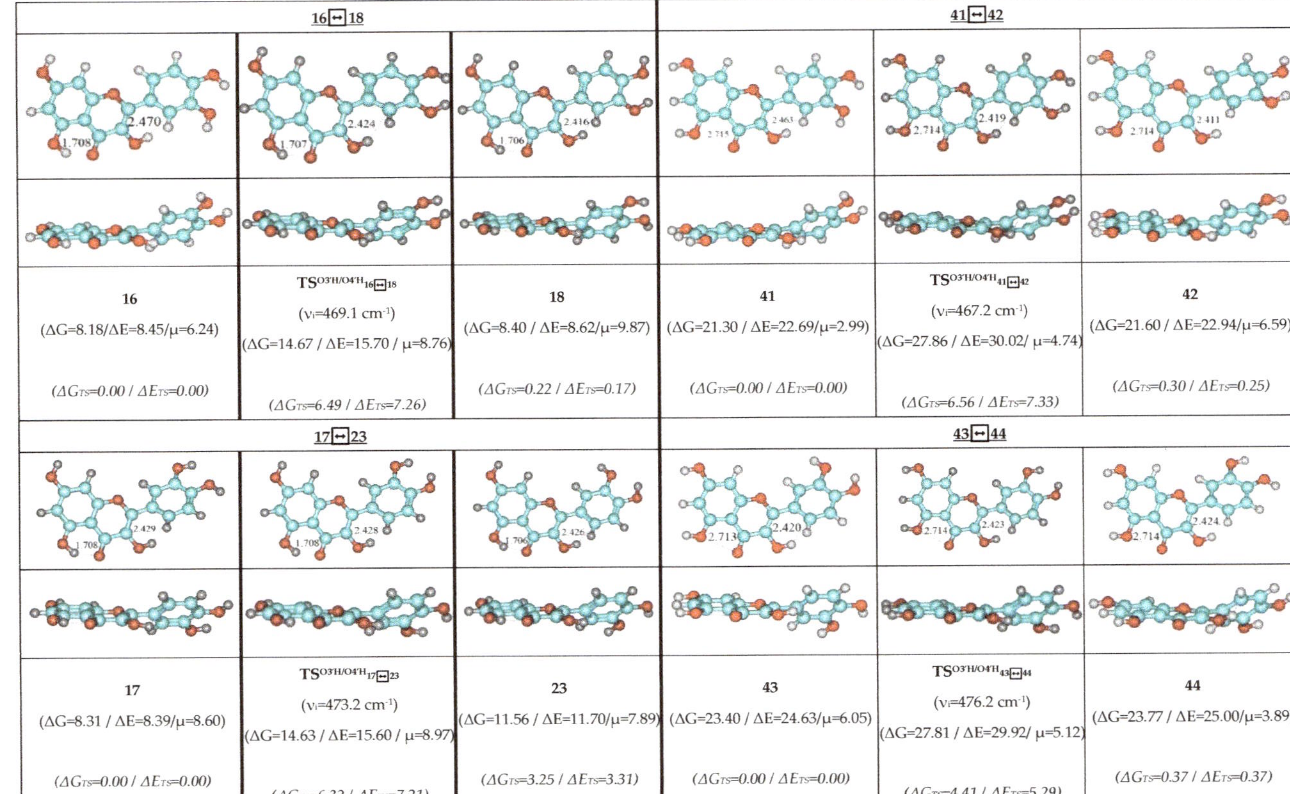

Figure 3. Geometrical structures of the quercetin molecule conformers and non-planar or locally non-planar TSs of their mutual concerted interconversions via the mirror-symmetrical rotation of the O3′H and O4′H hydroxyl groups around the C3′-O3′ and C4′-O4′ bonds, obtained at the MP2/6-311++G(2df,pd)//B3LYP/6-311++G(d,p) level of QM theory under normal conditions. For detailed designations see Figure 1.

In the case of the quercetin molecule, both of these tasks are overcomplicated. The reason is that the conformational mobility of this molecule is closely connected with its prototropic tautomerism [40,41]. It is known that the quercetin molecule has 202 molecular prototropic tautomers [40]. By contrast, now it is not known for sure all possible targets and their structure, despite all reasons to think that the range of this information would continuously grow together with the growing of the progress in bioinformatics and structural analysis. If also consider the conformationally-tautomeric variability of the targets, it would become clear that clarification of the structural grounds for the biological activity of the quercetin molecule is quite difficult task. We aimed to highlight this obstacle by the title of this paper.

4. Conclusions and Perspective for the Future Research

In this study, which is a logical continuation of our previous works on this topic [16–21], new pathways of the transformations of the conformers of the quercetin molecule into each other were found, which occurred due to the torsional mobility of the O3'H and O4'H hydroxyl groups.

It was established that the presence of only three degrees of freedom of the conformational mobility of the O3'H and O4'H hydroxyl groups is connected with their concerted behavior, which is controlled by the non-planar (in the case of the interconverting planar conformers) or locally non-planar (in other cases) $TSs^{O3'H/O4'H}$, in which O3'H and O4'H hydroxyl groups are oriented by the hydrogen atoms towards each other.

All these results assert that quercetin is a rather dynamical molecule, which is able to transform through the pathways into different conformers, forming complex networking.

We also shortly described the long-term perspectives for the investigation of the structural basis of the biological activity of quercetin.

Author Contributions: Setting of an idea of investigation, data preparation, analysis of the received results, writing and proofreading of the text of manuscript, references, Tables and Figures have been performed jointly by O.O.B. and D.M.H. All authors have read and agreed to the published version of the manuscript.

Funding: This research received no extra fuding.

References

1. Cianciosi, D.; Forbes-Hernández, T.Y.; Afrin, S.; Gasparrini, M.; Reboredo-Rodriguez, P.; Manna, P.P.; Zhang, J.; Lamas, L.B.; Flórez, S.M.; Toyos, P.A.; et al. Phenolic compounds in honey and their associated health benefits: A review. *Molecules* **2018**, *23*, 2322. [CrossRef] [PubMed]
2. Grytsenko, O.M.; Degtyarev, L.S.; Pilipchuck, L.B. Physico-chemical properties and electronic structure of quercetin. *Farmats. Zhurn.* **1992**, N2, 34–38.
3. Burda, S.; Oleszek, W. Antioxidant and antiradical activities of flavonoids. *J. Agric. Food Chem.* **2001**, *49*, 2774–2779. [CrossRef] [PubMed]
4. Grytsenko, O.M.; Pylypchuck, L.B.; Bogdan, T.V.; Trygubenko, S.A.; Hovorun, D.M.; Maksutina, N.P. Keto-enol prototropic tautomerism of quercetin molecule: Quantum-chemical calculations. *Farmats. Zhurn.* **2003**, N5, 62–65.
5. Olejniczak, S.; Potrzebowski, M.J. Solid state NMR studies and density functional theory (DFT) calculations of conformers of quercetin. *Org. Biomol. Chem.* **2004**, *2*, 2315–2322. [CrossRef]
6. Bentz, A.B. A review of quercetin: Chemistry, antioxidant properties, and bioavailability. *J. Young Investig.* **2009**, *19*, 1–14.
7. Tošović, J.; Marković, S.; Dimitrić Marković, J.M.; Mojović, M.; Milenković, D. Antioxidative mechanisms in chlorogenic acid. *Food Chem.* **2017**, *237*, 390–398. [CrossRef]
8. Nathiya, S.; Durga, M.; Devasena, T. Quercetin, encapsulated quercetin and its application—A review. *Int. J. Pharm. Pharm. Sci.* **2014**, *10*, 20–26.
9. David, A.V.A.; Arulmoli, R.; Parasuraman, S. Overviews of biological importance of quercetin: A bioactive flavonoid. *Pharmacogn. Rev.* **2016**, *10*, 84–89.

10. van Acker, S.A.; de Groot, M.J.; van den Berg, D.J.; Tromp, M.N.; Donné-Op den Kelder, G.; van der Vijgh, W.J.; Bast, A.A. A quantum chemical explanation of the antioxidant activity of flavonoids. *Chem. Res. Toxicol.* **1996**, *9*, 1305–1312. [CrossRef]

11. Bogdan, T.V.; Trygubenko, S.A.; Pylypchuck, L.B.; Potyahaylo, A.L.; Samijlenko, S.P.; Hovorun, D.M. Conformational analysis of the quercetin molecule. *Sci. Notes NaUKMA* **2001**, *19*, 456–460.

12. Trouillas, P.; Marsal, P.; Siri, D.; Lazzaroni, R.; Duroux, J.-C. A DFT study of the reactivity of OH groups in quercetin and taxifolin antioxidants: The specificity of the 3-OH site. *Food Chem.* **2006**, *97*, 679–688. [CrossRef]

13. Marković, Z.; Amić, D.; Milenković, D.; Dimitrić-Marković, J.M.; Marković, S. Examination of the chemical behavior of the quercetin radical cation towards some bases. *Phys. Chem. Chem. Phys.* **2013**, *15*, 7370–7378. [CrossRef] [PubMed]

14. Protsenko, I.O.; Hovorun, D.M. Conformational properties of quercetin: Quantum chemistry investigation. *Rep. Natl. Acad. Sci. Ukr.* **2014**, *N3*, 153–157. [CrossRef]

15. Vinnarasi, S.; Radhika, R.; Vijayakumar, S.; Shankar, R. Structural insights into the anti-cancer activity of quercetin on G-tetrad, mixed G-tetrad, and G-quadruplex DNA using quantum chemical and molecular dynamics simulations. *J. Biomol. Struct. Dyn.* **2019**. [CrossRef]

16. Brovarets', O.O.; Hovorun, D.M. Conformational diversity of the quercetin molecule: A quantum-chemical view. *J. Biomol. Struct. Dyn.* **2019**. [CrossRef]

17. Brovarets', O.O.; Hovorun, D.M. Conformational transitions of the quercetin molecule via the rotations of its rings: A comprehensive theoretical study. *J. Biomol. Struct. Dyn.* **2019**. [CrossRef]

18. Brovarets', O.O.; Hovorun, D.M. A hidden side of the conformational mobility of the quercetin molecule caused by the rotations of the O3H, O5H and O7H hydroxyl groups. In silico scrupulous study. *Symmetry* **2020**, *12*, 230.

19. Brovarets', O.O.; Protsenko, I.O.; Hovorun, D.M. Computational design of the conformational and tautomeric variability of the quercetin molecule. In Proceedings of the 6th Young Medicinal Chemist Symposium (EFMC-YMCS 2019, Athens, Greece, 5–6 September 2019; p. 50.

20. Brovarets', O.O.; Protsenko, I.O.; Hovorun, D.M. Comprehensive analysis of the potential energy surface of the quercetin molecule. In Proceedings of the "Bioheterocycles 2019" XVIII International Conference on Heterocycles in Bioorganic Chemistry, Ghent, Belgium, 17–20 June 2019; p. 84.

21. Brovarets', O.O.; Protsenko, I.O.; Zaychenko, G. Computational modeling of the tautomeric interconversions of the quercetin molecule. In Proceedings of the International Symposium EFMC-ACSMEDI Medicinal Chemistry Frontiers 2019, MedChemFrontiers 2019, Krakow, Poland, 10–13 June 2019; p. 114.

22. Filip, X.; Filip, C. Can the conformation of flexible hydroxyl groups be constrained by simple NMR crystallography approaches? The case of the quercetin solid forms. *Solid State Nucl. Magn. Reson.* **2015**, *65*, 21–28. [CrossRef]

23. Filip, X.; Miclaus, M.; Martin, F.; Filip, C.; Grosu, I.G. Optimized multi-step NMR-crystallography approach for structural characterization of a stable quercetin solvate. *J. Pharm. Biomed. Anal.* **2017**, *138*, 22–28. [CrossRef]

24. Tirado-Rives, J.; Jorgensen, W.L. Performance of B3LYP Density Functional Methods for a large set of organic molecules. *J. Chem. Theory Comput.* **2008**, *4*, 297–306. [CrossRef] [PubMed]

25. Parr, R.G.; Yang, W. *Density-Functional Theory of Atoms and Molecules*; Oxford University Press: Oxford, UK, 1989.

26. Lee, C.; Yang, W.; Parr, R.G. Development of the Colle-Salvetti correlation-energy formula into a functional of the electron density. *Phys. Rev. B* **1988**, *37*, 785–789. [CrossRef] [PubMed]

27. Frisch, M.J.; Trucks, G.W.; Schlegel, H.B.; Scuseria, G.E.; Robb, M.A.; Cheeseman, J.R.; Scalmani, G.; Barone, V.; Mennucci, B.; Petersson, G.A.; et al. *GAUSSIAN 09 (Revision B.01)*; Gaussian Inc.: Wallingford, CT, USA, 2010.

28. Brovarets', O.O.; Hovorun, D.M. Atomistic understanding of the C·T mismatched DNA base pair tautomerization via the DPT: QM and QTAIM computational approaches. *J. Comput. Chem.* **2013**, *34*, 2577–2590. [CrossRef] [PubMed]

29. Brovarets', O.O.; Tsiupa, K.S.; Hovorun, D.M. Non-dissociative structural transitions of the Watson-Crick and reverse Watson-Crick A·T DNA base pairs into the Hoogsteen and reverse Hoogsteen forms. *Sci. Rep.* **2018**, *8*, 10371.

30. Palafox, M.A. Molecular structure differences between the antiviral nucleoside analogue 5-iodo-2′-deoxyuridine and the natural nucleoside 2′-deoxythymidine using MP2 and DFT methods: Conformational analysis, crystal simulations, DNA pairs and possible behavior. *J. Biomol. Struct. Dyn.* **2014**, *32*, 831–851. [CrossRef]

31. Hratchian, H.P.; Schlegel, H.B. Finding minima, transition states, and following reaction pathways on *ab initio* potential energy surfaces. In *Theory and Applications of Computational Chemistry: The First 40 Years*; Dykstra, C.E., Frenking, G., Kim, K.S., Scuseria, G., Eds.; Elsevier: Amsterdam, The Netherlands, 2005; pp. 195–249.

32. Vásquez-Espinal, A.; Yañez, O.; Osorio, E.; Areche, C.; García-Beltrán, O.; Ruiz, L.M.; Cassels, B.K.; Tiznado, W. Theoretical study of the antioxidant activity of quercetin oxidation products. *Front. Chem.* **2019**, *7*, 818. [CrossRef]

33. Frisch, M.J.; Head-Gordon, M.; Pople, J.A. Semi-direct algorithms for the MP2 energy and gradient. *Chem. Phys. Lett.* **1990**, *166*, 281–289. [CrossRef]

34. Hariharan, P.C.; Pople, J.A. The influence of polarization functions on molecular orbital hydrogenation energies. *Theor. Chim. Acta* **1973**, *28*, 213–222. [CrossRef]

35. Krishnan, R.; Binkley, J.S.; Seeger, R.; Pople, J.A. Self-consistent molecular orbital methods. XX. A basis set for correlated wave functions. *J. Chem. Phys.* **1980**, *72*, 650–654. [CrossRef]

36. Atkins, P.W. *Physical Chemistry*; Oxford University Press: Oxford, UK, 1998.

37. Wigner, E. Über das Überschreiten von Potentialschwellen bei chemischen Reaktionen [Crossing of potential thresholds in chemical reactions]. *Zeits. Physik. Chem.* **1932**, *B19*, 203–216.

38. Keith, T.A. AIMAll (Version 10.07.01). 2010. Available online: aim.tkgristmill.com (accessed on 7 January 2020).

39. Matta, C.F.; Hernández-Trujillo, J. Bonding in polycyclic aromatic hydrocarbons in terms of the electron density and of electron delocalization. *J. Phys. Chem. A* **2003**, *107*, 7496–7504. [CrossRef]

40. Brovarets', O.O.; Hovorun, D.M. A new era of the prototropic tautomerism of the quercetin molecule: A QM/QTAIM computational advances. *J. Biomol. Struct. Dyn.* **2019**. [CrossRef]

41. Brovarets', O.O.; Hovorun, D.M. Intramolecular tautomerization of the quercetin molecule due to the proton transfer: QM computational study. *PLoS ONE* **2019**, *14*, e0224762. [CrossRef] [PubMed]

Article

Probing the Structure of [NiFeSe] Hydrogenase with QM/MM Computations

Samah Moubarak, N. Elghobashi-Meinhardt *, Daria Tombolelli and Maria Andrea Mroginski

Technische Universität Berlin, 135 Strasse-des-17, Juni, 10623 Berlin, Germany;
s.moubarak@campus.tu-berlin.de (S.M.); daria.tombolelli@campus.tu-berlin.de (D.T.);
andrea.mroginski@tu-berlin.de (M.A.M.)
* Correspondence: n.elghobashi-meinhardt@campus.tu-berlin.de

Received: 30 December 2019; Accepted: 18 January 2020; Published: 22 January 2020

Abstract: The geometry and vibrational behavior of selenocysteine [NiFeSe] hydrogenase isolated from *Desulfovibrio vulgaris* Hildenborough have been investigated using a hybrid quantum mechanical (QM)/ molecular mechanical (MM) approach. Structural models have been built based on the three conformers identified in the recent crystal structure resolved at 1.3 Å from X-ray crystallography. In the models, a diamagnetic Ni^{2+} atom was modeled in combination with both Fe^{2+} and Fe^{3+} to investigate the effect of iron oxidation on geometry and vibrational frequency of the nonproteic ligands, CO and CN-, coordinated to the Fe atom. Overall, the QM/MM optimized geometries are in good agreement with the experimentally resolved geometries. Analysis of computed vibrational frequencies, in comparison with experimental Fourier-transform infrared (FTIR) frequencies, suggests that a mixture of conformers as well as Fe^{2+} and Fe^{3+} oxidation states may be responsible for the acquired vibrational spectra.

Keywords: [NiFeSe] hydrogenase; quantum mechanics (QM)/molecular mechanics (MM), geometry optimizations; vibrational frequency analyses; Fourier transform infrared (FTIR) frequencies

---

1. Introduction

The present work addresses the structure of the active site of recombinant [NiFeSe] hydrogenase isolated from *Desulfovibrio vulgaris* Hildenborough [1]. The [NiFeSe] hydrogenase represents, in addition to the [NiFeS] hydrogenase, one of the two subclasses of [NiFe] hydrogenases. Hydrogenases are metalloenzymes with an active site that contains, under a reducing atmosphere, iron, nickel and sulfur as well as diatomic ligands like CN and CO, which catalyze the reversible oxidation of molecular hydrogen,

$$H_2 \leftrightarrow 2\,H^+ + 2\,e^-.$$

Unlike [FeFe] hydrogenases, which are irreversibly inactivated after exposure to O_2, [NiFe] hydrogenases can be reductively reactivated after exposure to O_2 [1]. Due to their potential of producing hydrogen, these enzymes represent a design template for molecular catalysts being applied in electrolyzers, which use electricity to split water into hydrogen and oxygen, as well as in fuel cells [2]. Nonetheless, as O_2 acts as an oxidant that competes with protons, the hydrogenase's protein scaffold protects the redox centers from high O_2 concentrations that would inhibit H_2 production at the active site [2].

Representing the predominant form, the [NiFeS] hydrogenase contains the catalytic Ni-Fe center, bound by two bridging cysteines and surrounded by two additional cysteine ligands on the Ni site and two CN- and one CO ligands coordinated to the Fe site. The [NiFeSe] hydrogenase is structured analogously, with the exception that a selenocysteine (Sec489) replaces a terminal cysteine (Figure 1A).

Three iron-sulfur clusters (proximal, medial, and distal) form an electron transfer path connecting the catalytic center to the protein surface (Figure 1B–E).

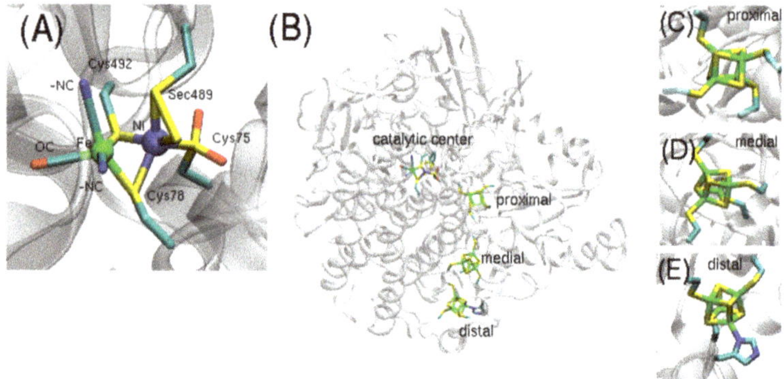

Figure 1. (**A**) The catalytic site of [NiFeSe] hydrogenase contains the Fe-Ni center (shown as green and blue spheres, respectively) coordinated to the side chains of Cys75, Cys78, and Cys492, and to Sec489. Three nonproteic ligands (two CN-ligands and one CO ligand) are coordinated to Fe. (**B**) Three [Fe₄S₄] clusters, (**C**) proximal, (**D**) medial, and (**E**) distal, form an electron transfer path connecting the catalytic center to the protein surface (protein depicted in white ribbon representation). Structure taken from PDB: 5JSH, resolution 1.3 Å [1].

In general, the activity of selenoproteins is usually higher than that of sulfur-containing protein homologues due to the lower pKa and higher nucleophilicity of the selenol group compared to that of the thiol group [3,4]. Selenium-containing cysteine (Se-Cys) is more acidic than sulfur-containing cysteine (S-Cys), and the high nucleophilicity of selenium leads to a higher O_2 tolerance of [NiFeSe] hydrogenase. The insertion of selenocysteine requires intricate biosynthetic machinery and is energetically more costly than cysteine insertion, so [NiFeSe] hydrogenases have evolved in only a few microorganisms, such as in *Desulfovibrio vulgaris* and *Methanococcus voltae* [2].

The enzymes belonging to the subfamily of [NiFeSe] hydrogenases are special because they have a fast H_2 production rate [5–7]. Compared to [NiFeS] hydrogenases, the Se enzyme records a higher H_2 production activity [8]. Indeed, if Se is available to the microorganism, a preferred expression of NiFeSe over NiFeS hydrogenase is observed [8]. Importantly, the recombinant expression system developed by Marques et al. is a step toward engineering large quantities of these valuable enzymes [1].

An earlier, lower-resolution (2.04 Å) structure of aerobically crystallized [NiFeSe] hydrogenase revealed the oxidized, "as-isolated" active site that lacks an oxide bridging ligand [9]. More recently, high-resolution (0.95–1.4 Å) crystal structures of [NiFeSe] hydrogenase have revealed three conformers of the active site that differ not only in the arrangement, but also in the number of sulfur atoms (Figure 2).

In the recent, high-resolution structures, conformers I and II contain a sulfur atom bound between the nickel atom and the selenocysteine unit (Figure 2A,B), while conformer III does not contain this additional sulfur atom (Figure 2C). In contrast, the nickel atom of conformer III is only bound to a selenocysteine. Conformer I and II, in turn, differ in their binding geometries of the Ni-S-Se-Cys unit. While in conformer II, the additional sulfur atom is directly inserted between the Ni-Se-Cys bond compared to conformer III, and in conformer I, the nickel atom forms a direct bond to the additional sulfur atom as well as to the selenocysteine unit. In addition, the sulfur and selenium atoms are also directly bound to each other.

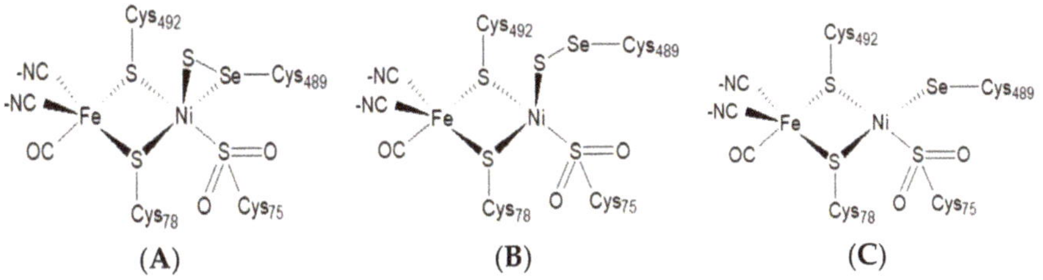

Figure 2. The three conformers of *D. vulgaris* [NiFeSe] hydrogenase (PDB: 5JSH) [1] differ in the connectivities of Cys489 with the Ni atom and in the number of sulfur atoms. Cys489 of conformer III does not contain a sulfur atom. (**A**) Conformer I, (**B**) conformer II, and (**C**) conformer III.

The structural and mechanistic details of [NiFeSe] hydrogenase have been investigated using a range of spectroscopic methods including X-ray diffraction (XRD) [2], Fourier transform infrared spectroscopy (FTIR) [2,10], resonance Raman (RR) [2], electron paramagnetic resonance (EPR) [1], circular dichroism (CD), UV/Vis [5], and gas chromatography-mass spectrometry GC-MS [1]. RR experiments indicate various redox dependent states, giving insight into hydrogen production as well as oxygen tolerance [2]. The UV/Vis spectrum of the [NiFeSe] hydrogenase is very similar to that of the [NiFe] hydrogenase, showing the characteristic broad absorption centered at 400 nm due to the oxidized FeS centers [5]. [NiFeSe] hydrogenase is nickel-EPR silent in the oxidized state [1], indicating that the enzyme is in a Ni(II) state that is contrary to most oxidized [NiFe] hydrogenases, in which nickel is in a paramagnetic Ni(III) spin state. GC experiments [1] indicate that H_2 production of [NiFeSe] hydrogenases is higher than that of [NiFeS] hydrogenases [5].

Particularly, FTIR spectroscopy has provided insight into the electronic structure and coordination pattern of the catalytic metal ions of hydrogenases. As the frequencies of the nonproteic ligands (CO, CN-) are highly sensitive to the changes in electronic environment, particularly at the Fe locus, FTIR spectra can help elucidate the electronic states governing the catalytic events. In general, for a given redox state, the frequencies of CO bands of [NiFeSe] are lower than those of equivalent redox states of standard sulfur-Ni-Fe hydrogenases [11]. This behavior is rationalized by the lower electronegativity of the Se atom relative to the S atom present in standard Ni-Fe hydrogenases; an increased π-donating effect from the *d* orbitals of Se to the $2p\pi^*$ orbital of the CO ligand is also present [12]. Also, the FTIR spectra of [NiFeSe] show CN-bands that are comparable in intensity with CO bands, suggesting a high π-mobility in the CN-ligands due to hydrogen bond coordination with atoms of surrounding amino acids [10].

In addition to a range of experimental studies, computational studies have also attempted to shed light on the structure and chemistry of [NiFe] hydrogenases with FeS active sites [13–15]. Hybrid quantum mechanical and molecular mechanical (QM/MM) applications have been particularly invaluable for probing the connection between the electronic structure and function in these complex enzymatic systems [16–18]. The combination of QM methodology, which describes the chemically active region involving charge-transfers or bond-formation, and MM methods to treat the bulk protein and solvent environment is a computationally efficient method for describing large biomolecules [16]. In the case of hydrogenase, this QM/MM approach has been shown to accurately describe experimentally measured vibrational frequencies [13–15].

The present work reports on QM/MM structural and vibrational frequency computations to analyze the three proposed [NiFeSe] conformers from the structures of Marques et al. [1]. For each of the three conformers, both Fe^{2+} and Fe^{3+} oxidization states have been investigated, providing for a total of six models. For the QM/MM optimized geometries of the six species, the IR spectra have been calculated and the vibrational frequencies for the CO and CN-ligands have been analyzed and compared to the experimental values. In the next section, the preparation of the computational models and the QM/MM methodology will be described, and then they will be followed by a discussion of the

QM/MM optimized geometries of each conformer. Finally, computed vibrational frequencies based on the QM/MM optimized geometries will be presented and analyzed.

2. Methods

Three conformers of [NiFeSe] hydrogenase from *D. vulgaris* Hildenborough were observed in the crystallographic data from PDB ID: 5JSH based on an oxidized species [1]. A model was therefore built for each of the three conformers by selecting the coordinates of the active site of each of the three conformers, which will henceforth be referred to as conformer I, II, and III (Figures 2 and 3). These conformers differ not only in the number of sulfur atoms and their bonding patterns at the Ni site, but also in the geometry of the selenocysteine.

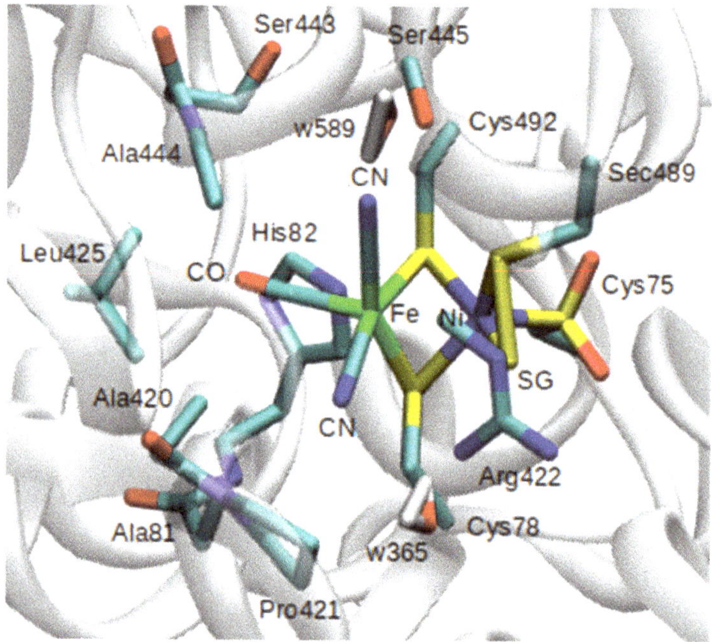

Figure 3. The positions of atoms optimized with quantum mechanical (QM) methodology include the Fe-Ni active site and the surrounding amino acids shown in color (carbon atoms in cyan, nitrogen atoms in blue, oxygen atoms in red, and sulfur and selenium atoms in yellow). Hydrogen atoms of protein are omitted for clarity.

Hydrogen atoms were modeled using CHARMM with the charmm36 force field [19,20]. His82 was protonated at position Nε due to possible interactions with the Fe[CN]$_2$CO unit. Unless otherwise specified, protonation of amino acid side chains was according to the standard assignment for pH 7. Missing atoms were modeled with CHARMM using the charmm36 force field [19,20].

The entire protein, including the catalytic center and three [Fe$_4$S$_4$] clusters, and crystal water molecules were then solvated in a water box with sodium chloride ions (0.1 M). Partial charges for the three [Fe$_4$S$_4$] clusters were adapted from previous work [13–15]. Energy minimizations were performed to optimize the hydrogen bond networks at the protein-solvent interface. In a final step, all protein atoms, crystal water, solvent water, and salt ions within a radius of 40 Å centered on the Ni atom were selected to comprise the complete model (Figure 4A).

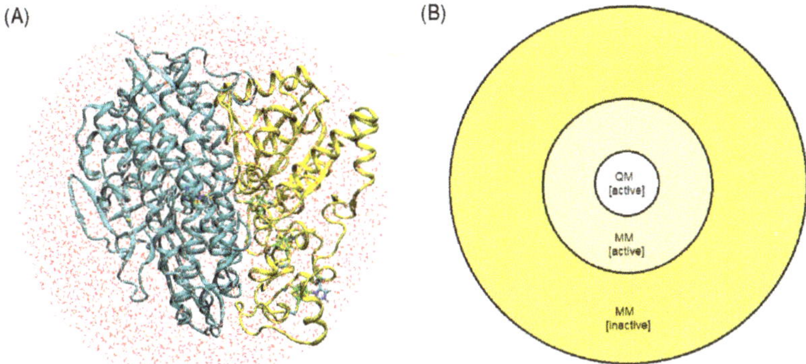

Figure 4. (**A**) The final model of *D. vulgaris* [NiFeSe] hydrogenase (PDB: 5JSH [1]) includes the two protein chains (blue and yellow ribbon representation), catalytic center, and three [Fe$_4$S$_4$] clusters, all solvated in an aqueous sphere (radius 40 Å) centered on the Ni atom of the active site. (**B**) Quantum mechanical (QM)/ molecular mechanical (MM) geometry optimizations are based on a division of the enzyme into three regions: the QM region that includes all atoms within a 6 Å radius around the central Ni atom, the MM active region that includes all non-QM atoms within a 15 Å radius around central Ni atom, and the MM inactive region that contains all remaining protein and water molecules from the crystal structure.

The QM/MM geometry optimization of each of the six models was carried out using a model that divides the enzyme into three regions: QM, MM active, and MM inactive (Figure 4B). Only the positions of atoms within the QM and MM active regions were optimized, whereas atoms in the MM inactive region were held fixed. Atoms included in the QM region are all atoms involved in the Fe-Ni catalytic center, as well as atoms belonging to amino acids Ala81, Ala420, Ala444, Arg422, His82, and Leu425.

Pro421, Ser443, and Ser445, two water molecules (w365 and w589) from the crystal structure, were also included in the QM region (Figure 3). Atoms belonging to protein and crystal water (oxygen) within a 15 Å radius of the Ni atom were considered as part of the MM active region.

Atoms in the QM region were treated with the Density Functional Theory (DFT) and the BP86 functional with the resolution of identity (RI) technique, as implemented in Turbomole6.2 [21–23]. A mixed basis set of def2-TZVP (Ni, Fe, S, and Se atoms) and 6-31G* (C, N, O, and H atoms) was used; this combination of basis sets has been tested and shown to be appropriate for geometry optimizations and vibrational frequency analyses on such enzymes [14].

Based on EPR studies indicating that [NiFeSe] hydrogenase is nickel-EPR silent in the oxidized state [1], the computations presented here treat Ni in the 2+ oxidation state. Furthermore, a common feature of hydrogenases is the presence in the active site of a low-spin Fe coordinated by CO- and CN-ligands [24,25]. Therefore, the electronic states considered in the models are based on [Ni^{2+}Fe^{2+}] and [Ni^{2+}Fe^{3+}] configurations. For Fe^{2+}, the total charge of the QM system is −1 for all three conformers and the total spin is S = 0 (multiplicity 1). For Fe^{3+}, the total charge of the system is 0 for all three conformers, giving a total spin of S = 1/2 and a spin multiplicity of 2.

The energetic coupling between the first and second layers (QM and mobile MM) was treated using electrostatic embedding with a charge-shift scheme as implemented in ChemShell [26–28]. Within the MM region, the charmm36 force field was used to describe all bonding and non-bonding interactions [19]. Any covalent bonds traversing the QM-MM border were cut and hydrogen link-atoms were used to satisfy valency. Vibrational frequency analyses and IR spectra were carried out with Turbomole6.2 [22,23] using the normal mode analysis approximation [29].

3. Results and Discussion

The QM/MM optimized geometries of each conformer are shown in Figure 5 for both Fe^{2+} (gray) and Fe^{3+} (orange) superimposed on the oxidized structure determined with X-ray crystallography at 1.3 Å resolution [1].

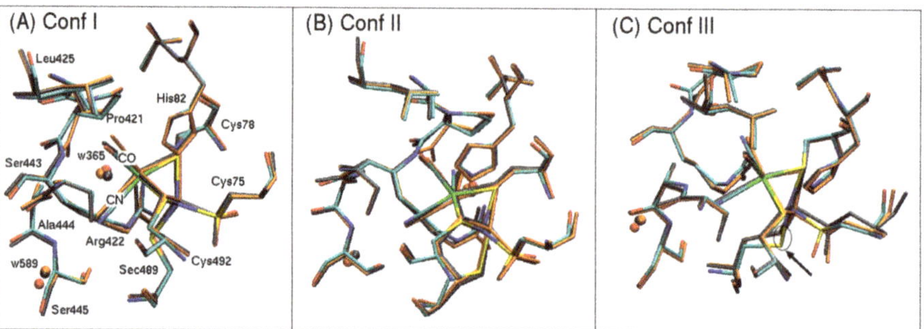

Figure 5. The QM/MM optimized geometry of the atoms included in the QM region of conformers I, II, and III for Fe^{2+} (atoms shown in gray) and Fe^{3+} (atoms shown in orange) are compared; the optimized geometries are superimposed on the geometries obtained from the crystal structure (PDB ID 5JSH [1]) (atoms shown in color: nickel atom is depicted in light blue, iron atom in green, sulfur and selenium atoms in yellow, carbon atoms in light blue, oxygen atoms in red, and nitrogen atoms in dark blue; hydrogen atoms are omitted for clarity).

In general, the QM/MM optimized geometries reproduce well the structures resolved with X-ray crystallography [1]. Bond lengths and relative orientations of atoms are well preserved. The positions of oxygen atoms from crystal water w365 and w589 are also preserved in each model, suggesting that the methodology is suited to describe the overall electronic environment. Bond lengths of atoms coordinated to a nickel or iron atom are reported in Tables 1 and 2 for $[Ni^{2+}Fe^{2+}Se]$ and $[Ni^{2+}Fe^{3+}Se]$, respectively; for comparison, the bond lengths obtained from the crystal structure (PDB ID: 5JSH from Marques et al. [1]) are listed in bold print. Ni-S lengths for the coordinated cysteine thiolates are well reproduced, but Fe-Ni distances in all three conformers are overestimated by 0.1–0.2 Å compared to the crystal structure. Particularly, conformer III demonstrates an overestimation in the Fe-Ni bond length for both Fe^{2+} and Fe^{3+} oxidation states (2.61 Å and 2.76 Å, respectively, versus the experimental value of 2.41 Å). This Fe-Ni separation is accompanied by a rotation of the side chain of Sec489, which leads to a 1 Å shift in the position of the selenium atom (see arrow pointing to Sec489 in Figure 5C). This overestimation in bond lengths is also observed for Ni-S and Ni-Se bonds for conformers I and II. For the oxidized species of $[Ni^{2+}Fe^{3+}Se]$, reasonable agreement is observed between the computed Fe-Ni bond length of conformer II (2.45 Å) and the experimentally reported value of 2.41 Å (Table 2). Conformers I and III overestimate the Fe-Ni distance by more than 0.1 Å and 0.34 Å, respectively, compared to the experimentally measured values [1].

The discrepancies between the computed and measured bond lengths suggest that either the assumed electronic state of the catalytic site may not accurately reflect the state what is captured in the crystal structure or that the assignment of atomic positions may be ambiguous due to insufficient experimental resolution. Here, we have treated the electronic state of nickel as 2+ since [NiFeSe] hydrogenases are known to be EPR silent, in contrast to standard Ni-Fe hydrogenases [8]. Another factor that may lead to discrepancy between computed and experimental geometries may be the existence of a bridging ligand (such as a hydride) that may not be resolved from electron density.

Table 1. Bond lengths (Å) from QM/MM-optimized geometries are listed for each of the three conformers I, II, and III for [$Ni^{2+}Fe^{2+}Se$] hydrogenase. For comparison, the bond lengths obtained from the crystal structure (PDB ID: 5JSH [1]) are listed in bold print.

	[$Ni^{2+}Fe^{2+}Se$] QM/MM Optimized Bond Lengths [Å]											
	Fe-Ni	Ni-S	Ni-Se	S-Se	Ni-S-Cys492	Ni-S-Cys78	Ni-S-Cys75	Fe-CN1	Fe-CN2	Fe-CO	Fe-S-Cys492	Fe-S-Cys78
I	2.52	2.21	2.37	2.17	2.19	2.25	2.22	1.85	1.87	1.74	2.28	2.26
	2.41	**1.98**	**2.34**	**2.20**	**2.22**	**2.31**	**2.19**	**1.82**	**1.82**	**1.82**	**2.34**	**2.26**
II	2.55	2.10	3.60	2.19	2.23	2.27	2.14	1.86	1.88	1.72	2.31	2.25
	2.41	**1.98**	**3.07**	**1.99**	**2.22**	**2.31**	**2.19**	**1.82**	**1.82**	**1.82**	**2.34**	**2.26**
III	2.61	-	2.24	-	2.24	2.27	2.18	1.85	1.88	1.73	2.20	2.32
	2.41	-	**2.39**	-	**2.22**	**2.31**	**2.19**	**1.82**	**1.82**	**1.82**	**2.34**	**2.26**

The bond lengths between nonproteic ligands (CO, CN1, and CN2) coordinated to Fe show reasonable agreement (0.05–0.14 Å). In our models, we have adopted the assignment of positions from the crystal structure of Marques et al. [1], namely CO pointing toward Leu425 and CN-ligands pointing toward Pro421 and Ser445. Nonetheless, the experimental resolution (1.3 Å) does not allow for unambiguous assignment of ligand position. In other words, if the CO ligand were exchanged with either of the CN-ligands, the electronic structure of the hydrogen bonding network to neighboring amino acid side chains and crystal water molecules would change (see Figure 3 for up-close view of active site). These alternate two scenarios have not been tested yet but would be worthwhile extensions of the current investigation.

Table 2. Bond lengths (Å) from QM/MM-optimized geometries are listed for each of the three conformers I, II, and III for [$Ni^{2+}Fe^{3+}Se$] hydrogenase. For comparison, the bond lengths obtained from the crystal structure (PDB ID: 5JSH from Marques et al. [1]) are listed in bold print.

	[$Ni^{2+}Fe^{3+}Se$] QM/MM Optimized Bond Lengths [Å]											
	Fe-Ni	Ni-S	Ni-Se	S-Se	Ni-S-Cys492	Ni-S-Cys78	Ni-S-Cys75	Fe-CN1	Fe-CN2	Fe-CO	Fe-S-Cys492	Fe-S-Cys78
I	2.54	2.17	2.38	2.17	2.21	2.25	2.21	1.85	1.88	1.72	2.27	2.27
	2.41	**1.98**	**2.34**	**2.20**	**2.22**	**2.31**	**2.19**	**1.82**	**1.82**	**1.82**	**2.34**	**2.26**
II	2.45	2.10	3.52	2.18	2.19	2.30	2.10	1.86	1.88	1.75	2.39	2.31
	2.41	**1.98**	**3.07**	**1.99**	**2.22**	**2.31**	**2.19**	**1.82**	**1.82**	**1.82**	**2.34**	**2.26**
III	2.76	-	2.24	-	2.25	2.32	2.15	1.85	1.88	1.72	2.29	2.20
	2.41	-	**2.39**	-	**2.22**	**2.31**	**2.19**	**1.82**	**1.82**	**1.82**	**2.34**	**2.26**

For each of the six QM/MM optimized structures, vibrational frequencies and IR intensities were computed; the results for Fe^{2+} and Fe^{3+} are listed in Table 3, respectively, and the spectra are shown in Figures 6 and 7, respectively. The results are compared with the FTIR spectra obtained from the oxidized form of native, wild-type (WT) [NiFeSe] hydrogenase as-isolated from *D. vulgaris* [10]. In their experiment, DeLacey et al. observed two sets of CO bands for different batches of purified enzymes [10]. These two bands were assigned to two different conformations, or isomers (IS), of the active site and termed Ni-IS$_I$ and Ni-IS$_{II}$; Ni-IS$_I$ is the conformer that shows a more intense CO band at 1904 cm^{-1} [10]. For sake of reference, the frequencies obtained from the [NiFeSe] species from the recombinant enzyme (r[NiFeSe]) of Marques et al., as-isolated (as is) and reduced (red) with H_2, are listed as well. An important note: the experimental values of DeLacey et al. listed in Table 3 were obtained from experiments that are not based on the same crystallized protein that is the basis of our computational study [1]. Nonetheless, as these experimental frequencies are for the native WT [NiFeSe] hydrogenase isolated from *D. vulgaris* under non-reducing conditions, we will refer to them as a reference with which to compare our calculated frequencies based on the oxidized, recombinant enzyme in Ref. [1].

Table 3. Computed vibrational frequencies of nonproteic ligands CO, CN1, and CN2 based on QM/MM-optimized geometries are listed for each of the three conformers I, II, and III for $[Ni^{2+}Fe^{2+}Se]$ and $[Ni^{2+}Fe^{3+}Se]$ hydrogenase. For the case of $[Ni^{2+}Fe^{2+}Se]$, frequencies of conformer II (frequencies shown in bold) show the best agreement with experimentally measured vibrational frequencies of Ni-IS$_I$, which are obtained from the oxidized form of native, wild-type [NiFeSe] hydrogenase as-isolated from *D. vulgaris* [10]; for the case of $[Ni^{2+}Fe^{3+}Se]$, frequencies of conformers I and III (frequencies shown in bold print) show best agreement with Ni-IS$_{II}$ frequencies. Vibrational frequencies obtained from the recombinant r[NiFeSe] species of Marques et al. [1] are listed as well for sake of comparison.

	v(CO) [cm^{-1}]	v(CN) [cm^{-1}] antisym.	v(CN) [cm^{-1}] sym.
Conf.	Calc. $[Ni^{2+}Fe^{2+}Se]$		
I	1896	2022	2063
II	**1906**	**2038**	**2089**
III	1889	2018	2065
	Calc. $[Ni^{2+}Fe^{3+}Se]$		
I	**1952**	**2058**	**2087**
II	1983	2074	2118
III	**1945**	**2060**	**2098**
	Experiment		
Ni-IS$_I$ [10]	1904	2076	2085
Ni-IS$_{II}$ [10]	1939	2079	2094
r[NiFeSe] as-isolated [1]	1904, 1940	2079	2085, 2094
r[NiFeSe] reduced [1]	1910,1926,1935	2059, 2063	2078, 2085

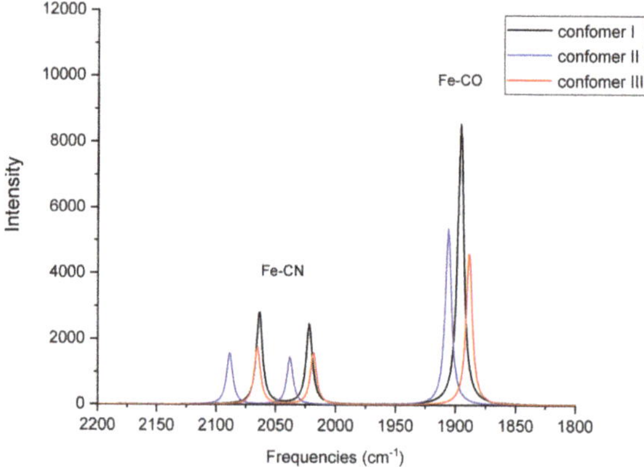

Figure 6. The computed vibrational infrared (IR) spectra for the QM/MM optimized geometries of conformer I (black line), II (blue line), and III (red line) are shown for $[Ni^{2+}Fe^{2+}Se]$ hydrogenase.

Comparison with the frequencies measured from an oxidized species isolated from the native, WT enzyme indicates that the computed frequencies are reasonable and are representative of the QM/MM optimized structures. Conformer II in the Fe^{2+} species demonstrates CO and symmetric CN-stretching frequencies that are in agreement with those of IS$_I$ (compare 1906 and 2089 cm^{-1} with 1904 and 2085 cm^{-1}, respectively). This agreement suggests the assumed Fe^{2+} oxidation state may be valid for the species that were measured in the experiments. Compared to the experimental antisymmetric CN-stretching frequencies for as-isolated ($\sim$2076–2079 cm^{-1}) or reduced ($\sim$2059–2063 cm^{-1}) species, the calculated antisymmetric CN-stretching frequencies for the Fe^{2+} species in all three conformers are red-shifted (2022 cm^{-1}, 2038 cm^{-1}, and 2018 cm^{-1}). Nonetheless, the antisymmetric CN-stretching frequency for conformer II (2038 cm^{-1}) is closest to the experimental values.

The discrepancy between computed and measured CN-antisymmetric stretching frequencies indicates that, although conformer II from Marques et al. may be present in the mixture of the oxidized species of the WT native enzyme measured in the experiment of DeLacey et al. [10], the antisymmetric stretching vibration may be sensitive to the protein environment in the model. A similar systematic red-shift in the computed CN-antisymmetric stretching frequency has been observed for membrane-bound [NiFe] hydrogenase from *Ralstonia eutropha* [13].

Based on the intense band that they observe at 1940 cm^{-1}, Marques et al. believe their as-isolated sample is more than 80% in the Ni-IS$_{II}$ state [1]; the computed vibrational frequencies for conformers I and III for Fe^{3+} (1952 cm^{-1} and 1945 cm^{-1}, respectively) are in best agreement with this frequency. Interestingly, based on electron density occupancy fitting, Marques et al. believe conformer I is dominant (51%), followed by conformer III (29%) and conformer II (20%) [1].

The isolated species of DeLacey et al. are believed to be diamagnetic Ni^{2+} states with one, or more than one, oxygen species in the active site [10]. An interesting suggestion has been made that the side chains of either cysteine or selenocysteine residues can be oxidized to sulfenates; this phenomenon has been used to explain data from X-ray absorption experiments [27]. In their experiments involving the Ni-Fe cofactor of the NAD-reducing soluble hydrogenase, Burgdorf et al. attribute the long measured Ni-S distance (3.6 Å) to thiol groups that may be in an oxidized form (Cys-SOH) [30]. Similar chemical modifications of Cys residues have been detected in a range of proteins [31]. DeLacey et al. also speculate that oxidation of their Ni-IS species leads to a Ni-OX species in which selenocysteine or cysteine ligands are oxidized [10].

In the case of Fe^{3+} (Table 3 and Figure 7), the computed CO and CN-symmetric stretching frequencies for conformers I (1952 cm^{-1} and 2087 cm^{-1}) and III (1945 cm^{-1} and 2098 cm^{-1}) agree well with the frequencies measured for both WT (1939 cm^{-1} and 2094 cm^{-1}) and recombinant enzymes (1940 cm^{-1}, 2085 cm^{-1}, and 2094 cm^{-1}), respectively. The computed CN-antisymmetric stretching frequencies for conformers I and III (2058 cm^{-1} and 2060 cm^{-1}, respectively) are red-shifted compared to the experimental frequencies, behavior that was also observed in the case of Fe^{2+}.

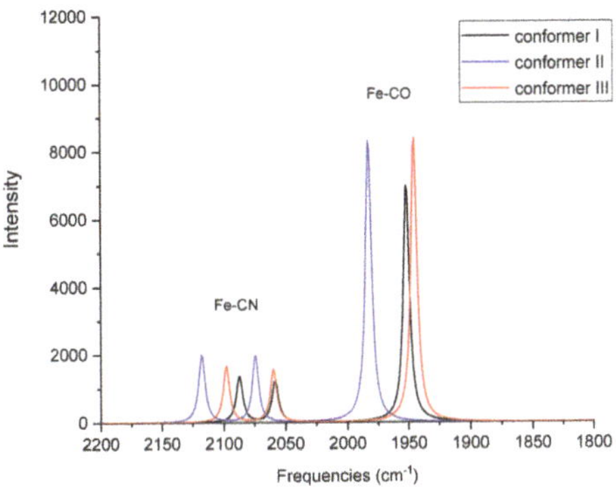

Figure 7. The computed vibrational IR spectra for the QM/MM optimized geometries of conformer I (black line), II (blue line), and III (red line) are shown for [Ni^{2+}Fe^{3+}Se] hydrogenase.

4. Conclusions and Outlook

Here we report for the first time a QM/MM computational investigation of the active site geometry of the [NiFeSe] hydrogenase based on the available structural data from X-ray crystallographic data collected at 1.3 Å resolution [1]. The IR vibrational frequencies of CN- and CO ligands coordinated to the Fe ion have been computed using a normal-mode analysis based on the QM/MM optimized geometries. Overall, the computed QM/MM optimized geometries are in good agreement with the experimentally resolved geometries, such that the computed vibrational frequencies may be assumed to reflect accurately the underlying structures. In comparison with experimental FTIR frequencies, the computed frequencies suggest that a mixture of structural isomers, as well as Fe oxidation states, may be present in a given protein sample, thus leading to the pattern of CO and CN-vibrational stretching bands observed. Based on the CO and CN-symmetric stretching frequencies, the Ni-IS$_I$ state can be assigned to conformer II in the case of Fe^{2+}, while the Ni-IS$_{II}$ state shows best agreement with conformers I and III in the case of Fe^{3+}. This pattern suggests that iron oxidation may be favored in conformers I and III. Independent of the iron oxidation state, conformers I and III show similar vibrational behavior, an interesting correlation that requires additional investigation. A systematic red-shift of 20–40 cm^{-1} in the antisymmetric stretching frequencies for both Fe^{2+} and Fe^{3+} is observed.

Here, the electronic state of nickel has been treated as 2+ as the oxidized states of [NiFeSe] hydrogenases are known to be EPR silent. Nonetheless, a worthwhile computational investigation would be a comparison of the optimized geometries considering additional ligand oxidization states, such as cysteines or selenocysteines oxidized to sulfenates. In addition, modeling bridging ligands at the Fe-Ni locus would provide some insight into the enzyme's catalytic mechanisms. Furthermore, as the assignment of CN- and CO ligands in the crystal structure may not necessarily be accurate, the present calculations should be extended to test for alternate assignments. Also, in the structure of Marques et al., the total Ni occupancy of the catalytic site could only be refined to 85%, while a minor Ni site could not be clearly resolved. Therefore, future models may also wish to address variable Ni positions.

Author Contributions: Methodology, S.M., N.E.-M., D.T.; Software, N.E.-M., D.T., S.M.; Formal Analysis, S.M., N.E.-M., D.T., M.A.M.; Investigation, S.M. Data Curation, S.M.; Writing—Original Draft Preparation, N.E.-M.; Writing—Review and Editing, N.E.-M., S.M., D.T., M.A.M.; Visualization, S.M., N.E.-M.; and Funding Acquisition, M.A.M. All authors have read and agreed to the published version of the manuscript.

Funding: Funded by the Deutsche Forschungsgemeinschaft (DFG, German Research Foundation) under Germany's Excellence Strategy—EXC 2008/1—390540038". Gefördert durch die Deutsche Forschungsgemeinschaft (DFG) im Rahmen der Exzellenzstrategie des Bundes und der Länder—EXC 2008/1—390540038". We acknowledge support by the German Research Foundation and the Open Access Publication Fund of TU Berlin. This work has been supported in part by the European Union's Horizon 2020 research and innovation programme under grant agreement No 810856.

Acknowledgments: The authors thank Christian Lorent for insightful discussions.

Conflicts of Interest: The authors declare no conflict of interest.

References

1. Marques, M.C.; Tapia, C.; Gutiérrez-Sanz, O.; Ramos, A.R.; Keller, K.L.; Wall, J.D.; Lacey, A.L.; Matias, P.M.; Pereira, I.A.C. The direct role of selenocysteine in [NiFeSe] hydrogenase maturation and catalysis. *Nat. Chem. Biol.* **2017**, *13*, 544–550. [CrossRef] [PubMed]

2. Yang, X.; Elrod, L.C.; Reibenspies, J.H.; Hall, M.B.; Darensbourgh, M.Y. Oxygen uptake in complexes related to [NiFeS]-and [NiFeSe]-hydrogenase active sites. *Chem. Sci.* **2019**, *10*, 1368–1373. [PubMed]

3. Axley, M.J.; Böck, A.; Stadtman, T.C. Catalytic properties of an Escherichia coli formate dehydrogenase mutant in which sulfur replaces selenium. *Proc. Nat. Acad. Sci. USA* **1991**, *88*, 8450–8454. [CrossRef] [PubMed]

4. Böck, A. Biosynthesis of selenoproteins–an overview. *Biofactors* **2005**, *11*, 77–78. [CrossRef]

5. Valente, F.M.A.; Oliveira, A.S.F.; Gnadt, N. Hydrogenases in *Desulfovibrio vulgaris* Hildenborough: Structural and physiologic characterisation of the membrane-bound [NiFeSe] hydrogenase. *J. Biol. Inorg. Chem.* **2005**, *10*, 667–682.

6. Rüdiger, O.; Gutiérrez-Sánchez, C.; Olea, D.; Pereira, I.A.C.; Vélez, M.; Fernández, V.M.; de Lacey, A.L. Enzymatic Anodes for Hydrogen Fuel Cells based on Covalent Attachment of Ni-Fe Hydrogenases and Direct Electron Transfer to SAM-Modified Gold Electrodes. *Electroanalysis* **2010**, *22*, 776–783. [CrossRef]

7. Riethausen, J.; Rüdiger, O.; Gärtner, W.; Lubitz, W.; Shafaat, H.S. Spectroscopic and Electrochemical Characterization of the [NiFeSe] Hydrogenase from *Desulfovibrio vulgaris* Miyazaki F: Reversible Redox Behavior and Interactions between Electron Transfer Centers. *ChemBioChem* **2013**, *14*, 1714–1719.

8. Valente, F.M.A.; Almeida, C.C.; Pacheco, I.; Carita, J.; Saraiva, L.M.; Pereira, I.A.C. Selenium Is Involved in Regulation of Periplasmic Hydrogenase Gene Expression in *Desulfovibrio vulgaris* Hildenborough. *J. Bacteriol.* **2006**, *188*, 3228–3235. [CrossRef]

9. Marques, M.C.; Coelho, R.; de Lacey, A.L.; Pereira, I.A.C.; Matia, P.M. The three-dimensional structure of [NiFeSe] Hydrogenase from *Desulfovibrio vulgaria* Hildenborough: A hydrogenase without a bridging ligand in the active site in its oxidised "as-isolated" state. *J. Mol. Biol.* **2010**, *396*, 893–907. [CrossRef]

10. DeLacey, A.L.; Gutiérrez-Sánchez, C.; Fernández, V.M.; Pacheco, I.; Pereira, I.A.C. FTIR spectroelectrochemical characterization of the Ni–Fe–Se hydrogenase from Desulfovibrio vulgaris Hildenborough. *J. Biol. Inorg. Chem.* **2008**, *13*, 1315–1320. [CrossRef]

11. DeLacey, A.L.; Fernández, V.M.; Rousset, M.; Cammack, R. Activation and inactivation of hydrogenase function and the catalytic cycle: Spectroelectrochemical studies. *Chem. Rev.* **2007**, *107*, 4304–4330.

12. Nakamoto, K. *Infrared and Raman Spectra of Inorganic and Coordination Compounds*; Part A; Wiley: New York, NY, USA, 1997; pp. 126–148.

13. Rippers, Y.; Horch, M.; Hildebrandt, P.; Zebger, I.; Mroginski, M.A. Revealing the Absolute Configuration of the CO and CN- Ligands at the Active Site of a [NiFe] Hydrogenase. *Chem. Phys. Chem.* **2012**, *13*, 3852–3856.

14. Siebert, E.; Rippers, Y.; Frielingsdorf, S.; Fritsch, J.; Schmidt, A.; Kalms, J.; Katz, S.; Lenz, O.; Scheerer, P.; Paasche, L.; et al. Resonance Raman Spectroscopic Analysis of the [NiFe] Active Site and the Proximal [4Fe-3S] Cluster of an O_2-Tolerant Membrane-Bound Hydrogenase in the Crystalline State. *J. Phys. Chem. B* **2015**, *119*, 13785–13796. [CrossRef]

15. Tombolelli, D.; Mroginski, M.A. Proton Transfer Pathways between active sites and proximal clusters in the membrane-bound [NiFe] hydrogenase. *J. Phys. Chem. B* **2019**, *123*, 3409–3420. [CrossRef]

16. Senn, H.M.; Thiel, W. QM/MM methods for biomolecular systems. *Angew. Chem.* **2009**, *48*, 1198–1229. [CrossRef]

17. Field, M.J.; Bash, P.A.; Karplus, M. A Combined Quantum Mechanical and Molecular Mechanical Potential for Molecular Dynamics Simulations. *J. Comput. Chem.* **1990**, *11*, 700–733. [CrossRef]

18. Warshel, A.; Levitt, M. Theoretical Studies of Enzymic Reactions: Dielectric, Electrostatic and Steric Stabilization of the Carbonium Ion in the Reaction of Lysozyme. *J. Mol. Biol.* **1976**, *103*, 227–249.

19. Brooks, B.R.; Brooks, C.L., III; Mackerell, A.D., Jr.; Nilsson, L.; Petrella, R.J.; Roux, B.; Won, Y.; Archontis, G.; Bartels, C.; Boresch, S.; et al. Charmm: The Biomolecular simulation Program. *J. Comp. Chem.* **2009**, *30*, 1545–1615.

20. Huang, J.; MacKerell, A.D., Jr. CHARMM36 all-atom additive protein force field: Validation based on comparison to NMR data. *J. Comput. Chem.* **2013**, *34*, 2135–2145. [CrossRef]

21. Eichkorn, K.; Treutler, O.; Öhm, H.; Häser, M.; Ahlrichs, R. Auxiliary basis sets to approximate Coulomb potentials. *Chem. Phys. Lett.* **1995**, *240*, 283–290. [CrossRef]

22. Ahlrichs, R.; Bär, M.; Häser, M.; Horn, C.K.H. Electronic structure calculations on work-station computers: The computer system Turbomole. *Chem. Phys. Lett.* **1989**, *162*, 165–169.

23. TURBOMOLE V7.0 2915. A Development of University of Karlsruhe and Forschungszentrum. Karlsruhe GmbH. (1989–2007). Available online: http://www.turbomole.com (accessed on 20 January 2020).

24. Fontecilla-Camps, J.C.; Volbeda, A.; Cavazza, C.; Nicolet, Y. Structure/Function Relationships of [NiFe]- and [FeFe]-Hydrogenases. *Chem. Rev.* **2007**, *107*, 4273–4303. [CrossRef]

25. Happe, R.P.; Roseboom, W.; Pierik, A.J.; Albracht, S.P.J.; Bagley, K.A. Biological activition of hydrogen. *Nature* **1997**, *385*, 126. [CrossRef]

26. Sherwood, P.; deVries, A.; Guest, M.; Schreckenbach, G.; Catlow, C.; French, S.; Sokol, A.; Bromley, S.; Thiel, W.; Turner, A.; et al. QUASI: A general purpose implementation of the QM/MM approach and its application to problems in catalysis. *J. Mol. Struct.* **2003**, *632*, 1–28.

27. Metz, S.; Kästner, J.; Sokol, A.; Keal, T.; Sherwood, P. ChemShell–a modular software package for QM/MM simulations. *Wiley Interdisc. Rev. Comp. Mol. Sci.* **2014**, *4*, 101–110. [CrossRef]

28. ChemShell. A Computational Chemistry Shell. Available online: www.chemshell.org (accessed on 20 January 2020).

29. Klapötke, T.M.; Schulz, A. *Quantenmechanische Methoden in der Hauptgruppenchemie*; Spektrum Akademie Verlag: Heidelberg, Germany, 1996; pp. 71–91.

30. Burgdorf, T.; Löscher, S.; Liebisch, P.; van der Linden, E.; Galander, M.; Lendzian, F.; Meyer-Klaucke, W.; Albracht, S.P.J.; Friedrich, B.; Dau, H.; et al. Structural and Oxidation-State Changes at Its Nonstandard Ni-Fe Site during Activation of the NAD-Reducing Hydrogenase from Ralstonia eutropha Detected by X-ray Absorption, EPR, and FTIR Spectroscopy. *J. Am. Chem. Soc.* **2005**, *127*, 576–592. [CrossRef] [PubMed]

31. Claiborne, A.; Yeh, J.I.; Mallett, T.C.; Luba, J.; Crane, E.J.; Charrier, V.; Parsonage, D. Protein-sulfenic acids: Diverse roles for an unlikely player in enzyme catalysis and redox regulation. *Biochemistry* **1999**, *38*, 15407–15416. [CrossRef]

Article

Major Depressive Disorder and Oxidative Stress: In Silico Investigation of Fluoxetine Activity against ROS

Cecilia Muraro [1], Marco Dalla Tiezza [1], Chiara Pavan [2], Giovanni Ribaudo [3], Giuseppe Zagotto [4] and Laura Orian [1,*]

[1] Department of Chemical Sciences, University of Padova, Via Marzolo 1, 35131 Padova, Italy
[2] Department of Medicine, University of Padova, Via Giustiniani 2, 35128 Padova, Italy
[3] Department of Molecular and Translational Medicine, University of Brescia, Viale Europa 11, 25123 Brescia, Italy
[4] Department of Pharmacological Sciences, University of Padova, Via Marzolo 5, 35131 Padova, Italy
* Correspondence: laura.orian@unipd.it; Tel.: +39-049-8275140

Received: 5 August 2019; Accepted: 23 August 2019; Published: 3 September 2019

Featured Application: Due to the seriousness of depressive disorders and their social implications, any strategy to improve clinical symptoms are of considerable importance. It is recognized that oxidative stress negatively impacts on this and other psychiatric diseases. Fluoxetine, a well-known and largely prescribed antidepressant drug, has antioxidant capacity, but, based on our analysis, this is due to its efficiency in increasing the concentration of free serotonin, rather than its direct ROS scavenging activity.

Abstract: Major depressive disorder is a psychiatric disease having approximately a 20% lifetime prevalence in adults in the United States (U.S.), as reported by Hasin et al. in *JAMA Psichiatry* 2018 75, 336–346. Symptoms include low mood, anhedonia, decreased energy, alteration in appetite and weight, irritability, sleep disturbances, and cognitive deficits. Comorbidity is frequent, and patients show decreased social functioning and a high mortality rate. Environmental and genetic factors favor the development of depression, but the mechanisms by which stress negatively impacts on the brain are still not fully understood. Several recent works, mainly published during the last five years, aim at investigating the correlation between treatment with fluoxetine, a non-tricyclic antidepressant drug, and the amelioration of oxidative stress. In this work, the antioxidant activity of fluoxetine was investigated using a computational protocol based on the density functional theory approach. Particularly, the scavenging of five radicals ($HO^\bullet$, $HOO^\bullet$, $CH_3OO^\bullet$, $CH_2{=}CHOO^\bullet$, and $CH_3O^\bullet$) was considered, focusing on hydrogen atom transfer (HAT) and radical adduct formation (RAF) mechanisms. Thermodynamic as well as kinetic aspects are discussed, and, for completeness, two metabolites of fluoxetine and serotonin, whose extracellular concentration is enhanced by fluoxetine, are included in our analysis. Indeed, fluoxetine may act as a radical scavenger, and exhibits selectivity for $HO^\bullet$ and $CH_3O^\bullet$, but is inefficient toward peroxyl radicals. In contrast, the radical scavenging efficiency of serotonin, which has been demonstrated in vitro, is significant, and this supports the idea of an indirect antioxidant efficiency of fluoxetine.

Keywords: free radical scavengers; antioxidants; fluoxetine; depressive disorder; major; oxidative stress; DFT calculations; reactive oxygen species

---

1. Introduction

1.1. Background

Fluoxetine hydrochloride (3-(p-trifluoromethylphenoxy)-N-methyl-3-phenylpropylamine HCl) (**1**, Figure 1) was presented to the community of physicians and medicinal chemists in the early 1970s, when it was described in a scientific journal as a selective serotonin-reuptake inhibitor [1]. Nevertheless, several years were required to fully develop this compound, which was later approved for the treatment of depression by the Food and Drug Administration (FDA, 1987) and marketed as Prozac (Eli Lilly) [2]. Fluoxetine is the prototype of selective serotonin-reuptake inhibitors (SSRIs), the first member of a new class of antidepressants initially introduced in the United States (U.S.) [2]. Moreover, it was also the first SSRI made available for clinical use in most countries besides the U.S. [3]. Despite early skepticism, Prozac became a true symbol of successful "blockbuster" drugs through the years. In fact, a fluoxetine capsule was depicted in the cover of *Newsweek*, celebrated as "a breakthrough drug for depression" by the same magazine (1990), as one of the "Pharmaceutical Products of the Century" by Fortune (1999) and, more in general, as the "happiness pill" in the pop culture [2].

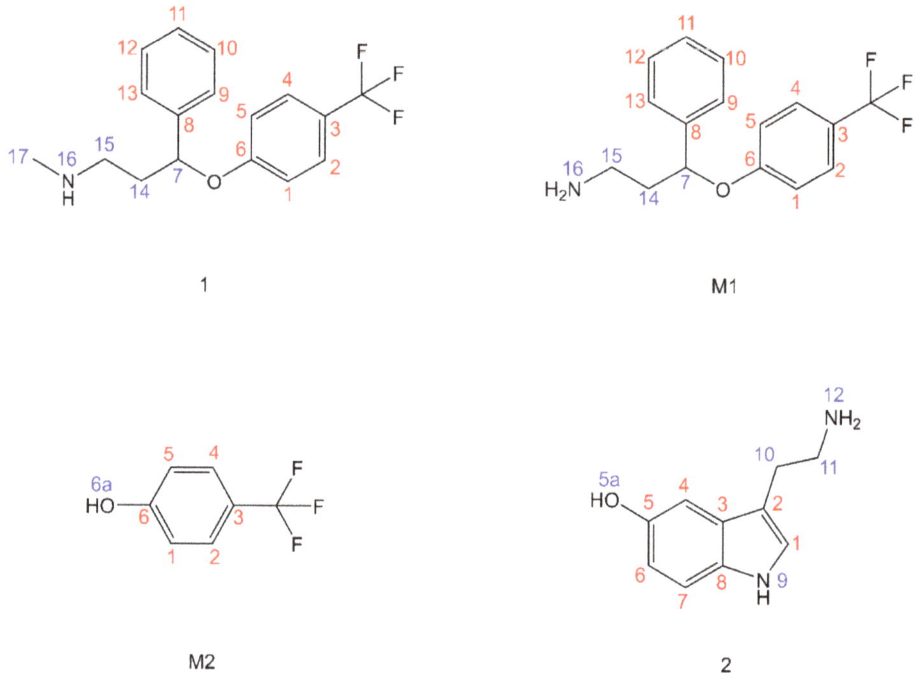

Figure 1. Fluoxetine (**1**) and its metabolites (**M1** and **M2**); serotonin (**2**).

By blocking the reuptake of serotonin (**2**, Figure 1) binding the neuronal presynaptic serotonin transporter SLC6A4 (Solute Carrier Family 6 Member 4), fluoxetine promotes an increase of the concentration of the neurotransmitter in the synaptic cleft. This leads to enhanced postsynaptic neuronal activity of the serotoninergic neurotransmission with limited effects on catecholaminergic neurons, differently from tricyclic antidepressants (TCAs) [4–6]. The observed therapeutic results are mainly due to the effect on the 5-HT$_{2C}$ receptors [7].

Fluoxetine is structurally related to diphenhydramine, which is an antihistaminic that had been developed 30 years before [5]. It is a racemic mixture of two enantiomers, R(−)-fluoxetine and S(+)-fluoxetine, the latter being 1.5 times more potent in inhibiting serotonin reuptake [3,8].

The molecular basis and the binding mode justifying this difference were studied from a structural point of view [9]. After the patent on racemic fluoxetine expired, attempts to develop S(+)-fluoxetine formulations were pursued [10]. Fluoxetine has been marketed in capsules (the typical dose is 20 mg for the treatment of depression), but formulations for prolonged drug release were also studied [5,11]. It has been approved worldwide for the treatment of major depression, but it is also effective against several other syndromes [1,5], with a clinical efficacy similar to that of TCAs and fewer cardiovascular and anticholinergic side effects [5]. Fluoxetine effectively acts on a wide spectrum of mood disorders and protects against the adverse effects of different types of stressors by decreasing some effects of stress on the immune system and by protecting against oxidative damage. Different molecular mechanisms have been proposed to explain its neuroprotective effects, including BDNF (Brain-derived Neurotrophic Factor) release, antagonism on NMDA (N-methyl-D-aspartate) receptors, the inhibition of NF-kB (Nuclear Factor Kappa B Subunit 1) activity, and inhibition of the release of pro-inflammatory factors (TNF-α, Tumor Necrosis Factor-Alpha, IL-1β, Interleukin 1 Beta) from microglial cells [12]. However, the underlying mechanisms of its therapeutic efficacy remain unclear, particularly those related to its antioxidant activity. The brain is very susceptible to oxidative stress, since it has a high energy requirement, and oxidative stress has been implicated in the pathogenesis of many psychiatric and degenerative disorders. Moreover, aversive stimuli promote peripheral oxidative stress, which leads to an increase in the generation of reactive oxygen species (ROS) in peripheral blood lymphocytes, granulocytes, and monocytes, and a meta-analysis showed an increase in oxidative stress markers such as 8-oxo-2′deoxyguanosine and F2-isoprostanes in depressed patients [13]. High level of proteins and lipid peroxidation and an imbalance between superoxide dismutase and catalase were found in the brain and in the submitochondrial particles into the brain, and the antidepressant treatment improves these oxidative stress parameters in patients with depression [14]. Higher serum total oxidant status and a lower serum total antioxidant capacity in depressed patients were reversed after treatment with antidepressants in animal models of depression as well as in patients [15–17].

It has been postulated that the capacity of fluoxetine to ameliorate the oxidative stress damage may be either due to a direct role of the molecule or achieved through the stimulation of some antioxidant enzymes. Oxidative stress can damage the cell through lipid peroxidation, DNA or protein oxidation, and mitochondrial damage [18]. In this connection, it must be pointed out that the central nervous system is composed of a high percentage of phospholipids, which can undergo peroxidation and generate ROS, and may eventually lead to potentially harmful conditions for cellular structures [19]. It has been demonstrated, mainly in preclinical but also in clinical studies, that fluoxetine provides a beneficial antioxidant effect through a combination of mechanisms: the inhibition of lipid peroxidation, an increase of glutaminergic transmission, the restoration of the normal metabolism of monoamines, the influence on ion balance, and a reduction of inflammation, which is connected to ROS production [18,20,21]. Kolla et al. showed that amitriptyline and fluoxetine protect against oxidative stress-induced damage in rat pheochromocytoma (PC12) cells, contrasting the effects of H_2O_2 [22]. Caiaffo et al. recently reviewed the evidences of the anti-inflammatory, antiapoptotic, and antioxidant activity of fluoxetine [19]. Particularly, several contributions highlighted that fluoxetine may positively influence, to different degrees, the expression and functioning of "endogenous components of the antioxidant defense system" (superoxide dismutase (SOD), catalase (CAT), glutathione peroxidase (GPx) as well as "non-enzymatic antioxidant components". Moreover, the same authors suggested that, based on the reported findings, fluoxetine provides its antioxidant effect only under conditions of oxidative damage, which can be connected to other diseases (diabetes, depression, ischemia) and/or polypharmacological treatments [19,23]. It must be also pointed out that very recently, Dalmizrak et al. studied the inhibitory potential of fluoxetine on glutathione reductase activity [24].

Kalogiannis et al. investigated the role of fluoxetine in ameliorating the condition of a mice model of cerebral inflammation. According to the working hypothesis of the authors, this SSRI may enhance the cerebral antioxidant capacity of serotonin by increasing its extracellular concentration in the brain [25]. In fact, serotonin is known to possess antioxidant capacity, and has been previously studied in silico as well as in vitro for its potentially protective role for the cell [26,27]. At the same time, the connection between the levels of this neurotransmitter and the antioxidant defense has been clearly established [28], and evidences of the in vitro antioxidant and radical scavenging properties of serotonin have been demonstrated using a combination of experimental techniques [29].

1.2. Aim of This Work

Computational methodologies are nowadays largely employed to rationalize chemical and biochemical phenomena. Particularly, accurate mechanistic studies on the control of oxidative stress by endogenous defenses, including enzymes such as glutathione peroxidase [30–32], and molecular drugs that act as mimics of these enzymes [33–37] or as radical scavengers [38] have been reported by some of us. In this work, we investigate in silico the antioxidant activity of fluoxetine via HAT (hydrogen atom transfer) and RAF (radical adduct formation) mechanisms. The purpose is to assess its radical scavenging efficiency; for completeness, two metabolites of fluoxetine (**M1** and **M2**, Figure 1) are included. Finally, the analysis is performed also on serotonin, the extracellular concentration of which is enhanced in patients treated with fluoxetine.

2. Materials and Methods

2.1. Calculation of $\Delta G°_{HAT/RAF}$

A conformational analysis of **1** was carried out using the semi-empirical quantum mechanical method GFN2-xTB [39–41]. Conversely, the most stable conformer of **2** was taken from [27]. For all the studied reactions of HAT and RAF, geometry optimizations of the reactants and products were performed in the gas phase without any constraint, using the M06-2X functional [42] combined with the 6-31G(d) basis set, as implemented in Gaussian 16 [43]. Spin contamination was checked for the doublet ground state species to assess the reliability of the wavefunction. Frequency calculations were carried out at the same level of theory (M06-2X/6-31G(d)) in order to confirm the stationary points (all positive frequencies) and to obtain the thermodynamic corrections at 1 atm and 298 K. Afterwards, single-point energy calculations were performed at M06-2X/6-311+G(d,p) in the gas phase in order to obtain more accurate energy values, and subsequently, at the same level of theory, in benzene and water using the continuum Solvation Model based on Density (SMD) [44,45]. This level of theory is denoted in the text (SMD)-M06-2X/6-311+G(d,p)//M06-2X/6-31G(d,p). Benzene and water mimic an apolar and a polar environment, respectively [46]. The calculation of the HAT/RAF Gibbs free energies for all the available sites was streamlined using an in-house Python script.

Natural spin densities and atomic charges were calculated with the natural population analysis (NPA) for C4 sites in 1 and its metabolites (Scheme 3) [47]. The electron spin density surfaces related to the localized natural bond orbitals (NBO) were also drawn for selected structures with Multiwfn [48] with a high-quality grid and the isodensity value of 0.003.

2.2. Calculation of $\Delta G^{\ddagger}_{HAT}$

On the basis of $\Delta G°_{HAT}$, the most reactive sites were identified, and the energy barriers were computed (level of theory: (SMD)-M06-2X/6-311+G(d,p)//M06-2X/6-31G(d,p)). The single imaginary frequency of all the transition states was analyzed to ensure that it corresponded to the correct vibrational mode. In gas phase, a reactant complex (RC) and a product complex (PC) exist on the potential energy surface, before and after the transition state (TS), respectively. In all cases, these species are located at higher energy than the free reactants and products; thus, energy barriers were calculated, referring to the free reactants in the gas phase as well as in the solvent [49].

3. Results

Radical scavenging occurs via different mechanisms, among which hydrogen atom transfer (HAT) and radical adduct formation (RAF) have been identified as mostly efficient in several organic molecules displaying antioxidant properties, including Trolox [50], melatonin [51], and capsaicin [52], but also psychotropic drugs such as zolpidem [38]. These mechanisms are summarized in Figure 2.

$$1 + R^\bullet \longrightarrow (1\text{-H})^\bullet + RH$$

$$1 + R^\bullet \longrightarrow (1R)^\bullet$$

$$R^\bullet = HO^\bullet,\ HOO^\bullet,\ CH_3O^\bullet,\ CH_3OO^\bullet,\ CH_2{=}CHOO^\bullet$$

Figure 2. Hydrogen atom transfer (HAT) and radical adduct formation (RAF) in fluoxetine (**1**).

HAT implies the transfer of $H^\bullet$ from one site of the scavenger to the ROS. Conversely, RAF consists in the addition of the ROS to a double bond site. The calculations of $\Delta G^\circ_{HAT/RAF}$ have been streamlined using an in-house Python script, in order to automatically screen all the available sites of fluoxetine (**1**), its metabolites, and serotonin (**2**). Activation energies $\Delta G^\ddagger_{HAT/RAF}$ were computed only for the most exergonic reactions. HAT results are presented in the following subsections, while the RAF mechanism and the related energetics are included in the Discussion.

3.1. Fluoxetine

HAT Gibbs free energy (ΔG°_{HAT}) was computed for all the possible sites of fluoxetine (Figure 1) considering five ROSs, i.e., $HO^\bullet$, $HOO^\bullet$, $CH_3OO^\bullet$, $CH_2{=}CHOO^\bullet$, and $CH_3O^\bullet$ (Figure 2). The results are shown in Figure 3 and reported in Table S1. $HO^\bullet$ is the most reactive and electrophilic oxygen-centered radical; the peroxyl radicals $HOO^\bullet$ and $CH_3OO^\bullet$ are much less reactive, and thus can reach remote cellular locations; $CH_2{=}CHOO^\bullet$ was chosen to mimic larger unsaturated peroxyl radicals; finally, $CH_3O^\bullet$ has an intermediate reactivity between $HO^\bullet$ and the peroxyl radicals.

The thermodynamic results, i.e., reaction energies are discussed first. Referring to the gas-phase values, it emerges that fluoxetine is selective for $HO^\bullet$ and alkoxyl radicals, excluding—in this latter case—the sites on the phenyl rings; it is not selective for peroxyl radicals. Focusing on the scavenging of $HO^\bullet$, HAT is thermodynamically strongly favored from N site 16 and from C sites 7, 14–15 and 17, among which 7 and 15 are associated to the largest (negative) values, i.e., −26.06 and −26.88 kcal mol^{-1}. A similar ΔG°_{HAT} is computed for HAT from N site 16 (−24.24 kcal mol^{-1}) and from C site 17 (−24.87 kcal mol^{-1}). In solvent, the same trends are maintained, the ΔG°_{HAT} becoming more negative with increasing polarity, and the largest negative value is found for HAT from N 16 in water (−31.41 kcal mol^{-1}). When considering $CH_3O^\bullet$, in the gas phase, HAT is thermodynamically favored only from sites 7 and 14–17. Meanwhile, ΔG°_{HAT} values are approximately half (in absolute value) those computed for $HO^\bullet$, except for those associated to C 14 and N 16, which are drastically reduced, changing from −18.31 and −24.24 kcal mol^{-1} to −3.81 and −9.73 kcal mol^{-1}, respectively. Also, for $CH_3O^\bullet$, increasing the solvent polarity results in thermodynamically more favored reactions. The reactivity of C sites 14 and 15 of fluoxetine recalls the reactivity of the topologically similar C site 4 in melatonin [46] and zolpidem [38]. By comparing the transfer of the methylene hydrogens to $HO^\bullet$ in these compounds, which were computed at the same level of theory, [38] the order is zolpidem (−33.3 kcal mol^{-1}) > melatonin (−28.9 kcal mol^{-1}) > fluoxetine (−26.88 kcal mol^{-1}). The value of ΔG°_{HAT} involving $HO^\bullet$ of Trolox is −38.7 kcal mol^{-1} [38]. As mentioned above, fluoxetine is not selective for peroxyl radicals: a negative value of ΔG°_{HAT}, i.e., −1.23 kcal mol^{-1}, is computed only for $CH_2{=}CHOO^\bullet$ in water.

The transition states and thus the activation energies were computed only for the exergonic reactions; they are reported in Table 1.

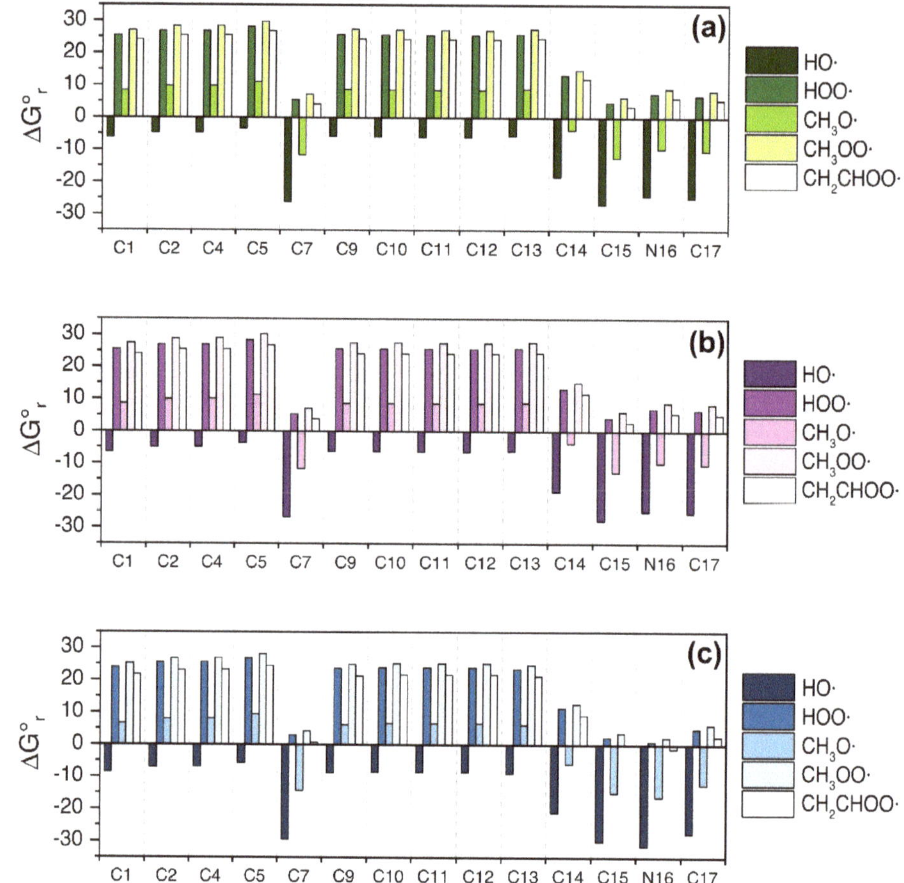

Figure 3. ΔG°_{HAT} (kcal mol^{-1}) in the gas phase (**a**), in benzene (**b**), and in water (**c**) for the scavenging of HO$^{\bullet}$, CH$_3$O$^{\bullet}$, HOO$^{\bullet}$, CH$_3$OO$^{\bullet}$, and CH$_2$=CHOO$^{\bullet}$ through HAT from all the available sites of **1**. Level of theory: (SMD)-M06-2X/6-311+G(d,p)//M06-2X/6-31G(d).

Table 1. $\Delta G^{\ddagger}_{HAT}$ (kcal mol^{-1}) in the gas phase, benzene, and water, for the scavenging of HO$^{\bullet}$ and CH$_3$O$^{\bullet}$ through HAT in **1**. Level of theory: (SMD)-M06-2X/6-311+G(d,p)//M06-2X/6-31G(d).

Site	Gas Phase	Benzene	Water	Gas Phase	Benzene	Water
		HO$^{\bullet}$			CH$_3$O$^{\bullet}$	
7	6.57	8.58	10.03	13.69	16.07	16.45
14	7.48	8.91	9.54	14.95	16.72	16.26
15	5.39	5.95	4.24	11.67	12.35	9.74
16	3.68	4.01	0.85	8.34	9.01	6.18
17	7.84	8.35	5.50	14.43	15.28	12.19

Considering the HAT from C sites 7 and 14, the energy barriers increase almost in all cases when going from the gas phase to benzene and water. Conversely, for the HAT from C sites 15 and 17 and N site 16, the lowest activation energies are computed in water. Kinetically, the easiest scavenging involves HO$^{\bullet}$ in water from N 16, which is also the most thermodynamically favored reaction.

3.2. Fluoxetine Metabolites

The pharmacokinetic and metabolic aspects of fluoxetine in the human body have been described in detail by Hiemke and Hartter and by Mandrioli et al. [3,5]. Fluoxetine is a lipophilic molecule that is almost completely absorbed after oral administration. Oral bioavailability is lower than 90% of the dose, due to hepatic first-pass extraction [53]. Fluoxetine highly accumulates in lungs, mainly due to lysosomal trapping, and it is characterized by a large volume of distribution (Vd), similar to that of TCAs [3]. Accumulation in the brain is lower than that of other SSRIs, with a brain to plasma ratio of 2.6:1 compared with 24:1 for fluvoxamine [54]. Fluoxetine is also characterized by a long half-life ($t_{1/2}$), ranging from 24 h to 4 days, which implies that between 1–22 months may be required to reach steady-state conditions during therapy [3,8]. Moreover, it has been observed that fluoxetine follows non-linear kinetics, since there is a disproportion between increase in blood concentration after dose escalation [3]. With a 20–60 mg/die dose, plasma levels range from 50 to 480 ng/mL [5].

R(−)-fluoxetine and S(+)-fluoxetine have different metabolic rates, since a four times greater clearance has been reported for the first, which means that the $t_{1/2}$ of the more active S(+) enantiomer is shorter [4]. Hepatic cytocrome P450 enzymes are responsible for the biotransformation in humans. Particularly, fluoxetine undergoes metabolic transformation mediated by CYP isoenzymes, and less than 10% is excreted unchanged or as fluoxetine N-glucuronide [55]. CYP2D6, and other isoforms such as CYP2C9 and CYP3A4, mediate N-demethylation to the active metabolite norfluoxetine (**M1**). Concerning this metabolite, the pharmacological difference between enantiomers is even stronger: S(+)-norfluoxetine is 20 times more potent in inhibiting serotonin uptake than the R(−) isomer [3]. Norfluoxetine has a longer $t_{1/2}$ (4–16 days), and it can also be converted to its glucuronide form [3,5]. CYP2D6 also mediates the further degradation of R(−)-fluoxetine, S(+)-fluoxetine, and S(+)-norfluoxetine, without influencing R(−)-norfluoxetine concentration. Other studies report that CYP2C9 is more selective for the R(−) enantiomer, while the remaining isoforms catalyze N-demethylation on both isomers with similar rates [56]. The oxidative O-dealkylation reaction of fluoxetine can also lead to the formation of another metabolite, p-trifluoromethylphenol (**M2**). This reaction is mediated by CYP2C19 and CYP3A4 [5].

Several analytical methods have been developed through the years for the identification and quantification of fluoxetine and its metabolites in biological samples. HPLC-UV, mass spectrometry, and gas chromatography are some of the tools included in the methods currently available in the literature [5,57–59]. The choice of applying one or more of such analytical techniques should also take into consideration the stereochemical aspects [6]. For the sake of completeness, we have investigated the scavenging activity of **M1** and **M2**.

$\Delta G°_{HAT}$ are shown in Figure 4 and reported in Table S2.

The results of norfluoxetine are very similar to those of **1**. Scavenging via HAT is thermodynamically most favored from C sites 7, 14–16. Particularly, in the gas phase, the largest (negative) $\Delta G°_{HAT}$ values are found for C 7 (−30.94 and −16.43 kcal mol^{-1} for HO$^\bullet$ and CH$_3$O$^\bullet$), which become even more negative with increasing solvent polarity, i.e., −31.48 and −16.41 kcal mol^{-1} in benzene and −32.61 and −17.53 kcal mol^{-1} in water, respectively. In general, the reactivity of **M1** from these sites toward the hydroxyl and methoxyl radicals via HAT is enhanced when compared to that of **1**. Particularly, focusing on site C 7, negative values of $\Delta G°_{HAT}$ are computed also for CH$_2$=CHOO$^\bullet$, ranging from −0.64 kcal mol^{-1} in the gas phase to −0.90 and −2.44 kcal mol^{-1} in benzene and water, respectively. The activation energies, which were computed for the exergonic reactions, are reported in Table 2.

When considering **M2**, the only strongly reactive site via HAT is 6a, i.e., the alcohol functional group. The selectivity is limited to the HO$^\bullet$ and CH$_3$O$^\bullet$. $\Delta G°_{HAT}$ values range from −11.24 kcal mol^{-1} for the alkoxyl radical in the gas phase to −27.09 kcal mol^{-1} for the hydroxyl radical in water. The activation energies increase with solvent polarity and are higher for this latter radical, with a value of 12.40 kcal mol^{-1} in water to be compared with 5.38 kcal mol^{-1} computed for the alkoxyl radical in the same conditions.

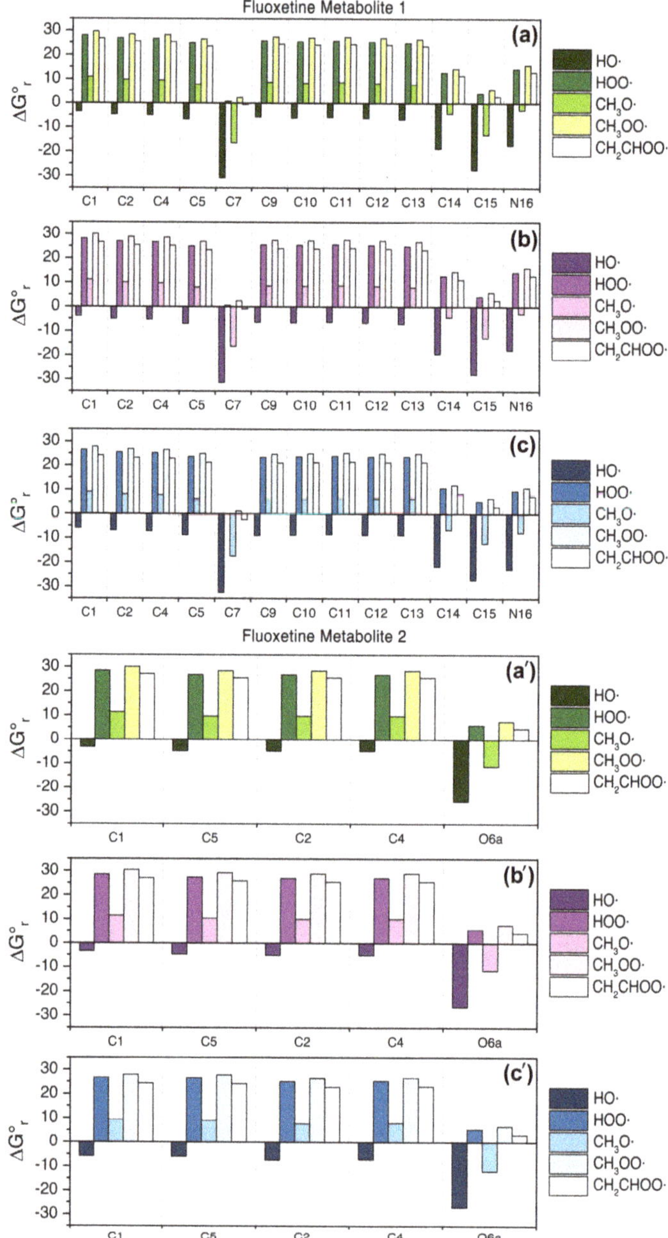

Figure 4. ΔG°_{HAT} (kcal mol^{-1}) in gas-phase for fluoxetine metabolite 1 (**a**) and fluoxetine metabolite 2 (**a′**), in benzene for fluoxetine metabolite 1 (**b**) and fluoxetine metabolite 2 (**b′**), and in water for fluoxetine metabolite 1 (**c**) and fluoxetine metabolite 2 (**c′**) for the scavenging of HO$^{\bullet}$, CH$_3$O$^{\bullet}$, HOO$^{\bullet}$, CH$_3$OO$^{\bullet}$, and CH$_2$=CHOO$^{\bullet}$ through HAT from all the available sites of fluoxetine metabolite 1 (**M1**) and metabolite 2 (**M2**). Level of theory: (SMD)-M06-2X/6-311+G(d, p)//M06-2X/6-31G(d).

Table 2. $\Delta G^{\ddagger}_{HAT}$ (kcal mol^{-1}) in the gas phase, benzene, and water, for the scavenging of HO$^{\bullet}$ and CH$_3$O$^{\bullet}$ through HAT in **M1** and **M2**. Level of theory: (SMD)-M06-2X/6-311+G(d,p)//M06-2X/6-31G(d).

Site	Gas Phase	Benzene	Water	Gas Phase	Benzene	Water
		HO$^{\bullet}$			CH$_3$O$^{\bullet}$	
7	5.40	7.99	11.58	12.23	15.19	17.63
14	5.36	7.33	10.44	13.64	15.96	18.05
15	4.35	5.42	6.69	9.79	11.07	11.55
16	3.84	4.87	5.71	11.37	12.61	13.07
6a [1]	6.21	8.26	12.40	1.83	3.11	5.38

[1] This is the only strongly reactive site (large negative ΔG°_{HAT}) of **M2** toward both HO$^{\bullet}$ and CH$_3$O$^{\bullet}$.

3.3. Serotonin

The radical scavenging activity of serotonin (**2**) was also systematically investigated, and the results are shown in Figure 5 and Table S3. The scavenging activity of **2** was expected for its well-known antioxidant capacity, which was likely related to its structural similarity to melatonin.

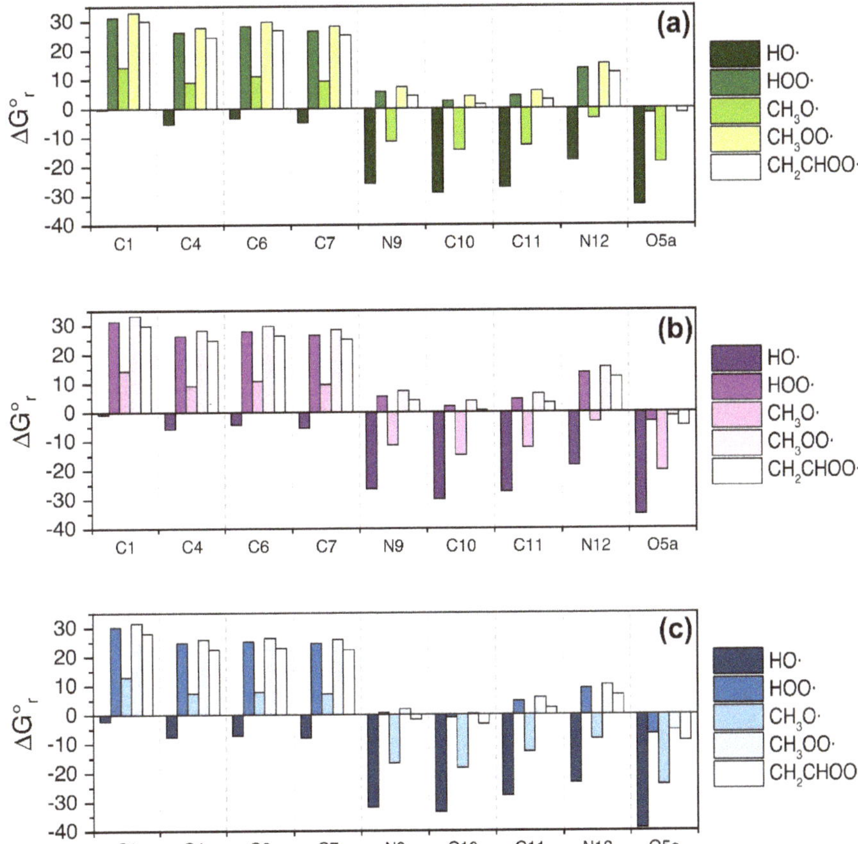

Figure 5. ΔG°_{HAT} (kcal mol^{-1}) in the gas phase (**a**), in water (**b**), and in benzene (**c**) for the scavenging of HO$^{\bullet}$, CH$_3$O$^{\bullet}$, HOO$^{\bullet}$, CH$_3$OO$^{\bullet}$, and CH$_2$=CHOO$^{\bullet}$ through HAT from all the available sites of **2**. Level of theory: (SMD)-M06-2X/6-311+G(d,p)//M06-2X/6-31G (d).

In the gas phase, the largest (negative) ΔG°_{HAT} values are computed for HAT to HO$^\bullet$ from sites 9, 10–12, and 5a, with values ranging from -18.02 kcal mol^{-1} (N of the primary amine) to -33.19 kcal mol^{-1} (O). The trend is maintained in the condensed phase, where the reaction energies become more negative with increasing solvent polarity, reaching the values of -23.50 kcal mol^{-1} (N of the primary amine) and -39.23 kcal mol^{-1} (O). Differently from **1** and its metabolites, serotonin also displays some modest activity toward peroxyl radicals, in particular CH2=CHOO$^\bullet$, when HAT occurs from the O site, with ΔG°_{HAT} going from -1.60 kcal mol^{-1} in the gas phase to -4.75 and -9.05 kcal mol^{-1} in benzene and water, respectively.

The transition states and thus the activation energies were computed only for the exergonic reactions involving the hydroxyl and the methoxyl radical; they are reported in Table 3. In fact, the activation energies for HAT from the most thermodynamically favored site are much higher when considering the other radicals, even when HAT occurs from site 5a (Table S4). By inspecting the values shown in Table 3, in the gas phase and benzene, the activation energies for the amines (sites 9 and 12) are almost identical; conversely, in water, $\Delta G^\ddagger_{HAT}$ from the N of the secondary amine becomes almost negligible when the hydroxyl radical is considered, i.e., 0.08 kcal mol^{-1}, a value which, combined with a large negative ΔG°_{HAT}, i.e., -31.84 kcal mol^{-1}, reveals that the HAT from this site is very efficient. Conversely, the thermodynamically most favored process, which is HAT from 5a, has the highest activation energy in water, i.e., 11.23 kcal mol^{-1}.

Table 3. $\Delta G^\ddagger_{HAT}$ (kcal mol^{-1}) in the gas phase, benzene, and water, for the scavenging of HO$^\bullet$ and CH$_3$O$^\bullet$ through HAT in **2**. Level of theory: (SMD)-M06-2X/6-311+G(d,p)//M06-2X/6-31G(d).

Site	Gas Phase	Benzene	Water	Gas Phase	Benzene	Water
		HO$^\bullet$			CH$_3$O$^\bullet$	
9	5.61	6.87	0.08	11.97	14.14	11.12
10	6.86	8.02	7.40	14.43	15.73	14.08
11	6.35	6.91	6.46	12.12	12.92	11.44
12	5.40	6.06	5.14	13.13	14.10	13.10
5a	5.02	6.14	11.23	7.29	8.50	5.36

4. Discussion

The data on HAT thermodynamics and kinetics for fluoxetine and its two metabolites show a selectivity for hydroxyl and alkoxyl radicals, in analogy to what has been reported for melatonin [46]. The most reactive sites are the hydroxyl group and the amine groups. Rather large negative ΔG°_{HAT} values are also computed from the C atoms of the hydrocarbon chain, which is in agreement with the reactivity described for melatonin and zolpidem [38]. In order to verify whether there are more efficient scavenging channels for fluoxetine, we have also investigated the RAF mechanism involving the aromatic C sites. ΔG°_{RAF} values for HO$^\bullet$ were computed for **1** as well as for **2**, and are reported in Tables S5 and S6. Focusing on HO$^\bullet$, it is worth noticing that the scavenging reactions of **1** that are less exergonic via HAT are significantly more exergonic via RAF, as it emerges for when comparing sites 1, 5, and 13, for which the ΔG°_{HAT} values are -6.14, -3.40, and -5.68 kcal mol^{-1}, and the ΔG°_{RAF} values are -7.98, -9.20, and -8.29 kcal mol^{-1}, respectively. Small negative Gibbs free reaction energies imply a larger stabilization of the reactants than the products. When comparing HAT and RAF, since the reactants are the same, it is useful to inspect the radical product. In Figure 6, the example of the spin densities computed for the radical products obtained from HAT and RAF involving site 1 and HO$^\bullet$ is shown. RAF leads to the formation of a radical product in which the spin density is significantly more delocalized on the benzene ring and the adjacent oxygen atom.

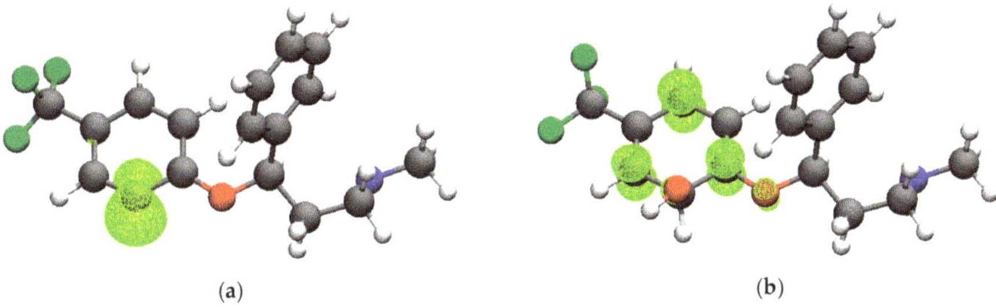

Figure 6. Spin densities on (1-H) $^{\bullet}$ when HAT occurs from C1 (**a**) (1OH) $^{\bullet}$ when RAF occurs at C1 (**b**). Level of theory: (SMD)-M06-2X/6-311+G(d,p)//M06-2X/6-31G(d).

There is no clear trend for RAF exergonicity when going from the gas phase to benzene and water. In general, the reactants should be more stabilized with increasing solvent polarity, and so the reaction Gibbs free energies should become more negative from benzene to water. However, this is not observed for sites 5, 6, and 8, which are located in a sterically crowded region of fluoxetine between the two orthogonal phenyl rings. In **2**, both hydroxyl and methoxyl radicals were tested, and we found that in gas-phase serotonin is selective via RAF for HO$^{\bullet}$ except from site 3. The process is weakly exergonic from site 8, too; these sites correspond to the junction carbon atoms. When CH$_3$O$^{\bullet}$ is considered, the HAT from these sites becomes strongly endergonic. Conversely, exergonic RAF occurs only on sites 1 and 4. The trend in solvent shows that the exergonicity increases with increasing solvent polarity, with the exceptions of sites 2–4 (HO$^{\bullet}$) and 4 (CH$_3$O$^{\bullet}$).

5. Conclusions

The major outcomes of our analysis can be summarized in the following. (i) Fluoxetine has antioxidant capacity acting as ROS scavenging via the HAT mechanism, and is selective for hydroxyl and methoxyl radicals. (ii) Scavenging may occur also via the RAF mechanism involving the C sites of the benzene rings. (iii) Serotonin is overall a better scavenger of hydroxyl and peroxyl radicals and, thanks to the presence of the alcoholic group, as in **M2**, which is one identified metabolite of fluoxetine, displays antioxidant activity also toward peroxyl radicals. These observations support the hypothesis that the clinical evidence of improvement of oxidative damage associated to fluoxetine therapy is likely due to the enhanced concentration of free serotonin, i.e., is an indirect effect.

Due to the seriousness of depressive disorders and their social implication, strategies to improve clinical symptoms are of considerable importance; particularly, the antioxidant capacity combined with the main therapeutic action of existing drugs is an idea that should be further investigated, and likely screened in silico as well as tested in vitro/vivo.

Supplementary Materials: The following are available online at http://www.mdpi.com/2076-3417/9/17/3631/s1: Tables S1–S6 (Gibbs free reaction and activation energies in the gas phase and solvents of the HAT and RAF processes here described); Cartesian coordinates of all the intermediates and transition states reported in this work.

Author Contributions: Conceptualization, L.O., G.Z. and C.P.; methodology, L.O.; investigation, C.M. and M.D.T.; resources, L.O.; data curation, C.M., M.D.T. and G.R.; writing—original draft preparation, L.O., C.P. and G.R.; writing—review and editing, L.O. and G.Z.; visualization, M.D.T.; supervision, L.O.; funding acquisition, L.O.

Funding: This research was funded by the Università degli Studi di Padova, thanks to the P-DiSC (BIRD2018-UNIPD) project MAD^{3}S (Modeling Antioxidant Drugs: Design and Development of computer-aided molecular Systems); P.I. L.O. All the calculations were carried out on Marconi (CINECA: Casalecchio di Reno, Italy) thanks to the ISCRA Grant REBEL2 (REdox state role in Bio-inspired ELementary reactions 2), P.I.: L.O.; M.D.T. is grateful to Fondazione CARIPARO for financial support (Ph.D. grant).

Conflicts of Interest: The authors declare no conflict of interest.

References

1. Wong, D.T.; Horng, J.S.; Bymaster, F.P.; Hauser, K.L.; Molloy, B.B. A selective inhibitor of serotonin uptake: Lilly 110140, 3-(p-Trifluoromethylphenoxy)-n-methyl-3-phenylpropylamine. *Life Sci.* **1974**, *15*, 471–479. [CrossRef]
2. Wong, D.T.; Perry, K.W.; Bymaster, F.P. The Discovery of Fluoxetine Hydrochloride (Prozac). *Nat. Rev. Drug Discov.* **2005**, *4*, 764–774. [CrossRef] [PubMed]
3. Hiemke, C.; Härtter, S. Pharmacokinetics of selective serotonin reuptake inhibitors. *Pharmacol. Ther.* **2000**, *85*, 11–28. [CrossRef]
4. Fuller, R.W.; Beasley, C.M. Fluoxetine mechanism of action. *J. Am. Acad. Child Adolesc. Psychiatry* **1991**, *30*, 849. [PubMed]
5. Mandrioli, R.; Forti, G.; Raggi, M. Fluoxetine metabolism and pharmacological interactions: The role of cytochrome P450. *Curr. Drug Metab.* **2006**, *7*, 127–133. [CrossRef] [PubMed]
6. Hancu, G.; Cârcu-Dobrin, M.; Budău, M.; Rusu, A. Analytical methodologies for the stereoselective determination of fluoxetine: An overview. *Biomed. Chromatogr.* **2018**, *32*, e4040. [CrossRef] [PubMed]
7. Ni, Y.G.; Miledi, R. Blockage of 5HT2C serotonin receptors by fluoxetine (Prozac). *Proc. Natl. Acad. Sci. USA* **1997**, *94*, 2036–2040. [CrossRef]
8. Wood, A.J.J.; Gram, L.F. Fluoxetine. *N. Engl. J. Med.* **1994**, *331*, 1354–1361. [CrossRef]
9. Zhou, Z.; Zhen, J.; Karpowich, N.K.; Law, C.J.; Reith, M.E.A.; Wang, D.-N. Antidepressant specificity of serotonin transporter suggested by three LeuT–SSRI structures. *Nat. Struct. Mol. Biol.* **2009**, *16*, 652–657. [CrossRef]
10. Spinks, D.; Spinks, G. Serotonin reuptake inhibition: An update on current research strategies. *Curr. Med. Chem.* **2002**, *9*, 799–810. [CrossRef]
11. Ciribassi, J.; Luescher, A.; Pasloske, K.S.; Robertson-Plouch, C.; Zimmerman, A.; Kaloostian-Whittymore, L. Comparative bioavailability of fluoxetine after transdermal and oral administration to healthy cats. *Am. J. Vet. Res.* **2003**, *64*, 994–998. [CrossRef] [PubMed]
12. Caraci, F.; Tascedda, F.; Merlo, S.; Benatti, C.; Spampinato, S.F.; Munafò, A.; Leggio, G.M.; Nicoletti, F.; Brunello, N.; Drago, F.; et al. Fluoxetine prevents Aβ1-42-induced toxicity via a paracrine signaling mediated by transforming-growth-factor-β1. *Front. Pharmacol.* **2016**, *7*, 389. [CrossRef] [PubMed]
13. Black, C.N.; Bot, M.; Scheffer, P.G.; Cuijpers, P.; Penninx, B.W.J.H. Is depression associated with increased oxidative stress? A systematic review and meta-analysis. *Psychoneuroendocrinology* **2015**, *51*, 164–175. [CrossRef] [PubMed]
14. Allen, J.; Romay-Tallon, R.; Brymer, K.J.; Caruncho, H.J.; Kalynchuk, L.E. Mitochondria and mood: Mitochondrial dysfunction as a key player in the manifestation of depression. *Front. Neurosci.* **2018**, *12*, 386. [CrossRef] [PubMed]
15. Novío, S.; Núñez, M.J.; Amigo, G.; Freire-Garabal, M. Effects of fluoxetine on the oxidative status of peripheral blood leucocytes of restraint-stressed mice. *Basic Clin. Pharmacol. Toxicol.* **2011**, *109*, 365–371. [CrossRef] [PubMed]
16. Réus, G.Z.; Matias, B.I.; Maciel, A.L.; Abelaira, H.M.; Ignácio, Z.M.; de Moura, A.B.; Matos, D.; Danielski, L.G.; Petronilho, F.; Carvalho, A.F.; et al. Mechanism of synergistic action on behavior, oxidative stress and inflammation following co-treatment with ketamine and different antidepressant classes. *Pharmacol. Rep.* **2017**, *69*, 1094–1102. [CrossRef] [PubMed]
17. Robertson, O.D.; Coronado, N.G.; Sethi, R.; Berk, M.; Dodd, S. Putative neuroprotective pharmacotherapies to target the staged progression of mental illness. *Early Interv. Psychiatry* **2019**. [CrossRef] [PubMed]
18. Herbet, M.; Gawrońska-Grzywacz, M.; Izdebska, M.; Piątkowska-Chmiel, I. Effect of the interaction between atorvastatin and selective serotonin reuptake inhibitors on the blood redox equilibrium. *Exp. Ther. Med.* **2016**, *12*, 3440–3444. [CrossRef] [PubMed]
19. Caiaffo, V.; Oliveira, B.D.R.; De Sá, F.B.; Evêncio Neto, J. Anti-inflammatory, antiapoptotic, and antioxidant activity of fluoxetine. *Pharmacol. Res. Perspect.* **2016**, *4*, e00231. [CrossRef] [PubMed]
20. Behr, G.A.; Moreira, J.C.F.; Frey, B.N. Preclinical and clinical evidence of antioxidant effects of antidepressant agents: Implications for the pathophysiology of major depressive disorder. *Oxid. Med. Cell. Longev.* **2012**, *2012*, 609421. [CrossRef]

21. Erman, H.; Guner, I.; Yaman, M.O.; Uzun, D.D.; Gelisgen, R.; Aksu, U.; Yelmen, N.; Sahin, G.; Uzun, H. The effects of fluoxetine on circulating oxidative damage parameters in rats exposed to aortic ischemia–reperfusion. *Eur. J. Pharmacol.* **2015**, *749*, 56–61. [CrossRef] [PubMed]

22. Kolla, N.; Wei, Z.; Richardson, J.S.; Li, X.M. Amitriptyline and fluoxetine protect PC12 cells from cell death induced by hydrogen peroxide. *J. Psychiatry Neurosci.* **2005**, *30*, 196–201. [PubMed]

23. Safhi, M.M.; Qumayri, H.M.; Masmali, A.U.M.; Siddiqui, R.; Alam, M.F.; Khan, G.; Anwer, T. Thymoquinone and fluoxetine alleviate depression via attenuating oxidative damage and inflammatory markers in type-2 diabetic rats. *Arch. Physiol. Biochem.* **2019**, *125*, 150–155. [CrossRef] [PubMed]

24. Dalmizrak, O.; Teralı, K.; Asuquo, E.B.; Ogus, I.H.; Ozer, N. The relevance of glutathione reductase inhibition by fluoxetine to human health and disease: Insights derived from a combined kinetic and docking study. *Protein J.* **2019**, 1–10. [CrossRef] [PubMed]

25. Kalogiannis, M.; Delikatny, E.J.; Jeitner, T.M. Serotonin as a putative scavenger of hypohalous acid in the brain. *Biochim. Biophys. Acta Mol. Basis Dis.* **2016**, *1862*, 651–661. [CrossRef] [PubMed]

26. Azouzi, S.; Santuz, H.; Morandat, S.; Pereira, C.; Côté, F.; Hermine, O.; El Kirat, K.; Colin, Y.; Le Van Kim, C.; Etchebest, C.; et al. Antioxidant and membrane binding properties of serotonin protect lipids from oxidation. *Biophys. J.* **2017**, *112*, 1863–1873. [CrossRef] [PubMed]

27. Lobayan, R.M.; Schmit, M.C.P. Conformational and NBO studies of serotonin as a radical scavenger. Changes induced by the OH group. *J. Mol. Graph. Model.* **2018**, *80*, 224–237. [CrossRef] [PubMed]

28. Da Rocha, A.M.; Kist, L.W.; Almeida, E.A.; Silva, D.G.H.; Bonan, C.D.; Altenhofen, S.; Kaufmann, C.G.; Bogo, M.R.; Barros, D.M.; Oliveira, S.; et al. Neurotoxicity in zebrafish exposed to carbon nanotubes: Effects on neurotransmitters levels and antioxidant system. *Comp. Biochem. Physiol. Part C Toxicol. Pharmacol.* **2019**, *218*, 30–35. [CrossRef] [PubMed]

29. Sarikaya, S.B.O.; Gulcin, I. Radical scavenging and antioxidant capacity of serotonin. *Curr. Bioact. Compd.* **2013**, *9*, 143–152. [CrossRef]

30. Orian, L.; Mauri, P.; Roveri, A.; Toppo, S.; Benazzi, L.; Bosello-Travain, V.; De Palma, A.; Maiorino, M.; Miotto, G.; Zaccarin, M.; et al. Selenocysteine oxidation in glutathione peroxidase catalysis: An MS-supported quantum mechanics study. *Free Radic. Biol. Med.* **2015**, *87*, 1–14. [CrossRef] [PubMed]

31. Bortoli, M.; Torsello, M.; Bickelhaupt, F.M.; Orian, L. Role of the Chalcogen (S, Se, Te) in the oxidation mechanism of the glutathione peroxidase active site. *ChemPhysChem* **2017**, *18*, 2990–2998. [CrossRef] [PubMed]

32. Maiorino, M.; Bosello-Travain, V.; Cozza, G.; Miotto, G.; Orian, L.; Roveri, A.; Toppo, S.; Zaccarin, M.; Ursini, F. *Glutathione Peroxidase 4 from Selenium: Its Molecular Biology and Role in Human Health*, 4th ed.; Springer International Publisher: New York, NY, USA, 2016; ISBN 9783319412832.

33. Orian, L.; Toppo, S. Organochalcogen peroxidase mimetics as potential drugs: A long story of a promise still unfulfilled. *Free Radic. Biol. Med.* **2014**, *66*, 65–74. [CrossRef] [PubMed]

34. Wolters, L.P.; Orian, L. Peroxidase activity of organic selenides: Mechanistic insights from quantum chemistry. *Curr. Org. Chem.* **2016**, *20*, 189–197. [CrossRef]

35. Ribaudo, G.; Bellanda, M.; Menegazzo, I.; Wolters, L.P.; Bortoli, M.; Ferrer-Sueta, G.; Zagotto, G.; Orian, L. Mechanistic insight into the oxidation of organic phenylselenides by H_2O_2. *Chem. A Eur. J.* **2017**, *23*, 2405–2422. [CrossRef] [PubMed]

36. Bortoli, M.; Zaccaria, F.; Dalla Tiezza, M.; Bruschi, M.; Fonseca Guerra, C.; Bickelhaupt, F.M.; Orian, L. Oxidation of organic diselenides and ditellurides by H_2O_2 for bioinspired catalyst design. *Phys. Chem. Chem. Phys.* **2018**, *20*, 20874–20885. [CrossRef] [PubMed]

37. Dalla Tiezza, M.; Ribaudo, G.; Orian, L. Organodiselenides: Organic catalysis and drug design learning from glutathione peroxidase. *Curr. Org. Chem.* **2019**. [CrossRef]

38. Bortoli, M.; Dalla Tiezza, M.; Muraro, C.; Pavan, C.; Ribaudo, G.; Rodighiero, A.; Tubaro, C.; Zagotto, G.; Orian, L. Psychiatric disorders and oxidative injury: Antioxidant effects of zolpidem therapy disclosed in silico. *Comput. Struct. Biotechnol. J.* **2019**, *17*, 311–318. [CrossRef] [PubMed]

39. Grimme, S.; Bannwarth, C.; Shushkov, P. A robust and accurate tight-binding quantum chemical method for structures, vibrational frequencies, and noncovalent interactions of large molecular systems parametrized for all spd-block elements ($Z = 1$–86). *J. Chem. Theory Comput.* **2017**, *13*, 1989–2009. [CrossRef] [PubMed]

40. Bannwarth, C.; Ehlert, S.; Grimme, S. GFN2-xTB—An accurate and broadly parametrized self-consistent tight-binding quantum chemical method with multipole electrostatics and density-dependent dispersion contributions. *J. Chem. Theory Comput.* **2019**, *15*, 1652–1671. [CrossRef] [PubMed]

41. Grimme, S. Exploration of chemical compound, conformer, and reaction space with meta-dynamics simulations based on tight-binding quantum chemical calculations. *J. Chem. Theory Comput.* **2019**, *15*, 2847–2862. [CrossRef] [PubMed]

42. Zhao, Y.; Truhlar, D.G. The M06 suite of density functionals for main group thermochemistry, thermochemical kinetics, noncovalent interactions, excited states, and transition elements: Two new functionals and systematic testing of four M06-class functionals and 12 other function. *Theor. Chem. Acc.* **2008**, *120*, 215–241. [CrossRef]

43. *Gaussian 16*, Revision B.01. Frisch, M.J.; Trucks, G.W.; Schlegel, H.B.; Scuseria, G.E.; Robb, M.A.; Cheeseman, J.R.; Scalmani, G.; Barone, V.; Petersson, G.A.; Nakatsuji, H.; Li, X.; Caricato, M.; Marenich, A.V.; Bloino, J.; Janesko, B.G.; Gomperts, R.; Mennucci, B.; Hratchian, H.P.; Ortiz, J.V.; Izmaylov, A.F.; Sonnenberg, J.L.; Williams; Ding, F.; Lipparini, F.; Egidi, F.; Goings, J.; Peng, B.; Petrone, A.; Henderson, T.; Ranasinghe, D.; Zakrzewski, V.G.; Gao, J.; Rega, N.; Zheng, G.; Liang, W.; Hada, M.; Ehara, M.; Toyota, K.; Fukuda, R.; Hasegawa, J.; Ishida, M.; Nakajima, T.; Honda, Y.; Kitao, O.; Nakai, H.; Vreven, T.; Throssell, K.; Montgomery, J.A., Jr.; Peralta, J.E.; Ogliaro, F.; Bearpark, M.J.; Heyd, J.J.; Brothers, E.N.; Kudin, K.N.; Staroverov, V.N.; Keith, T.A.; Kobayashi, R.; Normand, J.; Raghavachari, K.; Rendell, A.P.; Burant, J.C.; Iyengar, S.S.; Tomasi, J.; Cossi, M.; Millam, J.M.; Klene, M.; Adamo, C.; Cammi, R.; Ochterski, J.W.; Martin, R.L.; Morokuma, K.; Farkas, O.; Foresman, J.B.; Fox, D.J. Gaussian Inc.: Wallingford, CT, USA, 2016.

44. Marenich, A.V.; Cramer, C.J.; Truhlar, D.G. Universal Solvation model based on solute electron density and on a continuum model of the solvent defined by the bulk dielectric constant and atomic surface tensions. *J. Phys. Chem. B* **2009**, *113*, 6378–6396. [CrossRef] [PubMed]

45. Antony, J.; Sure, R.; Grimme, S. Using dispersion-corrected density functional theory to understand supramolecular binding thermodynamics. *Chem. Commun.* **2015**, *51*, 1764–1774. [CrossRef] [PubMed]

46. Galano, A. On the direct scavenging activity of melatonin towards hydroxyl and a series of peroxyl radicals. *Phys. Chem. Chem. Phys.* **2011**, *13*, 7178. [CrossRef] [PubMed]

47. Glendening, E.D.; Landis, C.R.; Weinhold, F. *NBO 6.0*: Natural bond orbital analysis program. *J. Comput. Chem.* **2013**, *34*, 1429–1437. [CrossRef] [PubMed]

48. Lu, T.; Chen, F. Multiwfn: A multifunctional wavefunction analyzer. *J. Comput. Chem.* **2012**, *33*, 580–592. [CrossRef] [PubMed]

49. Bortoli, M.; Wolters, L.P.; Orian, L.; Bickelhaupt, F.M. Addition-elimination or nucleophilic substitution? Understanding the energy profiles for the reaction of chalcogenolates with dichalcogenides. *J. Chem. Theory Comput.* **2016**, *12*, 2752–2761. [CrossRef] [PubMed]

50. Alberto, M.E.; Russo, N.; Grand, A.; Galano, A. A physicochemical examination of the free radical scavenging activity of trolox: Mechanism, kinetics and influence of the environment. *Phys. Chem. Chem. Phys.* **2013**, *15*, 4642. [CrossRef] [PubMed]

51. Galano, A.; Tan, D.X.; Reiter, R.J. Melatonin as a natural ally against oxidative stress: A physicochemical examination. *J. Pineal Res.* **2011**, *51*, 1–16. [CrossRef]

52. Galano, A.; Martínez, A. Capsaicin, a tasty free radical scavenger: Mechanism of action and kinetics. *J. Phys. Chem. B* **2012**, *116*, 1200–1208. [CrossRef]

53. Van Harten, J. Clinical pharmacokinetics of selective serotonin reuptake inhibitors. *Clin. Pharmacokinet.* **1993**, *24*, 203–220. [CrossRef] [PubMed]

54. Strauss, W.L.; Layton, M.E.; Hayes, C.E.; Dager, S.R. 19F magnetic resonance spectroscopy investigation in vivo of acute and steady-state brain fluvoxamine levels in obsessive-compulsive disorder. *Am. J. Psychiatry* **1997**, *154*, 516–522. [PubMed]

55. Benfield, P.; Heel, R.C.; Lewis, S.P. Fluoxetine. *Drugs* **1986**, *32*, 481–508. [CrossRef] [PubMed]

56. Margolis, J.M.; O'Donnell, J.P.; Mankowski, D.C.; Ekins, S.; Obach, R.S. (R)-, (S)-, and Racemic fluoxetine n-demethylation by human cytochrome P450 enzymes. *Drug Metab. Dispos.* **2000**, *28*, 1187–1191. [PubMed]

57. Zarghi, A.; Kebriaeezadeh, A.; Ahmadkhaniha, R.; Akhgari, M.; Rastkari, N. Selective liquid chromatographic method for determination of fluoxetine in plasma. *J. AOAC Int.* **2001**, *84*, 1735–1737. [PubMed]

58. Gupta, R.N.; Steiner, M. Determination of fluoxetine and norfluoxetine in serum by liquid chromatography with fluorescence detection. *J. Liq. Chromatogr.* **1990**, *13*, 3785–3797. [CrossRef]

59. Rohrig, T.P.; Prouty, R.W. A nortriptyline death with unusually high tissue concentrations. *J. Anal. Toxicol.* **1989**, *13*, 303–304. [CrossRef] [PubMed]

Article

Phase Equalization, Charge Transfer, Information Flows and Electron Communications in Donor–Acceptor Systems

Roman F. Nalewajski

Department of Theoretical Chemistry, Jagiellonian University, Gronostajowa 2, 30-387 Cracow, Poland; nalewajs@chemia.uj.edu.pl

Received: 14 March 2020; Accepted: 23 April 2020; Published: 23 May 2020

Abstract: Subsystem phases and electronic flows involving the acidic and basic sites of the donor (B) and acceptor (A) substrates of chemical reactions are revisited. The emphasis is placed upon the phase–current relations, a coherence of elementary probability flows in the preferred reaction complex, and on phase-equalization in the equilibrium state of the whole reactive system. The overall and partial charge-transfer (CT) phenomena in alternative coordinations are qualitatively examined and electronic communications in A—B systems are discussed. The internal polarization (P) of reactants is examined, patterns of average electronic flows are explored, and energy changes associated with P/CT displacements are identified using the chemical potential and hardness descriptors of reactants and their active sites. The nonclassical (phase/current) contributions to resultant gradient information are investigated and the preferred current-coherence in such donor–acceptor systems is predicted. It is manifested by the equalization of equilibrium local phases in the entangled subsystems.

Keywords: chemical reactivity theory; coordination complexes; donor–acceptor systems; partial electronic flows; phase–current relations; subsystem phases

---

1. Introduction

The Information Theory (IT) [1–8] of Fisher [1] and Shannon [3] has been successfully applied in an entropic interpretation of the molecular electronic structure (e.g., [9–11]). Several information principles have been investigated [9–16] and pieces of molecular electron density attributed to Atoms-in-Molecules (AIM) have been approached [12,16–20], providing the IT basis for the intuitive stockholder division of Hirshfeld [21]. Patterns of entropic bond multiplicities have been extracted from electronic communications in molecules [9–11,22–32], information distributions in molecules have been explored [9–11,33,34], and the nonadditive Fisher (gradient) information [1,2,9–11,35,36] has been linked to Electron Localization Function (ELF) [37–39] of Density Functional Theory (DFT) [40–45]. This analysis has enabled a formulation of the novel Contragradience (CG) probe for localizing chemical bonds [9–11,46], while the Orbital Communication Theory (OCT) of the chemical bond using the "cascade" propagations in molecular information systems has identified the bridge interactions between AIM [11,47–52], realized through orbital intermediates.

In molecular Quantum Mechanics (QM), the wavefunction phase or its gradient determining the effective velocity of probability density and its current give rise to nonclassical information/entropy supplements to classical measures of Fisher [1] and Shannon [3]. In resultant IT descriptors of electronic states, the information/entropy content in the probability (wavefunction modulus) distribution is combined with the relevant complement due to the current density (wavefunction phase) [53–62]. The overall gradient information is then proportional to the expectation value of the system kinetic energy of electrons. Such combined descriptors are also required in the phase distinction between the bonded (entangled) and nonbonded (disentangled) states of molecular subsystems, e.g., the substrate fragments of reactive systems [63–65]. This generalized treatment then allows one to interpret the variational principle for electronic energy as equivalent information rule, and to use the molecular virial theorem [66] in general reactivity considerations [67–71].

The extremum principles for the global and local (gradient) measures of the state resultant entropy have determined the phase-transformed ("equilibrium") states of molecular fragments, identified by their local "thermodynamic" phases [53–57]. The minimum-energy principle of QM has also been interpreted as physically-equivalent variational rule for the resultant gradient information, proportional to the state average kinetic energy [67–71]. In the *grand*-ensemble framework, they both determine the same thermodynamic equilibrium in an externally-open molecular system. This equivalence parallels identical predictions resulting from the minimum-energy and maximum-entropy principles of ordinary thermodynamics [72].

Elsewhere [63,73,74], the potential use of the DFT construction by Harriman, Zumbach, and Maschke (HZM) [75,76], of wavefunctions yielding the prescribed electron distribution, in a description of reactive systems has been examined. In such density-constrained Slater determinants, the defining Equidensity Orbitals (EO), of the Macke/Gilbert [77,78] type, exhibit the same molecular probability density, with the orbital orthogonality being assured by the EO local phases alone. These orbital states define the constrained multicomponent system composed of the mutually-closed orbital units, with each subsystem being characterized by its own phase and separate chemical-potential descriptors. Their simultaneous opening onto a common electron reservoir, and hence also onto themselves, generates the externally- and mutually-open orbital system, in which the EO fragments are effectively "bonded" (entangled) [63]. They then exhibit a common (molecular) phase descriptor and equalize their chemical potentials at the global reservoir level. This (mixed) equilibrium state is determined by the density operator corresponding to thermodynamic (grand-ensemble) probabilities of EO related to their orbital energies and average electron occupations.

In the present analysis, we focus on the phase component of electronic wavefunctions and the related current descriptor of molecular states, with special emphasis on the donor–acceptor reactive systems [63,67–71]. We examine the reactant phases and electronic flows involving the substrate acidic and basic active sites. The mutual relations between the state phase and current descriptors is explored, the phase-equalization in the equilibrium state of the whole reactive system is conjectured, and a coherence of probability flows in the preferred reaction complex is established. The overall and partial charge-transfer (CT) phenomena in alternative coordinations of the acidic (A) and basic (B) reactants are investigated, the electronic communications in alternative A—B complexes are qualitatively examined, and nonclassical contributions to the resultant gradient information are introduced. We also tackle the (internal) *polarization* (P) of reactants, induced by the (external) CT displacements between both substrates, and patterns of the resultant electronic flows on their active sites are qualitatively explored. The energy changes associated with specific P/CT fluxes in A—B systems are estimated using the familiar descriptors of the Charge Sensitivity Analysis (CSA) [79–85]: the global or regional chemical-potential/electronegativity [86–90] and hardness/softness [91] or Fukui function [92] descriptors of reactants and their acidic and basic active sites. The preferred coordination of reactants in such donor–acceptor systems is shown to exhibit a substantial current-coherence, which is also manifested by the equalization of the equilibrium local phases in the entangled (bonded) subsystems [63].

2. Molecular States and Their Phases

In QM, the state $\Psi(N)$ of N-electrons is represented by its representative vector in the abstract (Hilbert) space,

$$|\Psi(N)\rangle \equiv M|D(N)\rangle, \tag{1}$$

where M stands for its modulus ("length"),

$$M = \langle\Psi(N)|\Psi(N)\rangle^{1/2} \equiv |\Psi(N)|, \tag{2}$$

and $|D(N)\rangle$ denotes the corresponding directional ("unit") vector:

$$|D(N)| = \langle D(N)|D(N)\rangle^{1/2} = 1. \tag{3}$$

A variety of quantum states is exhausted by all admissible *orientations* of the unit vector $|D(N)\rangle$ [93]. Thus, for the unity-normalized state vectors, when $M = 1$, $|\Psi(N)\rangle \equiv |D(N)\rangle$.

The molecular state can be also identified by the spin-position representation of $|\Psi(N)\rangle$, called the wavefunction:

$$\Psi[\mathbf{Q}^{(N)}] = \langle \mathbf{Q}^{(N)}|\Psi(N)\rangle \equiv \Psi(N) = D[\Psi(N)]\,\exp\{i\Phi[\Psi(N)]\} \equiv D(N)\,F(N). \tag{4}$$

The squared modulus component,

$$D[\Psi(N)] = [\Psi(N)\Psi^{*}(N)]^{1/2} \equiv D(N), \tag{5}$$

determines the normalized probability distribution of N electrons:

$$P(N) = D(N)^2 = \langle \mathbf{Q}^{(N)}|\Psi(N)\rangle\,\langle \Psi(N)|\mathbf{Q}^{(N)}\rangle \equiv \langle \mathbf{Q}^{(N)}|P_\Psi(N)|\mathbf{Q}^{(N)}\rangle$$
$$= \langle \Psi(N)|\mathbf{Q}^{(N)}\rangle\,\langle \mathbf{Q}^{(N)}|\Psi(N)\rangle \equiv \langle \Psi(N)|P_\mathbf{Q}(N)|\Psi(N)\rangle,$$

$$\int P(N)\,d\mathbf{Q}^{(N)} \equiv 1, \tag{6}$$

Here, $P_\Psi(N)$ and $P_\mathbf{Q}(N)$ stand for the state and basis-set projection operators, respectively, while $\int d\mathbf{Q}^{(N)}$ denotes the integrations over spatial positions $\{r_k\}$ and summations over spin variables $\{\sigma_k\}$ of all N electrons. The state exponential factor $F(N)$ involves the N-electron phase function $\Phi[\Psi(N)]\} \equiv \Phi(N)$, which generates the state current density.

The wave function $\Psi(N)$ thus reflects projections (directional "cosines") of the state vector $|\Psi^{(N)}\rangle$ onto the basis vectors $\{|\mathbf{Q}^{(N)}\rangle\}$ of the adopted representation. It contains all essential "orientation" information about $|\Psi(N)\rangle$. This specific representation corresponds to the vector basis of N-electron states,

$$|\mathbf{Q}^{(N)}\rangle = |q_1, q_2, \dots, q_N\rangle = \{|q_k\rangle = |\sigma_k, r_k\rangle\}, \qquad k = 1, 2, \dots, N, \tag{7}$$

identified by the spin ($\{\sigma_k\}$) and position ($\{r_k\}$) variables of all N electrons. It includes the eigenvectors $\{|q_k\rangle\}$ of the electronic spin ($\mathbf{s}_k = i s_{k,x} + j s_{k,y} + k s_{k,z}$, $s_k = -|\mathbf{s}_k|$) and position ($\mathbf{r}_k$) operators:

$$\mathbf{s}_k{}^2|q_k\rangle = \frac{3}{4}\,\hbar^2|q_k\rangle \equiv s_k{}^2|q_k\rangle, \quad s_{k.z}|q_k\rangle = \sigma_k\,\hbar|q_k\rangle \equiv s_z|q_k\rangle, \qquad \sigma_k = \pm\frac{1}{2}, \quad \text{and}$$

$$\mathbf{r}_k|q_k\rangle = r_k|q_k\rangle. \tag{8}$$

In what follows, we also examine (normalized) molecular orbital (MO) states $\{|\psi_w\rangle\}$ of a single electron, $|\psi_w| = 1$. The state vector $|\psi_w\rangle$ is then given by the product of the spin ($|\xi_w\rangle$) and spatial ($|\varphi_w\rangle$) states:

$$|\psi_w\rangle \equiv |\xi_w\rangle|\varphi_w\rangle. \tag{9}$$

It generates the associated spin-orbital (SO) function in q-representation,

$$\psi_w(q) = \langle q|\psi_w\rangle = \langle \sigma|\xi_w\rangle\,\langle r|\varphi_w\rangle \equiv \xi_w(\sigma)\,\varphi_w(r), \tag{10}$$

the product of its normalized spin-function of an electron,

$$\xi_w(\sigma) = \langle \sigma|\xi_w\rangle, \qquad |\xi_w| = (\Sigma_\sigma\,|\xi_w(\sigma)|^2)^{1/2} = 1, \tag{11}$$

$$\xi_w(\sigma) \in \{\alpha(\sigma), \text{ spin-up}; \quad \beta(\sigma), \text{ spin-down}\}, \tag{12}$$

and the normalized (complex) spatial MO component,

$$\varphi_w(\mathbf{r}) = \langle \mathbf{r}|\varphi_w\rangle = d_w(\mathbf{r}) \exp[i\phi_w(\mathbf{r})] \equiv d_w(\mathbf{r}) f_w(\mathbf{r}), \tag{13}$$

$$\int \varphi_w(\mathbf{r})^* \varphi_w(\mathbf{r}) \, d\mathbf{r} = 1 \qquad \text{or} \qquad |\varphi_w| = 1, \tag{14}$$

with $p_w(\mathbf{r}) = d_w(\mathbf{r})^2$ defining the MO probability distribution. One customarily requires the spin-orbitals defining the N-electron configuration in a molecule to form the independent (orthonormal) set:

$$\langle \varphi_v|\varphi_w\rangle = \int \langle \varphi_v|\mathbf{r}\rangle\langle \mathbf{r}|\varphi_w\rangle \, d\mathbf{r} = \int \varphi_v(\mathbf{r})^* \varphi_w(\mathbf{r}) \, d\mathbf{r} = \delta_{k,l}. \tag{15}$$

To summarize, the complex MO wavefunction $\varphi_w(\mathbf{r})$ fully reflects the orientation properties of $|\varphi_w\rangle$ in the position representation. The square of its modulus $d_w(\mathbf{r}) = |\varphi_w(\mathbf{r})|$ determines the (normalized) spatial probability distribution in $|\psi_w\rangle$,

$$p_w(\mathbf{r}) = |\varphi_w(\mathbf{r})|^2 = d_w(\mathbf{r})^2 \geq 0, \qquad \int p_w(\mathbf{r}) \, d\mathbf{r} = 1, \tag{16}$$

while $|\xi_w(\sigma)|^2$ similarly determines the probability density of observing the specified spin component $s_z = \sigma\,\hbar$. The phase factor $f_w(\mathbf{r}) = \exp[i\phi_w(\mathbf{r})]$ identifies the orientation of the (normalized) state vector $|\varphi_w\rangle$ in the complex plane. It constitutes the $\mathbf{r}$-representation of the directional (unit) vector

$$|\mathrm{d}_w\rangle = |\varphi_w\rangle/|\varphi_w| = |\varphi_w\rangle, \qquad |\varphi_w| = \left[\int |\varphi_w(\mathbf{r})|^2 \, d\mathbf{r} \right]^{1/2} \equiv 1, \tag{17}$$

$$d_w(\mathbf{r}) = \langle \mathbf{r}|\mathrm{d}_w\rangle = \exp[i\phi_w(\mathbf{r})] = \cos\phi_w(\mathbf{r}) + i \sin\phi_w(\mathbf{r}). \tag{18}$$

The MO phase gradient ultimately determines the orbital current

$$j_w(\mathbf{r}) = [\hbar/(2mi)] \, [\varphi_w(\mathbf{r})^* \, \nabla\varphi_w(\mathbf{r}) - \varphi_w(\mathbf{r}) \, \nabla\varphi_w(\mathbf{r})^*] = (\hbar/m)] \, p_w(\mathbf{r}) \, \nabla\phi_w(\mathbf{r}). \tag{19}$$

As an illustration, we summarize in Appendix A the components of the quantum state describing a single electron, explore its probability/current descriptors, and summarize the relevant continuity relations. The latter result directly from the Schrödinger equation (SE) of molecular QM (see Appendix B).

One finally recalls that in the *one*-electron (MO) approximation the N-electron wavefunctions are defined as Slater determinants constructed from the configuration occupied SO, $\psi = (\psi_1, \psi_2, \dots, \psi_N)$,

$$\Psi(N) = \langle \mathbf{Q}^{(N)}|\Psi\rangle = (N!)^{-1/2} \, |\psi_1(\Psi)\psi_2(\Psi) \dots \psi_N(\Psi)| \equiv \det[\psi(\Psi)]. \tag{20}$$

3. Electronic Communications

In OCT of the chemical bond [9–11,22–32,62], one explores the entropy/information descriptors of molecular information channels, e.g., the electron probability networks in the orthonormal Atomic Orbital (AO) resolution $|\chi\rangle = \{|\chi_j\rangle, \quad j = 1, 2, \dots, t\}$,

$$\langle \chi_i|\chi_j\rangle = \int \langle \chi_i|\mathbf{r}\rangle \, \langle \mathbf{r}|\chi_j\rangle \, d\mathbf{r} = \int \chi_i(\mathbf{r})^* \, \chi_j(\mathbf{r}) \, d\mathbf{r} = \delta_{i,j}. \tag{21}$$

Such electron communications are reflected by the conditional probabilities of observing in the molecular bond system the monitored "output" orbitals, given the specified "input" orbitals.

In this standard LCAO MO approach, the optimum MO $|\varphi\rangle = \{|\varphi_w\rangle, \quad w = 1, 2, \ldots, t\}$ are represented as Linear Combinations of the AO basis functions:

$$\varphi(r) = \{\varphi_w(r) = \Sigma_j \chi_j(r)\, C_{j,w}\} = \chi(r)\, \mathbf{C}, \qquad \chi(r) = \langle r|\chi\rangle, \tag{22}$$

where the square matrix of the expansion coefficients, $\mathbf{C} = \langle \chi|\varphi\rangle = \{C_{i,w}\}$, satisfies the matrix orthonormality relations:

$$\mathbf{C}\mathbf{C}^{\dagger} = \{\delta_{i,j}\} = \mathbf{C}^{\dagger}\mathbf{C} = \{\delta_{w,w'}\} \equiv \mathbf{I}. \tag{23}$$

The AO-resolved communication theory explores the entropy/information descriptors of the electronic-signal propagation network between "input" and "output" AO states.

These scattering connections in the given electron configuration of Equation (20) are determined by its occupied MO, $\psi(\Psi)$. The representative probability of observing in Ψ the specified output AO $\chi_j(r) = \langle r|\chi_j\rangle = m_j(r)\, \exp[i\phi_j(r)]$, given the input AO $\chi_i(r) = \langle r|\chi_i\rangle = m_i(r)\, \exp[i\phi_i(r)]$,

$$P(\chi_j|\chi_i) = P(\chi_i, \chi_j)/P(\chi_i) \equiv P(j|i), \qquad \Sigma_j P(j|i) = 1, \tag{24}$$

is generated by the squared modulus of the associated conditional amplitude $A(\chi_j|\chi_i) \equiv A(j|i)$:

$$\{P(j|i) \equiv |A(j|i)|^2\}. \tag{25}$$

Here, $P(\chi_i) \equiv P(i)$ denotes the probability of detecting AO state $|\chi_i\rangle \equiv |i\rangle$ in the chemical bond system of the electron configuration in question, while $P(\chi_i, \chi_j) \equiv P(i, j)$ stands for the *joint*-probability of these two AO events, of simultaneously observing the two AO states $|i\rangle$ and $|j\rangle$ in Ψ. These probabilities satisfy the relevant normalizations:

$$\Sigma_i\, [\Sigma_j\, P(i, j)] = \Sigma_i\, P(i) = 1. \tag{26}$$

In accordance with the Superposition Principle (SP) of QM [93], the scattering amplitude $A(\zeta|\theta)$ between states $\theta(r) \equiv \langle r|\theta\rangle = m_\theta(r)\, \exp[i\phi_\theta(r)]$ and $\zeta(r) \equiv \langle r|\zeta\rangle = m_\zeta(r)\, \exp[i\phi_\zeta(r)]$ is defined by the mutual projection of the two state-vectors involved:

$$\begin{aligned} A(\zeta|\theta) = \langle\theta|\zeta\rangle &\equiv \int \theta(r)^{*}\, \zeta(r)\, dr = \int m_\theta(r)\, m_\zeta(r)\, \exp\{i[\phi_\zeta(r)-\phi_\theta(r)]\}\, dr \\ &\equiv \int M(\zeta|\theta)\, \exp\{i\Phi(\zeta|\theta)\}\, dr. \end{aligned} \tag{27}$$

This conditional amplitude is thus generated by the (local) effective modulus function $M(\zeta|\theta) = m_\theta(r)\, m_\zeta(r)$ and the phase component

$$\Phi(\zeta|\theta) = \phi_\zeta(r) - \phi_\theta(r). \tag{28}$$

Therefore, the conditional probability $P(\theta|\zeta)$ can be interpreted as the expectation value in one state of the (idempotent) projection operator onto another state:

$$\begin{aligned} P(\theta|\zeta) \equiv |A(\theta|\zeta)|^2 = \langle\zeta|\theta\rangle\, \langle\theta|\zeta\rangle &\equiv \langle\zeta|\mathrm{P}_\theta|\zeta\rangle \\ &= \langle\theta|\zeta\rangle\, \langle\zeta|\theta\rangle \equiv \langle\theta|\mathrm{P}_\zeta|\theta\rangle, \\ \mathrm{P}_\vartheta = |\vartheta\rangle\, \langle\vartheta|, \qquad \mathrm{P}_\vartheta{}^2 = \mathrm{P}_\vartheta, &\qquad \vartheta = (\theta, \zeta). \end{aligned} \tag{29}$$

In molecular electron configuration of Equation (20), each AO event is additionally conditional on the molecular state $\Psi(N)$, identified either by the N-electron projection operator $\mathrm{P}_\Psi(N) = |\Psi(N)\rangle\langle\Psi(N)|$ or the configuration (idempotent) *bond*-projection $\mathrm{P}_b(N)$ involving only the occupied MO selected by their finite occupation numbers

$$\mathbf{n}(\Psi) = \{n_w(\Psi)\delta_{w,w'}\},$$

$$n_w(\Psi) = 1\ (\Psi\text{-occupied MO}) \text{ or } n_w(\Psi) = 0\ (\Psi\text{-virtual MO}),$$

$$P_b(\Psi) = |\Psi(\Psi)\rangle\, \mathbf{n}(\Psi)\, \langle\Psi(\Psi)| = \Sigma_w\, |\psi_w(\Psi)\rangle\, n_w(\Psi)\, \langle\psi_w(\Psi)|$$
$$= \Sigma_s\, |\Psi_s(\Psi)\rangle\, \langle\Psi_s(\Psi)| \equiv \Sigma_s\, P_s(\Psi), \tag{30}$$

where $w = 1, 2, \dots, t$ (all SO, occupied, and virtual) and $s = 1, 2, \dots, N$ (occupied SO only).

The conditional probability of observing $|j\rangle$ given $|i\rangle$ in electronic configuration $|\Psi\rangle$ thus involves the *doubly*-conditional (coherent) amplitude $A_\Psi(\chi_j|\chi_i) \equiv A(\chi_j|\chi_i\|\Psi) \equiv A_\Psi(j|i)$:

$$P_\Psi(\chi_j|\chi_i) = P(\chi_j|\chi_i\|\Psi) \equiv P_\Psi(j|i) = |A_\Psi(j|i)|^2. \tag{31}$$

Its amplitude

$$A_\Psi(j|i) = \langle i|P_b(\Psi)|j\rangle = \Sigma_w\, \langle i|\psi_w(\Psi)\rangle\, n_w(\Psi)\, \langle\psi_w(\Psi)|j\rangle$$
$$= \Sigma_s\, \langle i|\Psi_s(\Psi)\rangle\, \langle\Psi_s(\Psi)|j\rangle = \Sigma_s\, C_{i,s}(\Psi)\, C_{j,s}(\Psi)^* \equiv \gamma_{i,j}(\Psi) \tag{32}$$

thus defines the relevant element of the (idempotent) Charge-and-Bond–Order (CBO) matrix of LCAO MO theory:

$$\gamma(\Psi) = \langle\chi|\Psi(\Psi)\rangle\, \mathbf{n}(\Psi)\, \langle\Psi(\Psi)|\chi\rangle = \mathbf{C}(\Psi)\, \mathbf{n}(\Psi)\, \mathbf{C}(\Psi)^\dagger = \{\gamma_{i,j}(\Psi)\}. \tag{33}$$

This molecular communication amplitude is thus characterized by the effective modulus and phase components of $\gamma_{i,j}(\Psi)$ resulting from corresponding descriptors of (complex) LCAO MO coefficients

$$\mathbf{C}(\Psi) = \{C_{k,w}(\Psi) = \langle\chi_k|\psi_w(\Psi)\rangle \equiv \langle k|\psi_w(\Psi)\rangle.$$

For the given electron configuration $|\Psi(N)\rangle$, it determines the molecular conditional probabilities between the specified pair of AO states. These scattering probabilities are now defined by the AO expectation values of the (nonidempotent) basis set projections in the occupied MO subspace,

$$P_l^b(\Psi) = P_b(\Psi)\, |l\rangle\, \langle l|\, P_b(\Psi) = P_b(\Psi)\, P_l\, P_b(\Psi), \qquad [P_l^b(\Psi)]^2 \neq P_l^b(\Psi), \tag{34}$$

of the system chemical bonds in $|\Psi(N)\rangle$:

$$P_\Psi(j|i) \equiv |A_\Psi(j|i)|^2 = \gamma_{i,j}(\Psi)\, \gamma_{j,i}(\Psi)$$
$$= \langle i|P_b(\Psi)|j\rangle\, \langle j|P_b(\Psi)|i\rangle = \langle i|P_b(\Psi)P_j P_b(\Psi)|i\rangle \equiv \langle i|P_j^b(\Psi)|i\rangle \tag{35}$$
$$= \langle j|P_b(\Psi)|i\rangle\, \langle i|P_b(\Psi)|j\rangle = \langle j|P_b(\Psi)P_i P_b(\Psi)|j\rangle \equiv \langle j|P_i^b(\Psi)|j\rangle.$$

4. Phase–Current Relation

In this section, we briefly explore the mutual relation between the phase and current descriptors [see Equation (19)] of the quantum state of an electron,

$$\psi(q) = \varphi(r)\xi(\sigma),$$

where

$$\varphi(r) = m(r)\, \exp[i\phi(r)]. \tag{36}$$

Its probability distribution reflects the square of the MO modulus factor, $p(r) = m(r)^2$, while its phase component $\phi(r)$ determines the state probability current of Equation (19):

$$j(r) = [\hbar/(2mi)]\, [\varphi(r)^*\, \nabla\varphi(r) - \varphi(r)\, \nabla\varphi(r)^*]$$
$$= (\hbar/m)\, p(r)\, \nabla\phi(r) \equiv p(r)\, V(r). \tag{37}$$

Here, the effective velocity of the probability "fluid",

$$V(r) = \{V_u(r) \equiv V_u(u, \{v \neq u\}), \quad u = x, y, z\},$$

measures the local current-per-particle and reflects the state phase gradient:

$$V(r) = j(r)/p(r) = (\hbar/m) \, \nabla \phi(r). \tag{38}$$

Therefore, velocity components of the probability flux are determined by the corresponding partials of the state phase function:

$$V_u(r) = (\hbar/m) \, [\partial \phi(r)/\partial u], \quad u = x, y, z. \tag{39}$$

It follows from this equation that the phase of electronic state can be formally reconstructed by an indefinite integration of the corresponding components of the velocity field of the probability flux:

$$\phi(u, \{v \neq u\}) = (m/\hbar) \int_{-\infty}^{u} V_u(\vartheta, \{v \neq u\})d\vartheta, u = x, y, z. \tag{40}$$

As an illustrative example, consider the fixed-momentum (free-particle) state

$$\varphi_K(r) = A \, \exp(iK \cdot r) = A \, \exp[i(K_x x + K_y y + K_z z),$$

$$K = (m/\hbar)V, \, V = iV_x + jV_y + kV_z = const. \tag{41}$$

One indeed recovers its phase components,

$$\phi(r) = K \cdot r = (m/\hbar) \, (V_x x + V_y y + V_z z), \tag{42}$$

by straightforward unidimensional integrations of Equation (40).

Consider now the phase/current dependent term of the overall content of MO resultant gradient-information [28,53–62] in Quantum Information Theory (QIT) [62]:

$$\begin{aligned} I[\phi] &= 4 \int p(r) \, [\nabla \phi(r)]^2 dr \equiv \int p(r) \, I_\phi(r) \, dr \\ &= (2m/\hbar)^2 \int p(r) \, V(r)^2 \, dr = (2m/\hbar)^2 \int p(r)^{-1} \, j(r)^2 \, dr. \end{aligned} \tag{43}$$

By analogy to the free-electron wavefunction of Equation (41), one introduces the wave-vector distribution in molecular systems, defined by the local phase gradient

$$K(r) = (m/\hbar) \, V(r) = \nabla \phi(r). \tag{44}$$

In terms of this local momentum measure, the nonclassical information functional simplifies:

$$I[\phi] = 4 \int p(r) \, K(r)^2 dr \equiv 4\langle K^2 \rangle. \tag{45}$$

Therefore, in such an inhomogeneous (molecular) electron density, the nonclassical MO information reflects the state average square $\langle K^2 \rangle$ of the wave-vector $K(r)$. It thus follows from the preceding equation that this nonclassical information measure, related to the current contribution to the state resultant kinetic energy, depends only on the magnitude of the local probability flow, being independent of its direction.

5. Current Coherence in Donor–Acceptor Systems

Electron flows in molecules are determined by the *energy* conditions. The molecular energetics is reflected by SE of QM. Indeed, the entropy/information flows in molecular systems are carried by the state electronic flows. The QIT description thus offers only a supplementary, alternative framework for exploring and ultimately better understanding the rules and structures of chemistry. The energetics of a nondegenerate ground-state of the whole molecular system is devoid of any phase aspect, while subtle preferences involving reactants may already involve local phases of electronic states in the substrate subsystems, reflecting their entanglement in the whole reactive complex [63–71].

As an illustrative example let us now recall the energy-preferred *complementary* reactive complex A—B [67,70,94,95] shown in Figure 1, consisting of the basic subsystem

$$B = (a_B \mid \ldots \mid b_B) \equiv (a_B|b_B)$$

and its reaction companion—the acidic substrate

$$A = (a_A \mid \ldots \mid b_A) \equiv (a_A|b_A),$$

where a_X and b_X denote the active acidic and basic sites of X, respectively. The four molecular fragments, $\lambda \in \{(a_A, b_A), (a_B, b_B)\}$, define the active parts in the system charge reconstruction. The acidic (electron acceptor) part is relatively harder, i.e., less responsive to external perturbations, thus exhibiting lower values of the fragment Fukui function or chemical softness descriptor, while the basic (electron donor) fragment is relatively more polarizable, as indeed reflected by higher response descriptors of its electron density or site populations. The acidic part a_X exerts an electron-*accepting* (stabilizing) influence on the neighboring part of another reactant Y, while the basic fragment b_X produces an electron-*donor* (destabilizing) effect on the fragment Y in its vicinity.

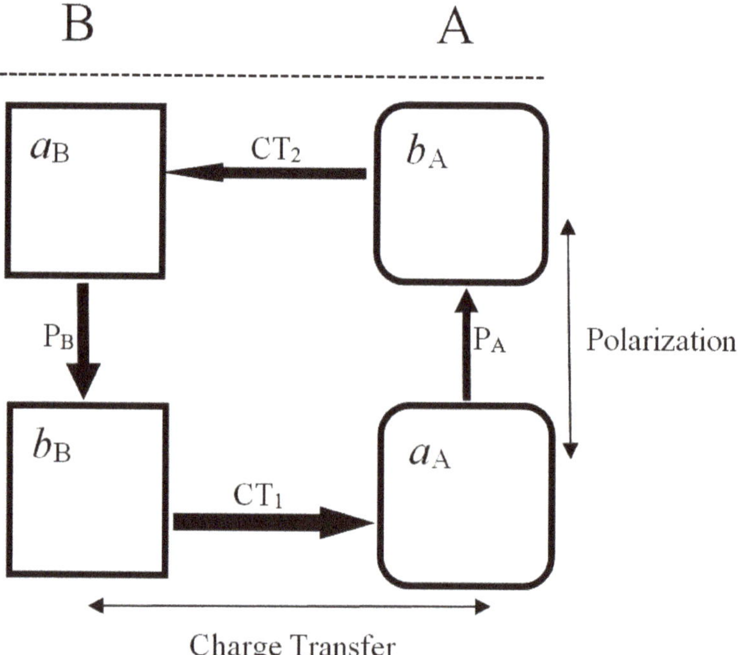

Figure 1. The Charge-Transfer (CT) $\{b_X \rightarrow a_Y\}$ and Polarizational (P$_X$) $\{a_X \rightarrow b_X\}$ electron flows involving acidic A = $(a_A|b_A)$ and basic B = $(a_B|b_B)$ reactants in the complementary arrangement R$_c$ of their acidic (a) and basic (b) sites in reactive complex R = (A|B). These electronic flows are seen to produce an effective concerted (circular) flux of electrons in the equilibrium reactive system R* = (A* | B*) = $(a_A{}^*|b_A{}^*|a_B{}^*|b_B{}^*)$ as a whole, with all fragments exhibiting the "flow-through" current pattern, which precludes an exaggerated depletion or accumulation of electrons on any site in reactive system.

In the most stable complementary (c) complex of Figure 1 [94,95], a geometrically accessible a-fragment of one reactant faces the geometrically accessible b-fragment of the other substrate

$$R_c \equiv \begin{bmatrix} a_A - b_B \\ b_A - a_B \end{bmatrix}, \tag{46}$$

while in a less favorable regional HSAB-type coordination the acidic (basic) fragment of one reactant faces the like-fragment of the other substrate:

$$R_{HSAB} \equiv \begin{bmatrix} a_A - a_B \\ b_A - b_B \end{bmatrix}. \tag{47}$$

A relative stability of R_c reflects an electrostatic preference: an electron-rich (repulsive, basic) fragment of one reactant indeed prefers to face an electron-deficient (attractive, acidic) part of the reaction partner. As shown in Figure 1, at finite separations between the two subsystems, spontaneous (primary) CT displacements between reactants trigger the induced (secondary) polarizational flows $\{P_X\}$ within each reactant, which restore the intra-substrate equilibria initially displaced by the presence of the other fragment and the inter-reactant CT:

$$R_c : \begin{array}{ccc} a_B & \leftarrow & b_A \\ \downarrow & & \uparrow \\ b_B & \rightarrow & a_A \end{array} , \quad R_{HSAB} : \begin{array}{ccc} a_B & \rightarrow & a_A \\ \uparrow & & \uparrow \\ b_B & \rightarrow & b_A \end{array} \tag{48}$$

It has been inferred from the (energetic) Electronegativity Equalization principle [95] that the electronic CT and P flows in R_c generate the *concerted* pattern shown in Figure 1 and Equation (48), which exhibits the maximum current (phase-gradient) coherence. It implies the least population activation on both reactants, which also energetically favors the complementary complex relative to the regional HSAB arrangement. Indeed, in R_{HSAB} coordination the energy-preferred *disconcerted* flow pattern implies a more exaggerated depletion of electrons on b_B and their more accentuated accumulation on a_A. In fact, these partial flows of electrons signify the "bridge" CT between the key ("diagonal") sites b_B and a_A, via the intermediate ("off-diagonal") sites a_B and b_A.

In the complementary arrangement of both reactants, the partial fluxes involving the four active sites indicate the "flow-through" behavior of all these fragments, which combines the external (CT) and internal (P) currents. One observes that the acidic sites accept electrons as a result of charge transfer between reactants and donate electrons through the substrate polarization. The opposite behavior of the basic sites is observed: they receive electrons due to P-mechanism and lose electrons as a result of CT. In the HSAB complex, one observes a similar flow-through pattern only on the a_B and b_A sites, b_B exclusively donates electrons both internally (P) and externally (CT), while a_A only accepts electrons from its complementary fragment b_A (P) and a_B (CT) part of the other reactant.

To summarize, in R_c, one observes the concerted flow of electrons involving all four active sites of both reactants, with small net changes in electron populations on these fragments, while the charge reconstruction in R_{HSAB} can be regarded as a transfer of electrons from b_B to a_A through the remaining (intermediate) sites a_B and b_A. The energetical preference of the complementary arrangement of reactants [94,95] also signifies the maximum *phase*-gradient coherence on each site, with its P or CT *inflow* part being accompanied by the associated outflow flux. Such a flow pattern thus corresponds to the least populational displacements on all constituent active sites. The above "activation" perspective then provides a natural physical explanation of the observed preference of the complementary coordination.

The energetically favored, concerted pattern of electronic fluxes in R_c can be realized only by an appropriate coherence of the site effective phase-gradients, of the pure or mixed states $\{\Psi_\lambda\}$ describing the mutually-closed fragments $\lambda \in \{a_A = 1, b_A = 2, a_B = 3, b_B = 4\}$ of both reactants. These states define the average resultant currents on each site, weighted by the site electron probability distribution $p_\lambda(r)$,

$$\langle j \rangle_\lambda = \int p_\lambda(r)\, j_\lambda(r)\, dr \equiv j(\lambda), \qquad \int p_\lambda(r)\, dr = 1, \tag{49}$$

already accounting for the average outflows-from and inflows-to the part in question, caused by the CT or P displacements in the system electronic structure.

The elementary P and CT currents of the flow-patterns in Equation (48) give rise to the corresponding vector sums $\{j(\lambda)\}$, the site *resultant* currents schematically drawn in Figure 2. They are seen to generate a "conrotatory" pattern in the complementary complex R_c and a collective "translational" pattern in R_{HSAB}. The magnitude $|\langle j \rangle_\lambda|$ of this average (directional) site descriptor of the probability-flow ultimately determines the size $|\Delta N_\lambda|$ of the net change in the fragment electron population in unit time,

$$\Delta N_\lambda = N_\lambda\, \langle j \rangle_\lambda, \tag{50}$$

where N_λ stands for the fragment average number of electrons in the separate reactant. The energy-favored (complementary) complex, which represents the least population-activation process, then corresponds to the lowest overall population displacement:

$$\Sigma_\lambda\, |\Delta N_\lambda| = \Sigma_\lambda\, N_\lambda\, |\langle j \rangle_\lambda| \equiv \Sigma_\lambda\, N_\lambda\, \langle j \rangle_\lambda \Rightarrow minimum. \tag{51}$$

(a)   (b)

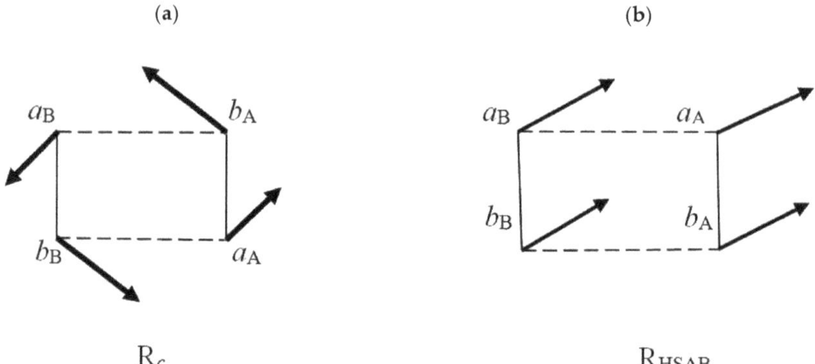

Figure 2. Qualitative diagrams of the *conrotatory* pattern of the resultant *site*-currents $\{j(\lambda)\}$ in the complementary complex R_c (a) and their *translational* pattern in the R_{HSAB} arrangement (b) of the acidic $A = (a_A|b_A)$ and basic $B = (a_B|b_B)$ reactants.

A reference to Equation (43) shows that the nonclassical gradient information of site λ also probes the average lengths of the directional phase-related properties:

$$\begin{aligned}
I_\lambda[\phi_\lambda] &= 4 \int p_\lambda(r)\, [\nabla\phi_\lambda(r)]^2 dr \\
&= (2m/\hbar)^2 \int p_\lambda(r)\, V_\lambda(r)^2\, dr \\
&= (2m/\hbar)^2 \int p_\lambda(r)^{-1}\, j_\lambda(r)^2\, dr.
\end{aligned} \tag{52}$$

Therefore, the least site-activation of Equation (51), which occurs for the current-coherence in reactive system, also implies the minimum of the (additive) overall nonclassical information content:

$$I[\phi] = \Sigma_\lambda\, I_\lambda[\phi_\lambda] \Rightarrow minimum. \tag{53}$$

Its absolute minimum is reached for the mutually-open (entangled) sites in the (bound) nondegenerate quantum state of the whole reactive system,

$$\mathrm{R}^* = (\mathrm{A}^* | \mathrm{B}^*) = (a_\mathrm{A}^* | b_\mathrm{A}^* | a_\mathrm{B}^* | b_\mathrm{B}^*),$$

when the local phase contribution identically vanishes. Moreover, in the information *equilibrium* state of the entangled molecular fragments, corresponding to the minimum of the system nonclassical gradient-information, this also implies the equalization of subsystem phases at the global phase of the wavefunction describing the reactive system as a whole (see also Appendix C).

The normalized probability distributions on sites $\{\kappa = (a, b)\}$ in reactants $\{X = (\mathrm{A}, \mathrm{B})\}$,

$$\{p_\lambda(\boldsymbol{r})\} = \{p_\kappa(\boldsymbol{r}; X) = \rho_\kappa(\boldsymbol{r}; X)/N_\kappa(X)\}, \qquad \int p_\kappa(\boldsymbol{r}; X)\, d\boldsymbol{r} = 1, \tag{54}$$

the shape factors of the fragment electronic densities $\{\rho_\kappa(\boldsymbol{r}; X)\}$, can be used as weights in determining the corresponding fragment (internal) averages of physical properties. For example, the site vector (directional) densities

$$\{k_\kappa(\boldsymbol{r}; X)\}, \qquad \{V_\kappa(\boldsymbol{r}; X)\} \qquad \text{and} \qquad \{j_\kappa(\boldsymbol{r}; X)\} \tag{55}$$

generate the associated average descriptors on each substrate:

$$\begin{aligned} \langle k(X)\rangle &= \Sigma_\kappa\, P_\kappa(X) \int p_\kappa(\boldsymbol{r}; X)\, k_\kappa(\boldsymbol{r}; X)\, d\boldsymbol{r} \equiv \Sigma_\kappa\, P_\kappa(X)\, \langle k(X)\rangle_\kappa \\ &\equiv \Sigma_\kappa \int w_\kappa(\boldsymbol{r}; X)\, k_\kappa(\boldsymbol{r}; X)\, d\boldsymbol{r} \equiv \int k(\boldsymbol{r}; X)\, d\boldsymbol{r}, \end{aligned} \tag{56}$$

$$\begin{aligned} \langle V(X)\rangle &= \Sigma_\kappa\, P_\kappa(X) \int p_\kappa(\boldsymbol{r}; X)\, V_\kappa(\boldsymbol{r}; X)\, d\boldsymbol{r} \equiv \Sigma_\kappa\, P_\kappa(X)\, \langle V(X)\rangle_\kappa \\ &\equiv \Sigma_\kappa \int w_\kappa(\boldsymbol{r}; X)\, V_\kappa(\boldsymbol{r}; X)\, d\boldsymbol{r} \equiv \int V(\boldsymbol{r}; X)\, d\boldsymbol{r}, \end{aligned} \tag{57}$$

$$\begin{aligned} \langle j(X)\rangle &= \Sigma_\kappa\, P_\kappa(X) \int p_\kappa(\boldsymbol{r}; X)\, j_\kappa(\boldsymbol{r}; X)\, d\boldsymbol{r} \equiv \Sigma_\kappa\, P_\kappa(X)\, \langle j(X)\rangle_\kappa \\ &\equiv \Sigma_\kappa \int w_\kappa(\boldsymbol{r})\, j_\kappa(\boldsymbol{r}; X)\, d\boldsymbol{r} \equiv \int j(\boldsymbol{r}; X)\, d\boldsymbol{r}. \end{aligned} \tag{58}$$

Here, $N(X) = \Sigma_\kappa\, N_\kappa(X)$ stands for the global number of electrons in reactant X,

$$P_\kappa(X) = N_\kappa(X)/N(X) \tag{59}$$

denotes the site probability in X, and the fragment and local weighting factors obey the usual (internal) normalizations for each reactant:

$$\Sigma_\kappa\, P_\kappa(X) = \Sigma_\kappa \int w_\kappa(\boldsymbol{r}; X)\, d\boldsymbol{r} = 1. \tag{60}$$

The corresponding quantities in the whole reactive system then result from weighting these substrate averages with their probabilities in R as a whole:

$$\{P_X(\mathrm{R}) = N(X)/N(\mathrm{R})\}, \qquad N(\mathrm{R}) = \Sigma_X\, N(X); \tag{61}$$

$$\langle k(\mathrm{R})\rangle = \Sigma_X\, P_X(\mathrm{R})\, \langle k(X)\rangle, \qquad \langle V(\mathrm{R})\rangle = \Sigma_X\, P_X(\mathrm{R})\, \langle V(X)\rangle, \qquad \langle j(\mathrm{R})\rangle = \Sigma_X\, P_X(\mathrm{R})\, \langle j(X)\rangle. \tag{62}$$

6. Overall Charge Transfer

Let us now briefly examine some energetic consequences of such displacements in electron populations on these functional sites of the acidic and basic subsystems [79–85]. The known (external) CT action of these substrates in the reactive complex R, the net inflow of electrons to A and an outflow from B, suggests the use of the "biased" estimates of the chemical potentials of the equilibrium reactant subsystems

$$\mathrm{A}^+ = (a_\mathrm{A}^+ | b_\mathrm{A}^+) \qquad \text{and} \qquad \mathrm{B}^+ = (a_\mathrm{B}^+ | b_\mathrm{B}^+)$$

in the polarized reactive complex $R_\alpha{}^+$, $\alpha = (c, \text{HSAB})$, defining its *substrate*-resolution. It combines the internally-open but mutually-closed reactants in presence of each other:

$$R_\alpha{}^+ = [A^+(\alpha)|B^+(\alpha)] = [a_A{}^+(\alpha) \mathbin{\vdots} b_A{}^+(\alpha) \mid a_B{}^+(\alpha) \mathbin{\vdots} b_B{}^+(\alpha)]. \tag{63}$$

Each coordination α defines its specific external potential due to the fixed nuclei in both subsystems,

$$v(\alpha) = v_A(\alpha) + v_B(\alpha), \tag{64}$$

which is constrained in definitions of the substrate populational derivatives of $R_\alpha{}^+$: the reactant chemical potentials and hardness descriptors.

The chemical potentials (negative electronegativities) of reactant subsystems in $R_\alpha{}^+$, $\mu(R_\alpha{}^+) = \{\mu_\alpha(X^+)\}$, represent partial derivatives of the system energy $E_\alpha{}^+[N(R_\alpha{}^+), v(\alpha)]$ with respect to the substrate electronic populations $N(R_\alpha{}^+) = \{N_\alpha(X^+)\}$ for the fixed external potential $v(\alpha)$, i.e., the "frozen" geometry of the whole reactive system,

$$\mu_\alpha(X^+) \equiv \partial E_\alpha{}^+[N(R_\alpha), v(\alpha)]/\partial N_\alpha(X^+)]_{v(\alpha)}, \qquad \alpha = (c, \text{HSAB}), \qquad X^+ \in (A^+, B^+). \tag{65}$$

These population "potentials" also reflect the subsystem electronegativities,

$$\chi_\alpha(X^+) \equiv \partial E_\alpha{}^+[Q(R_\alpha{}^+), v(\alpha)]/\partial Q_\alpha(X^+)]_{v(\alpha)} = -\mu_\alpha(X^+), \tag{66}$$

measuring the related partials of electronic energy $E_\alpha{}^+[Q(R_\alpha{}^+), v(\alpha)]$ with respect to the reactant net-charges $Q(R_\alpha{}^+) = \{Q_\alpha(X^+)\}$, $dQ_\alpha(X^+) = -dN_\alpha(X^+)$.

The internally-equalized chemical potentials

$$\mu(R_\alpha{}^+) = [\partial E_\alpha{}^+/\partial N(R_\alpha{}^+)]_{v(\alpha)} = \{\mu_\alpha(X^+) = \partial E_\alpha{}^+/\partial N_\alpha(X^+)\}$$

of the basic and acidic reactants, when acting as donor (B) and acceptor (A) of electrons, respectively, then read:

$$\begin{aligned}
\mu_\alpha(A^+) &= \mu_\alpha(a_A{}^+) = \mu_\alpha(b_A{}^+) = -I_A{}^+ \qquad \text{and} \\
\mu_\alpha(B^+) &= \mu_\alpha(a_B{}^+) = \mu_\alpha(b_B{}^+) = -A_B{}^+,
\end{aligned} \tag{67}$$

where $I_X{}^+$ and $A_X{}^+$ denote the ionization potential and electron affinity of X^+, respectively. These *biased* substrate descriptors apply to both $R_c{}^+$ and $R_{\text{HSAB}}{}^+$ complexes. Indeed, in the complementary arrangement, the amount of the first partial charge transfer dominates the second one (see Figure 1), $N(CT_1) > N(CT_2)$, so that B^+ net donates and A^+ accepts electrons.

The optimum amount of the resultant CT between A^+ and B^+ substrates in $R_\alpha{}^+$,

$$N_\alpha(CT) = N_\alpha(A^*) - N_\alpha(A^0) = N_\alpha(B^0) - N_\alpha(B^*) > 0, \tag{68}$$

where $\{N_\alpha(X^0)\}$ denote electron numbers in separate reactants and $\{N(X^*)\}$ stand for the average electron populations in the coordination final, equilibrium reactive system with the mutually-open subsystems,

$$R_\alpha{}^* = [A^*(\alpha) \mathbin{\vdots} B^*(\alpha)] = [a_A{}^*(\alpha) \mathbin{\vdots} b_A{}^*(\alpha) \mid a_B{}^*(\alpha) \mathbin{\vdots} b_B{}^*(\alpha)], \tag{69}$$

is determined by the corresponding in situ CT "force", the effective chemical-potential for this process (populational "gradient"), measuring the difference between chemical potentials of the polarized acidic and basic reactants,

$$\mu_\alpha(CT) = \partial E_\alpha{}^+[N_\alpha(CT), v(\alpha)]/\partial N_\alpha(CT) = \mu_\alpha(A^+) - \mu_\alpha(B^+) < 0, \tag{70}$$

and the coordination in situ hardness descriptor $\eta(\text{CT})$ for this electron transfer,

$$\eta_\alpha(\text{CT}) = \partial\mu_\alpha(\text{CT})/\partial N_\alpha(\text{CT})$$
$$= \eta_\alpha(\text{A}^+,\text{A}^+) - \eta_\alpha(\text{A}^+,\text{B}^+) + \eta_\alpha(\text{B}^+,\text{B}^+) - \eta_\alpha(\text{B}^+,\text{A}^+) > 0, \tag{71}$$

representing the effective CT-hardness (populational "Hessian"). Here, the elements of the hardness tensor in reactant resolution measure the second populational partials

$$\eta(\text{R}_\alpha{}^+) = [\partial^2 E_\alpha{}^+/\partial N(\text{R}_\alpha{}^+)\,\partial N(\text{R}_\alpha{}^+)]_{v(\alpha)} = [\partial\mu(\text{R}_\alpha{}^+)/\partial N(\text{R}_\alpha{}^+)]_{v(\alpha)} = \{\eta_\alpha(\text{X}^+,\text{Y}^+)\},$$

$$\eta_\alpha(\text{X}^+,\text{Y}^+) \equiv \{\partial^2 E_\alpha{}^+[\{N(\text{R}_\alpha{}^+)\},\,v(\alpha)]/\partial N_\alpha(\text{X}^+)\,\partial N_\alpha(\text{Y}^+)\}_{v(\alpha)}$$
$$= [\partial\mu_\alpha(\text{X}^+)/\partial N_\alpha(\text{Y}^+)]_{v(\alpha)}, \qquad \text{X}^+, \text{Y}^+ \in (\text{A}^+, \text{B}^+). \tag{72}$$

Since the acidic (basic) reactants are identified by their chemically hard (soft) character, as substrates of relatively small (high) polarizability, their chemical potentials obey the following inequality: $\mu_\alpha(\text{A}^+) < \mu_\alpha(\text{B}^+) < 0$.

Finally, the coordination equilibrium amount of the *inter*-reactant CT [79,80,82],

$$N_\alpha(\text{CT}) = -\mu_\alpha(\text{CT})/\eta_\alpha(\text{CT}), \tag{73}$$

generates the associated 2nd-order stabilization energy due to CT,

$$E_\alpha(\text{CT}) = \mu_\alpha(\text{CT})\,N_\alpha(\text{CT})/2 = -[\mu_\alpha(\text{CT})]^2/[2\eta_\alpha(\text{CT})] < 0. \tag{74}$$

It should be emphasized that this energy estimate already contains the implicit polarization contributions, due to internal charge adjustments within each internally-open reactant. Indeed, all *spontaneous* electron flows, of both of CT and P origins, stabilize the system.

7. Partial Electronic Flows

It is also of interest to establish energy contributions due to all partial electronic flows shown in the flux patterns in Equation (48), of both P and CT origins. As observed above, all such spontaneous responses of the reactive system contribute the stabilizing (negative) contributions to the reaction energy. The P-currents represent the *intra*-reactant responses created by a presence of the other substrate and the external CT.

The P/CT resolved flow patterns in Equation (48) and the associated energy terms call for the *site*-resolution of the polarized reactive system, now corresponding to the mutually-closed reaction sites on the internally- and mutually-closed reactants $\text{A} = (a_\text{A} \mid b_\text{A})$ and $\text{B} = (a_\text{B} \mid b_\text{B})$ (see Figure 1), for the external potential $v(\alpha)$ of Equation (64),

$$\text{R}_\alpha = [\text{A}(\alpha) \mid \text{B}(\alpha)] = [a_\text{A}(\alpha) \mid b_\text{A}(\alpha) \mid a_\text{B}(\alpha) \mid b_\text{B}(\alpha)]. \tag{75}$$

Their functional fragments, the acidic and basic sites in both reactants,

$$\lambda(\text{R}_\alpha) = \{\lambda_\alpha(\text{X}) \equiv [a_\text{X}(\alpha),\, b_\text{X}(\alpha)]\}$$
$$\equiv \{a_\text{A}(\alpha),\, b_\text{A}(\alpha),\, a_\text{B}(\alpha),\, b_\text{B}(\alpha)\} \equiv \{\lambda_1,\, \lambda_2,\, \lambda_3,\, \lambda_4\}, \tag{76}$$

which contain $n(\text{R}_\alpha) = \{n_\lambda(\alpha)\}$ electrons, are then characterized by different levels of the site chemical potentials

$$u(\text{R}_\alpha) = \partial E_\alpha[n(\text{R}_\alpha);\, v(\alpha)]/\partial n = \{u_\alpha(\lambda) = \partial E_\alpha/\partial n_\lambda \equiv u_\lambda\}, \tag{77}$$

measured by the partial energy derivatives with respect to average electron populations:

$$n(\text{R}_\alpha) = \{n_\alpha(\lambda) \equiv n_\lambda\} = \{n_\alpha(\text{X}) = [n_\alpha(a_\text{X}),\, n_\alpha(b_\text{X})]\}. \tag{78}$$

The site-hardness descriptors are similarly defined by the corresponding matrix of the second populational derivatives of the energy in this resolution level:

$$\mathbf{h}(R_\alpha) = \{h_{\lambda,\lambda'} = \partial^2 E_\alpha/\partial n_\lambda\, \partial n_{\lambda'} = \partial u_\alpha(\lambda)/\partial n_{\lambda'}\}. \tag{79}$$

In this more resolved perspective, one also applies the biased estimates of the fragment chemical potentials, measured either by the site negative ionization potential ($u_\lambda = -I_\lambda$) or its negative electron affinity ($u_\lambda = -A_\lambda$), when this fragment acts as an external electron donor or acceptor, respectively.

A reference to Figure 1 and Equation (48) shows that the given site λ can act in the following three ways:

(1) As internal (P) and external (CT) *donor* of electrons, e.g., b_B site in R_{HSAB};
(2) As internal and external *acceptor*, e.g., a_A site in R_{HSAB};
(3) As *flow*-through fragment, e.g., all sites in R_c and (a_B, b_A) parts of R_{HSAB}.

$$\tag{80}$$

In the latter category the inflow/outflow currents, either of CT or P origins, compete with one another, thus minimizing the site net electron accumulation or depletion. The corresponding chemical potentials for these types of behavior, the *biased* measures for Groups (1) and (2) and the *unbiased* estimates for Group (3), then read:

$$u_\alpha(1) = -I_\lambda, \qquad u_\alpha(2) = -A_\lambda, \qquad \text{and} \qquad u_\alpha(3) = -(I_\lambda + A_\lambda)/2. \tag{81}$$

These chemical potentials are expected to display the following hierarchy reflecting the site softnesses (polarizabilities):

$$u_\alpha(a_A) < u_\alpha(a_B) < u_\alpha(b_A) < u_\alpha(b_B) < 0. \tag{82}$$

The sum of the associated first-order changes in electronic energy, $\{\Delta E_\alpha^{(1)}(\lambda)\}$, of the site energies $\{E_\alpha(\lambda)\}$ following displacements $\{\Delta n_\lambda\}$ in their electron populations, then generates the following overall energy displacement:

$$\Delta E_\alpha^{(1)} = \Sigma_\lambda\, u_\alpha(\lambda)\, \Delta n_\lambda \equiv \Sigma_\lambda\, \Delta E_\alpha^{(1)}(\lambda). \tag{83}$$

One next observes that the population displacements are large in Groups (1) and (2) of the list (80), where the P and CT displacement *enhance* one another,

$$N_\alpha^{P}(\lambda) + N_\alpha^{CT}(\lambda) \equiv \Delta_\alpha(\lambda), \tag{84}$$

and small in Group (3), when they *cancel* one another:

$$N_\alpha^{P}(\lambda) + N_\alpha^{CT}(\lambda) \equiv \delta_\alpha(\lambda). \tag{85}$$

Therefore, the energy change in R_{HSAB} is determined by two large population displacements $\Delta_{HSAB}(\lambda)$, due to charge activations of the key sites a_A (strongly acidic) and b_B (strongly basic), and two remaining (small) flow-through displacements $\delta_{HSAB}(\lambda)$ of the mixed-character fragments a_B and b_A. In R_c, the collective charge displacement includes four flow-through $\delta_c(\lambda)$ contributions reflecting the charge activation on all sites. This observation further justifies the complementary preference in such donor-acceptor coordinations [94,95].

Consider now the second-order energetic consequences of the partial *charge transfer* (CT) flows, shown in the diagrams of Equation (48). A reference to Equations (70)–(74), (77), and (79) indicates that in R_c the dominating CT_1 process $b_B(\lambda_4) \to a_A(\lambda_1)$ is described by the following in situ gradient and Hessian descriptors:

$$u_c(CT_1) = u_c(a_A) - u_c(b_B) = u_1 - u_4 \qquad \text{and}$$
$$h_c(CT_1) = h_{1,1} - h_{1,4} + h_{4,4} - h_{4,1}. \tag{86}$$

They predict the optimum amount of this partial inter-reactant population displacement,

$$N_c(CT_1) = -u_c(CT_1)/h_c(CT_1),\tag{87}$$

and the associated stabilization energy:

$$E_c(CT_1) = -[u_c(CT_1)]^2/[2h_c(CT_1)].\tag{88}$$

The associated chemical potential and hardness descriptors for the CT_2 process $b_A(\lambda_2)\rightarrow a_B(\lambda_3)$ in R_c accordingly read:

$$\begin{aligned}u_c(CT_2) &= u_c(a_B) - u_c(b_A) = u_3(c) - u_2(c) \qquad \text{and}\\ h_c(CT_2) &= h_{3,3}(c) - h_{2,3}(c) + h_{2,2}(c) - h_{3,2}(c).\end{aligned}\tag{89}$$

They determine the optimum amount of CT_2 and the associated stabilization energy:

$$N_c(CT_2) = -u_c(CT_2)/h_c(CT_2), \qquad E_c(CT_2) = -[u_c(CT_2)]^2/[2h_c(CT_2)].\tag{90}$$

Consider now the associated inter-reactant fluxes in R_{HSAB}:

$$CT_1: \quad b_B(\lambda_4)\rightarrow b_A(\lambda_2) \qquad \text{and} \qquad CT_2: \quad a_B(\lambda_3)\rightarrow a_A(\lambda_1).$$

The first of these transfers of electrons generates the following in situ descriptors:

$$\begin{aligned}u_{HSAB}(CT_1) &= u_{HSAB}(b_A) - u_{HSAB}(b_B) \equiv u_2(HSAB) - u_4(HSAB) \qquad \text{and}\\ h_{HSAB}(CT_1) &= h_{2,2}(HSAB) - h_{2,4}(HSAB) + h_{4,4}(HSAB) - h_{4,2}(HSAB).\end{aligned}\tag{91}$$

They predict the following population displacement and energy change:

$$\begin{aligned}N_{HSAB}(CT_1) &= -u_{HSAB}(CT_1)/h_{HSAB}(CT_1) \qquad \text{and}\\ E_{HSAB}(CT_1) &= -[u_{HSAB}(CT_1)]^2/[2h_{HSAB}(CT_1)].\end{aligned}\tag{92}$$

For the second CT_2 displacement in R_{HSAB}, one similarly finds the relevant chemical in situ gradient and Hessian descriptors,

$$u_{HSAB}(CT_2) = u_{HSAB}(a_A) - u_{HSAB}(a_B) \equiv u_1(HSAB) - u_3(HSAB),$$

$$h_{HSAB}(CT_2) = h_{1,1}(HSAB) - h_{1,3}(HSAB) + h_{3,3}(HSAB) - h_{3,1}(HSAB),\tag{93}$$

and the resulting population and energy displacements:

$$N_{HSAB}(CT_2) = -u_{HSAB}(CT_2)/h_{HSAB}(CT_2), \qquad E_{HSAB}(CT_2) = -[u_{HSAB}(CT_2)]^2/[2h_{HSAB}(CT_2)].\tag{94}$$

These (primary) CT displacements in reactive complexes can be regarded as perturbations creating conditions for subsequent (secondary) polarization (P) responses in reactants.

In R_c, the initial charge transfers modify the site chemical potentials, initially equalized in the equilibrium reactants [see Equation (67)],

$$\begin{aligned}\lambda_1: &\quad \delta u_c(a_A) = [h_{1,1}(c) - h_{1,4}(c)]\,N_c(CT_1) + [h_{1,3}(c) - h_{1,2}(c)]\,N_c(CT_2),\\ \lambda_2: &\quad \delta u_c(b_A) = [h_{2,1}(c) - h_{2,4}(c)]\,N_c(CT_1) + [h_{2,3}(c) - h_{2,2}(c)]\,N_c(CT_2),\\ \lambda_3: &\quad \delta u_c(a_B) = [h_{3,1}(c) - h_{3,4}(c)]\,N_c(CT_1) + [h_{3,3}(c) - h_{3,2}(c)]\,N_c(CT_2),\\ \lambda_4: &\quad \delta u_c(b_B) = [h_{4,1}(c) - h_{4,4}(c)]\,N_c(CT_1) + [h_{4,3}(c) - h_{4,2}(c)]\,N_c(CT_2).\end{aligned}\tag{95}$$

The corresponding displacements in the regional HSAB coordination read:

$$
\begin{aligned}
\lambda_1: \quad \delta u_{\mathrm{HSAB}}(a_A) &= [h_{1,2}(\mathrm{HSAB}) - h_{1,4}(\mathrm{HSAB})]\,N_{\mathrm{HSAB}}(\mathrm{CT}_1) \\
&\quad + [h_{1,1}(\mathrm{HSAB}) - h_{1,3}(\mathrm{HSAB})]\,N_{\mathrm{HSAB}}(\mathrm{CT}_2), \\
\lambda_2: \quad \delta u_{\mathrm{HSAB}}(b_A) &= [h_{2,2}(\mathrm{HSAB}) - h_{2,4}(\mathrm{HSAB})]\,N_{\mathrm{HSAB}}(\mathrm{CT}_1) \\
&\quad + [h_{2,1}(\mathrm{HSAB}) - h_{2,3}(\mathrm{HSAB})]\,N_{\mathrm{HSAB}}(\mathrm{CT}_2), \\
\lambda_3: \quad \delta u_{\mathrm{HSAB}}(a_B) &= [h_{3,2}(\mathrm{HSAB}) - h_{3,4}(\mathrm{HSAB})]\,N_{\mathrm{HSAB}}(\mathrm{CT}_1) \\
&\quad + [h_{3,1}(\mathrm{HSAB}) - h_{3,3}(\mathrm{HSAB})]\,N_{\mathrm{HSAB}}(\mathrm{CT}_2), \\
\lambda_4: \quad \delta u_{\mathrm{HSAB}}(b_B) &= [h_{4,2}(\mathrm{HSAB}) - h_{4,4}(\mathrm{HSAB})]\,N_{\mathrm{HSAB}}(\mathrm{CT}_1) \\
&\quad + [h_{4,1}(\mathrm{HSAB}) - h_{4,3}(\mathrm{HSAB})]\,N_{\mathrm{HSAB}}(\mathrm{CT}_2).
\end{aligned}
\tag{96}
$$

These shifts in site electronegativities generate the associated in situ populational gradients for the subsequent P relaxation of reactants. For R_c, these internal flows are defined in Figure 1 and the corresponding current pattern in Equation (48). In the complementary complex, one finds:

$$
R_c: u_c(P_A) = \delta u_c(b_A) - \delta u_c(a_B), \qquad u_c(P_B) = \delta u_c(b_B) - \delta u_c(a_B),
\tag{97}
$$

while, in the HSAB coordination, where the internal P_A and P_B flows define $\{b_X \to a_X\}$ flows in each reactant,

$$
\begin{aligned}
R_{\mathrm{HSAB}}: u_{\mathrm{HSAB}}(P_A) &= \delta u_{\mathrm{HSAB}}(a_A) - \delta u_{\mathrm{HSAB}}(b_A), \\
u_{\mathrm{HSAB}}(P_B) &= \delta u_{\mathrm{HSAB}}(a_B) - \delta u_{\mathrm{HSAB}}(b_B).
\end{aligned}
\tag{98}
$$

Finally, to estimate magnitudes of these polarizational relaxations, one applies the following in situ hardness descriptors for the polarizational flows in reactants:

$$
\begin{aligned}
h_\alpha(P_A) &= h_{1,1}(\alpha) - h_{1,2}(\alpha) + h_{2,2}(\alpha) - h_{2,1}(\alpha), \\
h_\alpha(P_B) &= h_{3,3}(\alpha) - h_{3,4}(\alpha) + h_{4,4}(\alpha) - h_{4,3}(\alpha), \qquad \alpha = (c, \mathrm{HSAB}).
\end{aligned}
\tag{99}
$$

They ultimately determine the optimum sizes of these polarization transfers in R_α,

$$
N_\alpha(P_X) = -u_\alpha(P_X)/h_\alpha(P_X),
\tag{100}
$$

and the associated polarization-energies:

$$
E_\alpha(P_X) = -[u_\alpha(P_X)]^2/[2h_\alpha(P_X)], \qquad \alpha = (c, \mathrm{HSAB}), \qquad X = A, B.
\tag{101}
$$

To summarize, the overall stabilization energy in the *reactant*-resolution [Equation (74)] contains all four partial P/CT contributions in the *site*-resolution:

$$
E_\alpha(\mathrm{CT}) = [E_\alpha(\mathrm{CT}_1) + E_\alpha(\mathrm{CT}_2)] + \Sigma_{X=A,B}\,E_\alpha(P_X), \qquad \alpha = (c, \mathrm{HSAB}).
\tag{102}
$$

8. Communication Considerations

These alternative coordinations in reactive systems can be also qualitatively approached within the communication theory of the chemical bond [9–11,22–32,62]. To directly connect to the discussion of the preceding section, one again adopts the *site*-resolution on both reactants [see Equation (76)], in which the system communication (information) channel is defined by the network of conditional probabilities of observing in R_α the "output" ("receiver") fragment λ', given the "input" ("source") site λ:

$$
P_\alpha(\lambda'|\lambda) = P_\alpha(\lambda', \lambda)/P_\alpha(\lambda) \equiv P_\alpha(\lambda \to \lambda'), \qquad \Sigma_{\lambda'}\,P_\alpha(\lambda'|\lambda) = 1.
\tag{103}
$$

Here, $P_\alpha(\lambda)$ denotes the *site*-probability and $P_\alpha(\lambda', \lambda)$ stands for the *joint*-probability of the occurrence of the *two*-site event in the bond system of R_α. They must satisfy the usual normalizations:

$$\Sigma_\lambda \left[\Sigma_{\lambda'} \, P_\alpha(\lambda', \lambda) \right] = \Sigma_\lambda \, P_\alpha(\lambda) = 1. \tag{104}$$

The key CT sites in both reactants determine the overall chemical character of these subsystems in reactive complexes: the *acceptor* (acidic) site in A, a_A, and the *donor* (basic) site in B, b_B. These crucial fragments are accentuated by the bold symbols in Figure 3, where the dominating communications between the nearest neighbors in both coordinations are shown. The R_c diagram in the Figure 3a shows that only the complementary arrangement exhibits the *direct* communication $b_B \rightarrow a_A$, reflected by a high conditional probability $P_c(b_B \rightarrow a_A)$, besides the *double*-cascade propagation $b_B \rightarrow [a_B \rightarrow b_A] \rightarrow a_A$ of a relatively low probability,

$$P_c(b_B \rightarrow [a_B \rightarrow b_A] \rightarrow a_A) = P_c(b_B \rightarrow a_B) \, P_c(a_B \rightarrow b_A) \, P_c(b_A \rightarrow a_A) << P_c(b_B \rightarrow a_A). \tag{105}$$

The R_{HSAB} coordination Figure 3b generates two indirect (*single*-cascade) scatterings between the crucial sites a_A and b_B:

$$\begin{aligned} P_{HSAB}(b_B \rightarrow a_B \rightarrow a_A) &= P_{HSAB}(b_B \rightarrow a_B) \, P_{HSAB}(a_B \rightarrow a_A) \quad \text{and} \\ P_{HSAB}(b_B \rightarrow b_A \rightarrow a_A) &= P_{HSAB}(b_B \rightarrow b_A) \, P_{HSAB}(b_A \rightarrow a_A). \end{aligned} \tag{106}$$

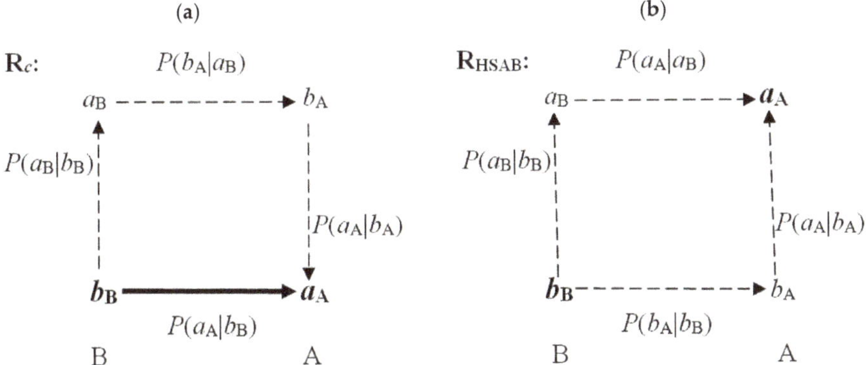

Figure 3. The dominating (nearest neighbor) communications between the chemically bonded sites of the Acid–Base complexes R_c (**a**) and R_{HSAB} (**b**), respectively. The former involves a strong (direct) $b_B \rightarrow a_A$ propagation between the key sites a_A and b_B, which determine the overall chemical behavior of reactants, and a weak (intermediate) *double*-bridge communication $b_B \rightarrow [a_B \rightarrow b_A] \rightarrow a_A$, while the latter exhibits only two indirect (*single*-bridge) communications: $b_B \rightarrow a_B \rightarrow a_A$ and $b_B \rightarrow b_A \rightarrow a_A$.

As indicated above, in communication theory the indirect scatterings involve products of the relevant direct propagations. Therefore, the conditional probability of the direct step $b_B \rightarrow a_A$ in R_c must dominate over all indirect communications between these sites. This provides additional rationale for the observed complementary preference in the acid-base coordination.

Moreover, the soft (basic) fragment is predicted to overlap with the neighboring sites more strongly compared to the hard (acidic) fragment, thus giving rise to stronger electron communications:

$$P_\alpha(b_X \rightarrow \lambda) > P_\alpha(a_X \rightarrow \lambda) \quad \text{and} \quad P_\alpha(\lambda \rightarrow b_X) > P_\alpha(\lambda \rightarrow a_X). \tag{107}$$

The strongest communications are thus expected for most *covalently*-bonded neighboring fragments b_A and b_B in the HSAB complex, while the weakest propagations are predicted between its two hard (acidic) sites a_A and a_B, which bind more *ionically*:

$$P(b_B \rightarrow b_A) \gg P(a_B \rightarrow a_A). \tag{108}$$

In R_{HSAB}, this qualitative analysis thus suggests a *covalent* character of the chemical bond between the basic sites of both reactants, and the *ionic* bond between their acidic groups. In R_c, one similarly predicts a strong *coordination* bond between a_A and b_B, and a predominantly *covalent* bond between a_B and b_A.

The current density $j(r)$ [see Equation (37)] reflects the flow of electronic probability density $p(r)$ with an effective velocity measuring the current-per-particle: $V(r) = j(r)/p(r)$. It implies the associated flux of the system resultant gradient-information, $J_I(r) = I(r)V(r)$. Here, $I(r) = I_p(r) + I_\phi(r)$ denotes the density-per-electron of the overall information combining the classical contribution $I_p(r) = [\nabla \ln p(r)]^2$ and its nonclassical supplement $I_\phi(r) = [2\nabla\phi(r)]^2$ due to the state phase component [see Equation (43)]. The information continuity equation then predicts the vanishing classical contribution to the resultant information source and a finite production of its nonclassical part [62–64,67,73].

The shifts in electron populations on active sites of donor–acceptor systems [Equation (48)] also imply the associated redistributions of the system resultant gradient-information. The concerted flows in R_c redistribute the information density more evenly, compared to a more localized redistribution in R_{HSAB}, where electronic flows net transport the information between the key sites of both reactants: from b_B to a_A. The complementary coordination thus corresponds to a lower level of the overall *determinicity*-information [see Equation (53)] compared to that in HSAB-type coordination. Therefore, the former reactive system represents more information uncertainty in comparison to the latter complex, thus exhibiting a higher resultant gradient-entropy (*indeterminicity*-information).

9. Conclusions

To paraphrase Progogine [96], the classical, *modulus* component of molecular wavefunction determines the electronic probability distribution—the state structure of "being"—while the gradient of the nonclassical *phase* variable generates the current density—the system structure of "becoming". In this work, we have explored in some detail the mutual relation between the phase component of electronic wavefunction and its current descriptor, expressing the former as an indefinite integral over the probability velocity field. The probability and current descriptors of electronic states are both related by the quantum continuity relations implied by SE of molecular QM, which have also been summarized. The optimum distribution of the state electronic density is determined by density variational principle for the system electronic energy, while the equilibrium current density results from the subsidiary extremum rule for the nonclassical entropy or information measure, which determines the state optimum phase for the given electron density. The elementary chemical processes have also been monitored using the classical entropy/information descriptors [97–100].

In phenomenological approaches to chemical reactivity, it is customary to distinguish the mutually bonded (open, entangled) and nonbonded (closed, disentangled) status of the of electronic distributions in subsystems. The former reflects the quantum state of the whole reactive system, while the latter refers to states of the mutually closed reactants. The bonded or nonbonded/frozen character of molecular fragments at such hypothetical reaction stages also delineates the allowed types of electronic communications between substrates, which are responsible for the interreactant chemical bond: the *promolecular* reference state $R^0 = (A^0 \mid B^0)$, describing the "frozen" (molecularly placed) electron distributions in the ground-states of separate reactants $\{X^0\}$, does not allow for any communications between the system constituent AIM or their basis functions; the *polarized* reactive system $R^+ = (A^+ \mid B^+)$ opens internal communications within each reactant, while the final *equilibrium*, fully "relaxed" molecular system $R^* = (A^* \mid B^*)$ already accounts for all *intra-* and *inter-*substrate propagations.

We have also demonstrated that the global equilibrium in R^* establishes a common phase component of the fragment states (see Appendix C). In other words, the bonded status of molecular fragments implies the phase equalization of the effective subsystem states, at the equilibrium phase related to the "molecular" electron distribution, in the interacting system as a whole. Therefore, the bonding (entangling) of molecular fragments represents the *phase*-phenomenon reflecting their common (molecular) electronic state, past or present (see Appendix D). This phase equalization is independent of the actual distance between subsystems, thus representing a long-range correlation effect.

The coherence of electronic currents has also been shown to play a crucial role in establishing the energetic preference of the least-activated complementary reactive complex combining the donor and acceptor substrates. Alternative coordinations of such reactants in R_c and R_{HSAB} complexes, respectively, imply different patterns of the dominating electronic currents and communications. The specific redistributions of the system electronic and information densities, via P and/or CT, channels, have been qualitatively discussed and the resultant pattern of the site currents have been established. We have also examined energetic implications of the overall and partial P/CT electron flows in A—B complexes using the chemical potential (electronegativity) and hardness/softness descriptors of reactants and their active sites, defined in the DFT based reactivity theory.

The phase equalization in the mutually open subsystems provides a consistent theoretical framework for distinguishing the mutually bonded and nonbonded states of their electron distributions. This IT phase criterion of electronic equilibria in molecular systems complements the familiar energetic principle of the chemical-potential (electronegativity) equalization. The complete set of requirements for the quantum equilibrium in the bonded reactive system thus calls for equalizations of both the modulus (probability) and phase (current) related (local) intensities, the chemical potentials and phases of molecular substrates, at the corresponding global descriptors characterizing the reactive system as a whole.

Thus, for the given probability distributions of subsystems, it is the equalized, "molecular" phase of the whole reactive system which marks the bonded (equilibrium) state in the reaction substrates. The phase component of molecular electronic states thus emerges as an important "association" fabric in chemical systems, which "glues" reactants in their "bonded" (entangled) condition. It keeps the "memory" of the present or past interactions between subsystems, and it shapes the probability fluxes between the substrate active sites, which effect the information flows in the reactive system.

Funding: This research received no external funding.

Conflicts of Interest: The author declares no conflict of interest.

Appendix A Continuity Relations Revisited

The dynamics of electronic states is determined by SE, which also implies specific rates of time evolutions of both components of complex wavefunctions, and of their physical descriptors: the state probability and current distributions. The time derivatives of the modulus and phase parts of electronic states reflect the relevant continuity equations in molecular QM. For simplicity, let us consider a single electron in state $|\Psi(t)\rangle$ at time t, or the associated (complex) wavefunction in position representation:

$$\psi(\mathbf{r}, t) = \langle \mathbf{r}|\Psi(t)\rangle = R(\mathbf{r}, t)\exp[i\phi(\mathbf{r}, t)], \qquad \phi(\mathbf{r}, t) \geq 0. \tag{A1}$$

Its modulus (R) and phase (ϕ) components determine the state physical attributes of the electron probability and current densities:

$$p(\mathbf{r}, t) = \psi(\mathbf{r}, t)^{*}\psi(\mathbf{r}, t) = R(\mathbf{r}, t)^{2}, \tag{A2}$$

$$j(r, t) = [\hbar/(2mi)] \, [\psi(r, t)^* \, \nabla\psi(r, t) - \psi(r, t)\nabla\psi(r, t)^*]$$
$$= (\hbar/m) \, p(r, t) \, \nabla\phi(r, t) \tag{A3}$$
$$\equiv p(r, t) \, V(r, t).$$

The effective velocity $V(r, t)$ of the probability "fluid" measures the current-per-particle and reflects the state *phase*-gradient:

$$V(r, t) = j(r, t)/p(r, t) = (\hbar/m) \, \nabla\phi(r, t). \tag{A4}$$

Equation (A1) also identifies the two (additive) components of the wavefunction logarithm:

$$\ln\psi \, (r, t) = \ln R(r, t) + i\phi(r, t), \tag{A5}$$

which determine resultant measures of the global and gradient content of the state entropy or information. For example, the complex entropy descriptor [59,62],

$$S[\psi] = -2\langle\psi|\ln\psi|\psi\rangle = -2\int p \, [\ln R + i \, \phi] \, dr$$
$$\equiv S[p] + i \, S[\phi], \tag{A6}$$

combines as its real part the Shannon entropy $S[p]$ in probability distribution p, and the nonclassical phase supplement $S[\phi]$, which determines its imaginary component.

The corresponding Fisher-type measure of the state resultant gradient information $I[\psi]$ or entropy $M[\psi]$ are defined by expectation values of the associated (Hermitian) operators [53–62]

$$I = \quad 4\Delta = 4[(\nabla\ln R)^2 \quad (i\nabla\phi)^2] = 4[(\nabla\ln R)^2 + (\nabla\phi)^2] = 4|\nabla\ln\psi| \qquad \text{and}$$
$$M = 4[(\nabla\ln R)^2 + (i\nabla\phi)^2], \tag{A7}$$

$$I[\psi] = \langle\psi|I|\psi\rangle = 4\int R^2[(\nabla\ln R)^2 + (\nabla\phi)^2] \, dr = 4\int[(\nabla R)^2 + (R\,\nabla\phi)^2] \, dr \equiv I[R] + I[\phi]$$
$$= \int p[(\nabla\ln p)^2 + 4(\nabla\phi)^2] \, dr = \int p^{-1}(\nabla p)^2 \, dr + 4\int p \, (\nabla\phi)^2 \, dr \equiv I[p] + I[\phi], \tag{A8}$$

$$M[\psi] = \langle\psi|M|\psi\rangle = 4\{\langle\psi|(\nabla\ln R)^2|\psi\rangle + \langle\psi|(i\nabla\phi)^2|\psi\rangle\} = I[p] - I[\phi]$$
$$= 4\int[(\nabla R)^2 - (\nabla\phi)^2] \, dr \equiv M[R] + M[\phi] \tag{A9}$$
$$= \int p[(\nabla\ln p)^2 - 4(\nabla\phi)^2] \, dr = M[p] + M[\phi].$$

The former is thus related to the state average kinetic energy of electrons, $T[\psi] = \langle\psi|T|\psi\rangle$, determined by the quantum operator $T = -\hbar^2/(2m) \, \Delta = [\hbar^2/(8m)] \, I$:

$$I[\psi] = (8m/\hbar^2) \, T[\psi] \equiv \sigma T[\psi]. \tag{A10}$$

This proportionality allows one to use the molecular virial theorem [66] in general considerations on the information redistribution in the bond-formation process and during chemical reactions [67–70]. It follows from Equations (A6)–(A9) that the independent additive components of Equation (A5) indeed determine the average values of the resultant entropy/information descriptors of molecular electronic states, with the local IT densities being weighted by the state electron probability distribution.

In the molecular scenario, the electron is moving in the external potential $v(r)$ due to the "frozen" nuclear frame of the familiar Born–Oppenheimer approximation. The electronic Hamiltonian

$$H(r) = -[\hbar^2/(2m)] \, \Delta + v(r) \equiv T(r) + v(r), \tag{A11}$$

determines the quantum dynamics of electronic states expressed by SE,

$$i\hbar \, (\partial\psi/\partial t) = H\psi, \tag{A12}$$

which can be also formulated in terms of the (unitary) evolution operator

$$U(t - t_0) \equiv U(\tau) = \exp(-i\hbar^{-1}\tau\, H), \tag{A13}$$

$$\psi(t) = U(\tau)\, \psi(t_0). \tag{A14}$$

This equation and its complex conjugate then imply the associated temporal evolutions of the wavefunction components R and ϕ (see also Appendix B):

$$\partial R/\partial t = -\nabla R \cdot V, \tag{A15}$$

$$\partial\phi/\partial t = [\hbar/(2m)]\, [R^{-1}\Delta R - (\nabla\phi)^2] - v/\hbar. \tag{A16}$$

These dynamical equations can be ultimately recast as the corresponding continuity relations. Consider first the continuity of probability distribution:

$$\partial p/\partial t = 2R\,(\partial R/\partial t) = -(2R\,\nabla R)\cdot V = -\nabla p \cdot V = -\nabla\cdot j \qquad \text{or}$$
$$\sigma_p \equiv dp/dt = \partial p/\partial t + \nabla\cdot j = \partial p/\partial t + \nabla p \cdot V = 0. \tag{A17}$$

The divergence of probability flux of Equation (A3)

$$\nabla\cdot j = \nabla p \cdot V + p\,\nabla\cdot V = \nabla p \cdot V, \tag{A18}$$

further implies the vanishing divergence of the velocity field V related to $\nabla^2\phi = \Delta\phi$:

$$\nabla\cdot V = (\hbar/m)\,\Delta\phi = 0 \qquad \text{or} \qquad \Delta\phi = 0. \tag{A19}$$

The total *time*-derivative $dp(r)/dt$ determines the vanishing local probability "source": $\sigma_p(r) = 0$. It measures the time rate of change in an infinitesimal volume element of probability fluid *moving* with velocity $V = dr/dt$, while the partial derivative $\partial p[r(t), t]/\partial t$ refers to volume element around the *fixed* point in space. Indeed, separating the explicit time dependence of $p(r, t)$ from its implicit dependence through the particle position $r(t)$, $p(r, t) = p[r(t), t]$, gives:

$$\sigma_p(r, t) = \partial p[r(t), t]/\partial t + (dr/dt)\cdot\partial p(r, t)/\partial r$$
$$= \partial p(r, t)/\partial t + V(r, t)\cdot\nabla p(r, t) = \partial p(r, t)/\partial t + \nabla\cdot j(r, t) = 0. \tag{A20}$$

Turning now to the phase dynamics of Equation (A16), one first realizes that the effective velocity V of the *probability*-current $j = pV$ also determines the *phase*-flux and its divergence:

$$J = \phi\, V \qquad \text{and} \qquad \nabla\cdot J = \nabla\phi \cdot V = (\hbar/m)\,(\nabla\phi)^2. \tag{A21}$$

This complementary flow descriptor ultimately generates a finite *phase*-source:

$$\sigma_\phi \equiv d\phi/dt = \partial\phi/\partial t + \nabla\cdot J = \partial\phi/\partial t + V\cdot\nabla\phi \neq 0. \tag{A22}$$

Using Equation (A16) eventually gives:

$$\sigma_\phi = [\hbar/(2m)][R^{-1}\Delta R + (\nabla\phi)^2] - v/\hbar. \tag{A23}$$

To summarize, the effective velocity of probability-current also determines the phase-flux in molecules. The source (net production) of the classical *probability*-variable of electronic states identically vanishes, while that of their nonclassical, *phase*-part, determined by state components (p,ϕ) and the system external potential v, remains finite.

The local production of electronic current $j = p\,V$ is also of interest,

$$\sigma_j \equiv dj/dt = \sigma_p\, V + p\,(dV/dt) = p\,(dV/dt), \tag{A24}$$

where we have recognized Equation (A17). A reference to Equations (A3) and (A17) then gives:

$$\sigma_j = (\hbar/m)\, p\, d/dt(\nabla\phi) = (\hbar/m)\, p\, \nabla(d\phi/dt) = (\hbar/m)\, p\, \nabla\sigma_\phi. \tag{A25}$$

Hence, using Equation (A23) finally gives:

$$\sigma_j = [\hbar^2/(2m^2)]\, [R\, \nabla^3 R - (\Delta R)\, \nabla R] - (p/m)\, \nabla v. \tag{A26}$$

Appendix B Schrödinger Equation and Wavefunction Components

Let us now examine the implications of the (complex) SE [Equation (A12)] when expressed in terms of the modulus and phase components of wavefunction [see Equation (A1)]:

$$\begin{aligned} i\hbar\,(\partial\psi/\partial t) &= i\hbar\,[(\partial R/\partial t) + iR(\partial\phi/\partial t)]\,\exp(i\psi) \\ &= H\psi = \{-[\hbar^2/(2m)]\,[\Delta R + 2i\nabla R\cdot\nabla\phi - R(\nabla\phi)^2] + vR\}\,\exp(i\phi), \end{aligned}$$

where we use Equation (A19). Multiplying both sides of this equation by $\exp(-i\phi)$ and dividing by $\hbar R$ finally gives:

$$i\,(\partial\ln R/\partial t) - (\partial\phi/\partial l) = -[\hbar/(2m)]\,[R^{-1}\Delta R + 2i(\nabla\ln R)\cdot\nabla\phi - (\nabla\phi)^2] + v/\hbar. \tag{A27}$$

Comparing the *imaginary* parts of the preceding equation generates the dynamic equation for the time evolution of the modulus part of electronic state,

$$\partial\ln R/\partial t = -\,[(\hbar/m)\,\nabla\phi]\cdot\nabla\ln R = -\,V\cdot\nabla\ln R, \tag{A28}$$

which can be directly transformed into the probability continuity equation (A17):

$$\partial p/\partial t = -\nabla\cdot j \qquad \text{or} \qquad \sigma_p = dp/dt = 0. \tag{A29}$$

Equating the *real* parts of Equation (A27) similarly determines the phase-dynamics of Equation (A16):

$$\partial\phi/\partial t = [\hbar/(2m)]\,[R^{-1}\Delta R - (\nabla\phi)^2] - v/\hbar. \tag{A30}$$

The latter equation also determines the phase-production $\sigma_\phi = d\phi/dt$ [Equation (A23)] for its flux definition of Equation (A21), in the phase continuity relation:

$$\partial\phi/\partial t = -\nabla\cdot J + \sigma_\phi. \tag{A31}$$

The average electronic energy in state (A1) thus combines the following component contributions:

$$\begin{aligned} \langle E\rangle_\psi &= \langle\psi|H|\psi\rangle = -[\hbar^2/(2m)]\int[R\Delta R - R^2(\nabla\phi)^2]\,dr + \int R^2\,v\,dr \\ &= [\hbar^2/(2m)]\int[(\nabla R)^2 + R^2(\nabla\phi)^2]dr + \int R^2\,v\,dr \equiv \langle T\rangle_\psi + \langle V_{ne}\rangle_\psi, \end{aligned} \tag{A32}$$

where we use the relevant integration by parts and $\langle V_{ne}\rangle_\psi$ denotes the state average electron-nuclei attraction energy. The phase-dependent part of the average kinetic energy $\langle T\rangle_\psi$ identically vanishes in the stationary electronic state

$$\psi_s(r, t) = R_s(r)\,\exp[i\phi_s(t)], \tag{A33}$$

for the sharply specified electronic energy

$$E_s = R_s(r)^{-1}\,H(r)\,R_s(r) = -\,[\hbar^2/(2m)]\,R_s(r)^{-1}\Delta R_s(r) + v(r) = const., \tag{A34}$$

when $\phi_s(t) = -\,(E_s/\hbar)\,t \equiv -\,\omega_s t$ and hence $\nabla\phi_s(t) = 0$.

The open systems and their fragments are described by grand ensembles and exhibit continuously varying, fractional average numbers of electrons; the closed molecules and their parts similarly exhibit integer numbers of electrons. The pure state $\Psi_R^*(N, t_0)$ of the whole externally-closed but internally-open reactive complex $R^* = (A^* \mid B^*)$ at time $t_0 \equiv 0$, defined by its overall external potential $v = v_A + v_B$ and number of electrons $N = N_A + N_B$ (integer), can be expanded

$$\Psi_R^*(N, t_0) = \Sigma_s \, C_s(t_0) \, R_s(N), \qquad C_s(t_0) = \langle R_s(N) | \Psi_R^*(N, t_0) \rangle, \tag{A35}$$

in terms of the complete set of the system stationary states, the eigensolutions

$$H(N) \, R_s(N) = E_s \, R_s(N) \tag{A36}$$

of N-electron molecular Hamiltonian:

$$H(N) = T(N) + V_{ne}(N) + U_{ee}(N),$$
$$T(N) = \Sigma_k \, T(k), \qquad V_{ne}(N) = \Sigma_k \, v(k), \qquad U_{ee}(N) = \Sigma_{k<l} \, g(k, l). \tag{A37}$$

Here, $R_s(N)$ denotes the time-independent amplitude (modulus) function of N electrons, while $T(N)$, $V_{ne}(N)$ and $U_{ee}(N)$ stand for the quantum operators of electronic kinetic, attraction and repulsion energies, respectively. Given the state $\Psi(N, t_0)$ at the initial time $t_0 = 0$, its form at time τ [see Equations (A13) and (A14)] is determined by the interval evolution operator

$$U(\tau, N) = \exp[-i\hbar^{-1}\tau \, H(N)], \tag{A38}$$

$$\Psi(N, \tau) = U(\tau) \, \Psi(N, t_0) = \Sigma_s \, C_s(t_0) \, \{\exp[i\phi_s(\tau)] \, R_s(N)\}$$
$$\equiv \Sigma_s \, C_s(t_0) \, \Psi_s(N, \tau), \tag{A39}$$

where $\Psi_s(N, \tau) = R_s(N) \exp[i\phi_s(\tau)]$ is the full stationary state of N-electrons corresponding to energy E_s.

Such pure-state expansion is not available for the open systems in their final (mixed) equilibrium states, described by the fragment density operators defining the associated ensembles of Hamiltonians for different numbers of electrons. At the polarization stage $R^+ = (A^+ \mid B^+)$, however, involving the mutually- and externally-closed substrates, each subsystem X^+ conserves the initial (integer) number of electrons N_X^0 so that the interacting-fragment Hamiltonians $\{H_X(N_X^0)\}$ for the overall external potential v of the whole system are well defined, e.g.,

$$H_A(N_A^0) = T(N_A^0) + V_{ne}(N_A^0) + U_A(N_A^0) + U_{AB}(N_A^0, N_B^0),$$
$$U_A(N_A^0) = \Sigma_{(k<l)\in A} \, g(k, l), \qquad U_{AB}(N_A^0, N_B^0) = \Sigma_{k\in A}\Sigma_{l\in B} \, g(k, l), \qquad \text{etc.} \tag{A40}$$

They determine the stationary eigenproblems for subsystems containing the initial, integer numbers of electrons in separate reactants,

$$H_X(N_X^0) \, \varphi_u(N_X^0) = E_u(X^+) \, \varphi_u(N_X^0), \qquad X = A, B, \tag{A41}$$

and associated expansions at time t_0 of general states in the polarized subsystems:

$$\Psi_X^+(N_X^0, t_0) = \Sigma_u \, C_u(X^+, t_0) \, \varphi_u(N_X^0), \qquad C_u(X^+, t_0) = \langle \varphi_u(N_X^0) | \Psi_X^+(N_X^0, t_0) \rangle. \tag{A42}$$

Their evolution in time interval τ is determined by the reactant operators

$$U_X(\tau, N_X^0) = \exp[-i\hbar^{-1}\tau \, H_X(N_X^0)], \qquad X = A, B, \tag{A43}$$

$$\Psi_X^+(N_X^0, \tau) = U_X(\tau, N_X^0) \, \Psi_X^+(N_X^0, t_0)$$
$$= \Sigma_u \, C_u(X^+, t_0) \, \{\exp[i\phi_u(X^+, \tau)] \, \varphi_u(N_X^0)\}$$
$$\equiv \Sigma_u \, C_u(X^+, t_0) \, \Phi_u(N_X^0, \tau), \tag{A44}$$

where

$$\phi_u(X^+,\tau) = -\,[E_u(X^+)/\hbar]\,\tau \equiv -\,\omega_u(X^+)\tau \tag{A45}$$

denotes the stationary phase of subsystem X^+, while

$$\Phi_u(N_X^0,\,\tau) = \varphi_u(N_X^0)\,\exp[i\phi_u(X^+,\,\tau)] \tag{A46}$$

stands for the full stationary state of reactant X in presence of the other substrate, corresponding to the fragment energy $E_u(X^+)$.

Products of such stationary wavefunctions of both fragments, describing their distinguishable groups of electrons,

$$\begin{aligned}
\{\Theta_{u,w}(A^+,\,B^+;\,\tau) &= \Phi_u(N_A^0,\,\tau)\,\Phi_w(N_B^0,\,\tau) \\
&= \varphi_u(N_A^0)\,\varphi_w(N_B^0)\,\exp\{i[\phi_u(A^+,\tau)+\phi_w(B^+,\tau)]\} \\
&\equiv \vartheta_{u,w}(N_A^0,\,N_B^0)\,\exp[i\theta_{u,w}(\tau)]\},
\end{aligned}$$

$$\theta_{u,w}(\tau) = -\hbar^{-1}[E_u(A^+)+E_w(B^+)]\tau = -[\omega_u(A^+)+\omega_w(B^+)]\,\tau = -\omega_{u,w}\,\tau, \tag{A47}$$

thus constitute the complete basis for expanding general electronic states of the polarized reactive system R^+ [compare Equation (A39)]:

$$\begin{aligned}
\Psi_R^+(N,\,\tau) &= \Sigma_{u,w}\,D_{u,v}(t_0)\,\{\exp[i\theta_{u,w}(\tau)]\vartheta_{u,w}(N_A^0,\,N_B^0)\} \\
&= \Sigma_{u,w}\,D_{u,v}(t_0)\,\Phi_u(N_A^0,\,\tau)\,\Phi_w(N_B^0,\,\tau) \\
&= \Sigma_{u,w}\,D_{u,v}(t_0)\,\Theta_{u,w}(A^+,\,B^+;\,\tau), \\
D_{u,v}(t_0) &= \langle\vartheta_{u,w}(N_A^0,\,N_B^0)|\Psi_R^+(N,\,t_0)\rangle.
\end{aligned} \tag{A48}$$

While stationary states of the whole reactive system imply the purely time-dependent phase, the effective (mixed) states of reactants X^* in $R^* = (A^*|B^*)$, identified by their partial densities $\{\rho_X^* = N_X^*p_X^*,\,N_X^* = \int\rho_X^*dr\,\text{(fractional)}\}$, pieces of molecular electron density

$$\rho_R^* = \rho_A^* + \rho_B^*, \tag{A49}$$

exhibit the local phases generating nonvanishing phase gradients. Indeed, the HZM construction [73–76] of wavefunctions yielding the prescribed probability distributions $\{p_X^*\}$ on subsystems gives rise to finite electronic currents on reactants. This current pattern in both substrates or on their acidic and basic sites manifests the valence-state activation of such open fragments, which generates nonvanishing contributions to the associated (nonclassical) entropy/information descriptors.

Appendix C Information Principle

The partial currents $\{j_\alpha(r,\,\lambda)\}$ and probability distributions $\{p_\alpha(r,\,\lambda)\}$ in active fragments

$$\lambda \in (a_A = 1,\, b_A = 2,\, a_B = 3,\, b_B = 4) \tag{A50}$$

of the reactive system R_α, $\alpha = (c,\,HSAB)$, determine the additive site contributions to the resultant (nonclassical) gradient-information [see Equations (43) and (A8)] in R_α:

$$I_\alpha[\{j_\alpha(\lambda)\}] = \Sigma_\lambda\,I_\alpha^\lambda[j_\alpha(\lambda)], \qquad I_\alpha^\lambda[j_\alpha(\lambda)] = (2m/\hbar)^2\int p_\alpha(r,\,\lambda)^{-1}\,j_\alpha(r,\,\lambda)^2\,dr. \tag{A51}$$

The minimum of this current-*determinicity* measure $I_\alpha[\{j_\alpha(\lambda)\}]$ implies the maximum of the complementary nonclassical current-*indeterminicity* descriptor, of the site gradient "entropy"

[Equation (A9)] containing negative nonclassical contribution [62,63]. Consider such a nonclassical information principle subject to the local constraint of preserving the resultant local current in R_α,

$$j_\alpha(r) = \Sigma_\lambda \, j_\alpha(r, \lambda), \tag{A52}$$

$$\delta\{I_\alpha[\{j_\alpha(\lambda)\}] - \int \xi_\alpha(r) \cdot j_\alpha(r) \, dr\} = 0, \tag{A53}$$

where the vector Lagrange multiplier $\xi_\alpha(r)$ enforces the local constraint of Equation (A53). The Euler equations determining the optimum site-currents then read:

$$\delta I_\alpha[\{j_\alpha(\lambda)\}]/\delta j_\alpha(r, \lambda) = (8m^2/\hbar^2) \, [j_\alpha(r, \lambda)/p_\alpha(r, \lambda)] \equiv (8m^2/\hbar^2) \, V_\alpha(r, \lambda) = \xi_\alpha(r). \tag{A54}$$

Therefore, at the minimum of the current-information the fragment equilibrium (*eq.*) probability-velocity is λ-independent,

$$V_\alpha(r, \lambda) = [\hbar^2/(8m^2)] \, \xi_\alpha(r) \equiv V_\alpha^{eq.}(r). \tag{A55}$$

Moreover, since this effective velocity descriptor is determined by the phase gradients of Equations (37) and (44), one concludes from the preceding equation that the equilibrium phases $\{\phi_\alpha(r, \lambda)\}$ of the site "states" can differ only by a constant, irrelevant in QM, thus containing the same (site "equalized") local contribution. The minimization of the nonclassical gradient information thus gives rise to the site *phase*-equalization in the information-equilibrium state of the reactive system as a whole [63].

Appendix D Reactant Entanglement

The A and B subsystems of the reactive system R = A—B ultimately represent the bonded (entangled) subsystems in the equilibrium complex $R^* = (A^* \mid B^*)$ containing the mutually- and internally-open substrates $\{X^*\}$. Therefore, when brought into temporary interaction and then infinitely separated in $R^*(\infty) = A^* + B^*$, such dissociated fragments can no longer be described by the individual wavefunctions for each reactant, even after the interaction has utterly ceased and the "molecular" Hamiltonian is given by the sum of subsystem Hamiltonians:

$$H[R(\infty)] = H(A) + H(B). \tag{A56}$$

Indeed, QM deals with a *tensor product* of subsystem spaces, not with their *sum* [101]. However, for additive parts of the Hamiltonian, the "separation theorem" of QM predicts

$$\Psi[R(\infty)] = \Psi(A) \, \Psi(B), \tag{A57}$$

so that the following natural question arises: What is the molecular trace ("memory") of the past interaction left in the entangled separated fragments? As we have already argued elsewhere, it is the phase part of the system wavefunction which preserves a memory of the temporary interaction between subsystems (see also Appendix C).

The IT equilibrium phase of the given quantum system as a whole has been previously linked to its probability distribution $p(r)$ [62,63], $\phi_{eq.}(r) = - (\frac{1}{2}) \ln p(r)$, thus predicting the equilibrium current proportional to the negative gradient of probability density [see Equation (37)]:

$$j_{eq.}(r) = (\hbar/m) \, p(r) \, \nabla\phi_{eq.}(r) \equiv p(r) \, V_{eq.}(r) = - [\hbar/(2m)] \, \nabla p(r). \tag{A58}$$

Therefore, the product function in Equation (A57) of the separated "dissociated" subsystems still preserves the memory of the probability distribution $p(r)$ of the past (interacting) subsystems at a finite separation between reactants, contained in the equilibrium phase for this "molecular" separated reactive system.

The same phase criterion applies to diagnosing the entanglement distinction between the *bonded* and *nonbonded* status of reactants in the equilibrium, $R^* = (A^* \mid B^*)$, and polarized, $R^+ = (A \mid B)$, complexes, respectively. For the same probability distribution $p_R(r)$ in both these hypothetical states, the two subsystems are free to exchange electrons in the former, while in the latter this flow of electrons is forbidden. The equilibrium phase of $\Psi(R^*)$ then reflects the negative of $\ln p_R$, $\phi_{eq.}(R^*) = \phi_{eq.}[p_R]$, while at a finite separation R_{AB} between reactants $\phi_{eq.}(R^+)$ is proportional to the negative of $\ln[(p_A p_B)^{1/2}]$. Therefore, it is also the phase component which distinguishes the bonded (entangled) state of reactants in R^* from their nonbonded (disentangled) status in R^+, also at finite separations between the two reactants.

The mutual opening of the interacting reactants at their finite separation establishes the state global phase $\phi_{eq.}[R^*(x_{AB})]$, where x_{AB} denotes the separation coordinate, which is naturally transformed in the dissociation limit $R_{AB}\rightarrow\infty$ into the equilibrium phases of reactants,

$$\phi_{eq.}[R^*(R_{AB}\rightarrow\infty)]\rightarrow\{\phi_{eq.}[A^*(x_{AB} \rightarrow -\infty)] \text{ or } \phi_{eq.}[B^*(x_{AB} \rightarrow +\infty)]\}, \tag{A59}$$

upon the infinite separation of the mutually open subsystems, when

$$\rho_R[R^*(R_{AB}\rightarrow\infty)] \rightarrow \{\rho_A^*(x_{AB} \rightarrow -\infty) \text{ or } \rho_B^*(x_{AB} \rightarrow +\infty)\}. \tag{A60}$$

Appendix E Density Matrices for Interacting Subsystems

In describing the mixed states of interacting subsystems, it is useful to apply the density matrix formalism of QM [102,103]. Consider again the acid and base reactants in the polarized (interacting) system $R^+ = [A^+(x)|B^+(\xi)]$, where x and ξ denote their internal coordinates, respectively. For example, in the topological, *physical*-space partitioning [104] of the molecular electron density, into pieces belonging to separate basins of the physical space, these coordinates describe the disjoint sets of the position variables in these regions of space, while in the functional-space division schemes, e.g., the stockholder partition [21], each of these coordinates explores the whole physical space of the allowed positions of the separate groups of electrons.

The physical properties of subsystems are represented by quantum operators acting on the internal degrees-of-freedom of the fragment in question. Since R^+ as a whole is assumed to represent an isolated system, it is described by the specific wavefunction $\Psi(x, \xi)$, the *pure* quantum state of the reactive system. For example, the interacting-fragment Hamiltonians in R^+, $\{H_X(N_X^0)\}$ act on the position variables of N_X^0 electrons belonging to X^+. The stationary states of $H_A(N_A^0) \equiv H_A(x)$,

$$H_A(x)\,\varphi_s(x) = E_s(A^+)\,\varphi_s(x), \tag{A61}$$

form the complete set capable of expanding the state of the whole system:

$$\Psi(x, \xi) = \Sigma_s\, \Phi_s(\xi)\,\varphi_s(x), \qquad \Phi_s(\xi) = \int \varphi_s^*(x)\,\Psi(x, \xi)\,dx. \tag{A62}$$

Notice, however, that the simple product representation of $\Psi(x, \xi)$ is not available at finite separations, between interacting subsystems, so that the state of each reactant cannot be described by the substrate wavefunction dependent on its own internal coordinates.

The quantum operator of the physical property $L(A)$ of subsystem A^+ will act only on variables x: $L(A) = L_x$. Its average value in the general state of Equation (A62), of the reactive system as a whole, then reads:

$$\begin{aligned}
\langle L \rangle_\Psi &= \int\int \Psi^*(x, \xi)\, L_x\, \Psi(x, \xi)\, dx\, d\xi \\
&= \Sigma_s \Sigma_{s'}\, [\int \Phi_{s'}^*(\xi)\, \Phi_s(\xi)\, d\xi]\, [\int \varphi_{s'}^*(x)\, L_x\, \varphi_s(x)\, dx] \\
&\equiv \Sigma_s\, \Sigma_{s'}\, \rho_{s,s'}(A)\, L_{s',s}(A) = \text{tr}_A[\rho(A)\, L(A)].
\end{aligned} \tag{A63}$$

Here, $\rho(A) = \{\rho_{s',s}\}$ stands for the effective density matrix of subsystem $A^+(x)$, already integrated over coordinates of electrons in the complementary subsystem $B^+(\xi)$. The partial trace of the preceding equation thus enables one to calculate the ensemble average of the subsystem quantity $L(A)$ as if this part of R^+ were isolated, being in the effective *mixed* state defined by subsystem density matrix $\rho(A)$ in representation $\{\varphi_s(x)\}$.

Writing the preceding equation in the subsystem position representation,

$$\langle L \rangle_\Psi = \int\int \rho_A(x, x')\, \langle x'|L_x|x \rangle \, dx\, dx', \tag{A64}$$

where $\langle x'|L_x|x \rangle = L_x \delta(x' - x)$, one obtains the following expression for the subsystem density matrix:

$$\begin{aligned}
\rho_A(x, x') &= \int \Psi^*(x', \xi)\, \Psi(x, \xi)\, d\xi = \Sigma_s \Sigma_{s'} \, \rho_{s,s'}(A)\, \varphi_{s'}^{*}(x')\, \varphi_s(x) \\
&\equiv \Sigma_s \Sigma_{s'} \, \rho_{s,s'}(A)\, \Omega_{s',s}(x, x') = \mathrm{tr}_A[\rho(A)\, \Omega_A(x, x')].
\end{aligned} \tag{A65}$$

These expressions for the expectation values of subsystem operators in the mixed quantum states are independent of the applied representation. The corresponding dynamics of the reactant density matrices involved are also uniquely determined by molecular SE.

To summarize, also the polarized (interacting) reactants $\{X^+\}$ in R^+ cannot be described by a single wavefunction of the *pure* quantum state. They have to instead be characterized by the density matrix reflecting an incoherent mixture of subsystem states, weighted by the ensemble probability factors and corresponding to the substrate *mixed* quantum state.

References

1. Fisher, R.A. Theory of statistical estimation. *Proc. Camb. Phil. Soc.* **1925**, *22*, 700–725. [CrossRef]
2. Frieden, B.R. *Physics from the Fisher Information—A Unification*; Cambridge University Press: Cambridge, UK, 2004.
3. Shannon, C.E. The mathematical theory of communication. *Bell Syst. Tech. J.* **1948**, *27*, 379–493, 623–656. [CrossRef]
4. Shannon, C.E.; Weaver, W. *The Mathematical Theory of Communication*; University of Illinois: Urbana, IL, USA, 1949.
5. Kullback, S.; Leibler, R.A. On information and sufficiency. *Ann. Math. Stat.* **1951**, *22*, 79–86. [CrossRef]
6. Kullback, S. *Information Theory and Statistics*; Wiley: New York, NY, USA, 1959.
7. Abramson, N. *Information Theory and Coding*; McGraw-Hill: New York, NY, USA, 1963.
8. Pfeifer, P.E. *Concepts of Probability Theory*; Dover: New York, NY, USA, 1978.
9. Nalewajski, R.F. *Information Theory of Molecular Systems*; Elsevier: Amsterdam, The Netherlands, 2006.
10. Nalewajski, R.F. *Information Origins of the Chemical Bond*; Nova Science Publishers: New York, NY, USA, 2010.
11. Nalewajski, R.F. *Perspectives in Electronic Structure Theory*; Springer: Heidelberg, Germany, 2012.
12. Nalewajski, R.F.; Parr, R.G. Information theory, atoms-in-molecules and molecular similarity. *Proc. Natl. Acad. Sci. USA* **2000**, *97*, 8879–8882. [CrossRef] [PubMed]
13. Nalewajski, R.F. Information principles in the theory of electronic structure. *Chem. Phys. Lett.* **2003**, *272*, 28–34. [CrossRef]
14. Nalewajski, R.F. Information principles in the Loge Theory. *Chem. Phys. Lett.* **2003**, *375*, 196–203. [CrossRef]
15. Nalewajski, R.F.; Broniatowska, E. Information distance approach to Hammond Postulate. *Chem. Phys. Lett.* **2003**, *376*, 33–39. [CrossRef]
16. Nalewajski, R.F.; Parr, R.G. Information-theoretic thermodynamics of molecules and their Hirshfeld fragments. *J. Phys. Chem. A* **2001**, *105*, 7391–7400. [CrossRef]
17. Nalewajski, R.F. Hirshfeld analysis of molecular densities: Subsystem probabilities and charge sensitivities. *Phys. Chem. Chem. Phys.* **2002**, *4*, 1710–1721. [CrossRef]
18. Parr, R.G.; Ayers, P.W.; Nalewajski, R.F. What is an atom in a molecule? *J. Phys. Chem. A.* **2005**, *109*, 3957–3959. [CrossRef]
19. Nalewajski, R.F.; Broniatowska, E. Atoms-in-Molecules from the stockholder partition of molecular two-electron distribution. *Theoret. Chem. Acc.* **2007**, *117*, 7–27. [CrossRef]

20. Heidar-Zadeh, F.; Ayers, P.W.; Verstraelen, T.; Vinogradov, I.; Vöhringer-Martinez, E.; Bultinck, P. Information-theoretic approaches to Atoms-in-Molecules: Hirshfeld family of partitioning schemes. *J. Phys. Chem. A* **2018**, *122*, 4219–4245. [CrossRef] [PubMed]

21. Hirshfeld, F.L. Bonded-atom fragments for describing molecular charge densities. *Theoret. Chim. Acta* **1977**, *44*, 129–138. [CrossRef]

22. Nalewajski, R.F. Entropic measures of bond multiplicity from the information theory. *J. Phys. Chem. A* **2000**, *104*, 11940–11951. [CrossRef]

23. Nalewajski, R.F. Entropy descriptors of the chemical bond in Information Theory: I. Basic concepts and relations. Mol Phys 102:531-546; II. Application to simple orbital models. *Mol. Phys.* **2004**, *102*, 547–566. [CrossRef]

24. Nalewajski, R.F. Entropic and difference bond multiplicities from the two-electron probabilities in orbital resolution. *Chem. Phys. Lett.* **2004**, *386*, 265–271. [CrossRef]

25. Nalewajski, R.F. Reduced communication channels of molecular fragments and their entropy/information bond indices. *Theoret. Chem. Acc.* **2005**, *114*, 4–18. [CrossRef]

26. Nalewajski, R.F. Partial communication channels of molecular fragments and their entropy/information indices. *Mol. Phys.* **2005**, *103*, 451–470. [CrossRef]

27. Nalewajski, R.F. Entropy/information descriptors of the chemical bond revisited. *J. Math. Chem.* **2011**, *49*, 2308–2329. [CrossRef]

28. Nalewajski, R.F. Quantum information descriptors and communications in molecules. *J. Math. Chem.* **2014**, *52*, 1292–1323. [CrossRef]

29. Nalewajski, R.F. Multiple, localized and delocalized/conjugated bonds in the orbital-communication theory of molecular systems. *Adv. Quantum Chem.* **2009**, *56*, 217–250.

30. Nalewajski, R.F.; Szczepanik, D.; Mrozek, J. Bond differentiation and orbital decoupling in the orbital communication theory of the chemical bond. *Adv. Quantum Chem.* **2011**, *61*, 1–48.

31. Nalewajski, R.F.; Szczepanik, D.; Mrozek, J. Basis set dependence of molecular information channels and their entropic bond descriptors. *J. Math. Chem.* **2012**, *50*, 1437–1457. [CrossRef]

32. Nalewajski, R.F. Electron communications and chemical bonds. In *Frontiers of Quantum Chemistry*; Wójcik, M., Nakatsuji, H., Kirtman, B., Ozaki, Y., Eds.; Springer: Singapore, 2017; pp. 315–351.

33. Nalewajski, R.F.; Świtka, E.; Michalak, A. Information distance analysis of molecular electron densities. *Int. J. Quantum Chem.* **2002**, *87*, 198–213. [CrossRef]

34. Nalewajski, R.F.; Broniatowska, E. Entropy displacement analysis of electron distributions in molecules and their Hirshfeld atoms. *J. Phys. Chem. A* **2003**, *107*, 6270–6280. [CrossRef]

35. Nalewajski, R.F. Use of Fisher information in quantum chemistry. *Int. J. Quantum Chem.* **2008**, *108*, 2230–2252. [CrossRef]

36. Nalewajski, R.F.; Köster, A.M.; Escalante, S. Electron localization function as information measure. *J. Phys. Chem. A* **2005**, *109*, 10038–10043. [CrossRef]

37. Becke, A.D.; Edgecombe, K.E. A simple measure of electron localization in atomic and molecular systems. *J. Chem. Phys.* **1990**, *92*, 5397–5403. [CrossRef]

38. Silvi, B.; Savin, A. Classification of chemical bonds based on topological analysis of electron localization functions. *Nature* **1994**, *371*, 683–686. [CrossRef]

39. Savin, A.; Nesper, R.; Wengert, S.; Fässler, T.F. ELF: The electron localization function. *Angew. Chem. Int. Ed. Engl.* **1997**, *36*, 1808–1832. [CrossRef]

40. Hohenberg, P.; Kohn, W. Inhomogeneous electron gas. *Phys. Rev.* **1964**, *136B*, 864–971. [CrossRef]

41. Kohn, W.; Sham, L.J. Self-consistent equations including exchange and correlation effects. *Phys. Rev.* **1965**, *140A*, 133–1138. [CrossRef]

42. Levy, M. Universal variational functionals of electron densities, first-order density matrices, and natural spin-orbitals and solution of the v-representability problem. *Proc. Natl. Acad. Sci. USA* **1979**, *76*, 6062–6065. [CrossRef] [PubMed]

43. Parr, R.G.; Yang, W. *Density-Functional Theory of Atoms and Molecules*; Oxford University Press: New York, NY, USA, 1989.

44. Dreizler, R.M.; Gross, E.K.U. *Density Functional Theory: An Approach to the Quantum Many-Body Problem*; Springer: Berlin, Germany, 1990.

45. Nalewajski, R.F. (Ed.) *Density Functional Theory I-IV, Topics in Current Chemistry*; Springer: Berlin, Germany, 1966; Volume 180–183.
46. Nalewajski, R.F.; de Silva, P.; Mrozek, J. Use of nonadditive Fisher information in probing the chemical bonds. *J. Mol. Struct.* **2010**, *954*, 57–74. [CrossRef]
47. Nalewajski, R.F. Through-space and through-bridge components of chemical bonds. *J. Math. Chem.* **2011**, *49*, 371–392. [CrossRef]
48. Nalewajski, R.F. Chemical bonds from through-bridge orbital communica-tions in prototype molecular systems. *J. Math. Chem.* **2011**, *49*, 546–561. [CrossRef]
49. Nalewajski, R.F. On interference of orbital communications in molecular systems. *J. Math. Chem.* **2011**, *49*, 806–815. [CrossRef]
50. Nalewajski, R.F.; Gurdek, P. On the implicit bond-dependency origins of bridge interactions. *J. Math. Chem.* **2011**, *49*, 1226–1237. [CrossRef]
51. Nalewajski, R.F. Direct (through-space) and indirect (through-bridge) components of molecular bond multiplicities. *Int. J. Quantum Chem.* **2012**, *112*, 2355–2370. [CrossRef]
52. Nalewajski, R.F.; Gurdek, P. Bond-order and entropic probes of the chemical bonds. *Struct. Chem.* **2012**, *23*, 1383–1398. [CrossRef]
53. Nalewajski, R.F. Exploring molecular equilibria using quantum information measures. *Ann. Phys.* **2013**, *525*, 256–268. [CrossRef]
54. Nalewajski, R.F. On phase equilibria in molecules. *J. Math. Chem.* **2014**, *52*, 588–612. [CrossRef]
55. Nalewajski, R.F. Quantum information approach to electronic equilibria: Molecular fragments and elements of non-equilibrium thermodynamic description. *J. Math. Chem.* **2014**, *52*, 1921–1948. [CrossRef]
56. Nalewajski, R.F. Phase/current information descriptors and equilibrium states in molecules. *Int. J. Quantum Chem.* **2015**, *115*, 1274–1288. [CrossRef]
57. Nalewajski, R.F. Quantum information measures and molecular phase equilibria. In *Advances in Mathematics Research*; Baswell, A.R., Ed.; Nova Science Publishers: New York, NY, USA, 2015; Volume 19, pp. 53–86.
58. Nalewajski, R.F. On phase/current components of entropy/information descriptors of molecular states. *Mol. Phys.* **2014**, *112*, 2587–2601. [CrossRef]
59. Nalewajski, R.F. Complex entropy and resultant information measures. *J. Math. Chem.* **2016**, *54*, 1777–1782. [CrossRef]
60. Nalewajski, R.F. Phase description of reactive systems. In *Conceptual Density Functional Theory*; Islam, N., Kaya, S., Eds.; Apple Academic Press: Waretown, NJ, USA, 2018; pp. 217–249.
61. Nalewajski, R.F. Quantum information measures and their use in chemistry. *Curr. Phys. Chem.* **2017**, *7*, 94–117. [CrossRef]
62. Nalewajski, R.F. *Quantum Information Theory of Molecular States*; Nova Science Publishers: New York, NY, USA, 2016.
63. Nalewajski, R.F. Resultant information approach to donor-acceptor systems. In *An Introduction to Electronic Structure Theory*; Nova Science Publishers: New York, NY, USA, 2020; in press.
64. Nalewajski, R.F. Information equilibria, subsystem entanglement and dynamics of overall entropic descriptors of molecular electronic structure. *J. Mol. Model.* **2018**, *24*, 212–227. [CrossRef]
65. Nalewajski, R.F. On entangled states of molecular fragments. *Trends Phys. Chem.* **2016**, *16*, 71–85.
66. Nalewajski, R.F. Virial theorem implications for the minimum energy reaction paths. *Chem. Phys.* **1980**, *50*, 127–136. [CrossRef]
67. Nalewajski, R.F. Understanding electronic structure and chemical reactivity: Quantum-information perspective. *Appl. Sci.* **2019**, *9*, 1262–1292, In *The Application of Quantum Mechanics to the Reactivity of Molecules*. [CrossRef]
68. Nalewajski, R.F. *Overall Entropy/Information Descriptors of Electronic States and Chemical Reactivity*; Islam, N., Bir Singh, S., Ranjan, P., Haghi, A.K., Eds.; Apple Academic Press: Waretown, NJ, USA, 2020; in press.
69. Nalewajski, R.F. On entropy/information description of reactivity phenomena. In *Advances in Mathematics Research*; Baswell, A.R., Ed.; Nova Science Publishers: New York, NY, USA, 2019; Volume 26, pp. 97–157.
70. Nalewajski, R.F. Role of electronic kinetic energy (resultant gradient information) in chemical reactivity. *J. Mol. Model.* **2019**, *25*, 259–278, Berski, S., Sokalski, W.A. (Eds.). [CrossRef] [PubMed]

71. Nalewajski, R.F. On classical and quantum entropy/information descriptors of molecular electronic states. In *Research Methodologies and Practical Applications of Chemistry*; Pogliani, L., Haghi, A.K., Islam, N., Eds.; Apple Academic Press: Waretown, NJ, USA, 2019; in press.

72. Callen, H.B. *Thermodynamics: An Introduction to the Physical Theories of Equilibrium Thermostatics and Irreversible Thermodynamics*; Wiley: New York, NY, USA, 1962.

73. Nalewajski, R.F. Equidensity orbitals in resultant-information description of electronic states. *Theoret. Chem. Acc.* **2019**, *138*, 108–123, In *Chemical Concepts from Theory and Computation*; Liu, S. (Ed.). [CrossRef]

74. Nalewajski, R.F. Resultant information description of electronic states and chemical processes. *J. Phys. Chem. A* **2019**, *123*, 45–60. [CrossRef] [PubMed]

75. Harriman, J.E. Orthonormal orbitals for the representation of an arbitrary density. *Phys. Rev. A* **1980**, *24*, 680–682. [CrossRef]

76. Zumbach, G.; Maschke, K. New approach to the calculation of density functionals. *Phys. Rev. A* **1983**, *28*, 544–554, Erratum, *Phys Rev A* 29, 1585–1587. [CrossRef]

77. Macke, W. Zur wellermechanischen behandlung von vielkoeperproblemen. *Ann. Phys.* **1955**, *17*, 1–9. [CrossRef]

78. Gilbert, T.L. Hohenberg-Kohn theorem for nonlocal external potentials. *Phys. Rev. B* **1975**, *12*, 2111–2120. [CrossRef]

79. Nalewajski, R.F.; Korchowiec, J.; Michalak, A. Reactivity criteria in charge sensitivity analysis. *Top. Curr. Chem.* **1996**, *183*, 25–141, In *Density Functional Theory IV*; Nalewajski, R.F., Ed.

80. Nalewajski, R.F.; Korchowiec, J. *Charge Sensitivity Approach to Electronic Structure and Chemical Reactivity*; World Scientific: Singapore, 1997.

81. Geerlings, P.; De Proft, F.; Langenaeker, W. Conceptual density functional theory. *Chem. Rev.* **2003**, *103*, 1793–1873. [CrossRef]

82. Nalewajski, R.F. Sensitivity analysis of charge transfer systems: In situ quantities, intersecting state model and its implications. *Int. J. Quantum Chem.* **1994**, *49*, 675–703. [CrossRef]

83. Nalewajski, R.F. Charge sensitivity analysis as diagnostic tool for predicting trends in chemical reactivity. In *Proceedings of the NATO ASI on Density Functional Theory*; (Il Ciocco, 1993); Dreizler, R.M., Gross, E.K.U., Eds.; Plenum: New York, NY, USA, 1995; pp. 339–389.

84. Chattaraj, P.K. (Ed.) *Chemical Reactivity Theory: A Density Functional View*; CRC Press: Boca Raton, FL, USA, 2009.

85. Nalewajski, R.F. Chemical reactivity description in density-functional and information theories. *Acta Phys. Chim. Sin.* **2017**, *33*, 2491–2509, In *Chemical Concepts from Density Functional Theory*; Liu, S.(Ed.).

86. Gyftopoulos, E.P.; Hatsopoulos, G.N. Quantum-thermodynamic definition of electronegativity. *Proc. Natl. Acad. Sci. USA* **1965**, *60*, 786–793. [CrossRef] [PubMed]

87. Perdew, J.P.; Parr, R.G.; Levy, M.; Balduz, J.L. Density functional theory for fractional particle number: Derivative discontinuities of the energy. *Phys. Rev. Lett.* **1982**, *49*, 1691–1694. [CrossRef]

88. Mulliken, R.S. A new electronegativity scale: Together with data on valence states and on ionization potentials and electron affinities. *J. Chem. Phys.* **1934**, *2*, 782–793. [CrossRef]

89. Iczkowski, R.P.; Margrave, J.L. Electronegativity. *J. Am. Chem. Soc.* **1961**, *83*, 3547–3551. [CrossRef]

90. Parr, R.G.; Donnelly, R.A.; Levy, M.; Palke, W.E. Electronegativity: The density functional viewpoint. *J. Chem. Phys.* **1978**, *69*, 4431–4439. [CrossRef]

91. Parr, R.G.; Pearson, R.G. Absolute hardness: Companion parameter to absolute electronegativity. *J. Am. Chem. Soc.* **1983**, *105*, 7512–7516. [CrossRef]

92. Parr, R.G.; Yang, W. Density functional approach to the frontier-electron theory of chemical reactivity. *J. Am. Chem. Soc.* **1984**, *106*, 4049–4050. [CrossRef]

93. Dirac, P.A.M. *The Principles of Quantum Mechanics*; Oxford University Press: Oxford, UK, 1967.

94. Chandra, A.K.; Michalak, A.; Nguyen, M.T.; Nalewajski, R.F. On regional matching of atomic softnesses in chemical reactions: Two-reactant charge sensitivity study. *J. Phys. Chem. A* **1998**, *102*, 10182–10188. [CrossRef]

95. Nalewajski, R.F. Manifestations of the maximum complementarity principle for matching atomic softnesses in model chemisorption systems. *Top. Catal.* **2000**, *11*, 469–485. [CrossRef]

96. Prigogine, I. *From Being to Becoming: Time and Complexity in the Physical Sciences*; Freeman WH & Co.: San Francisco, CA, USA, 1980.

97. Flores-Gallegos, N.; Esquivel, R.O. Von Neumann entropies analysis in Hilbert space for the dissociation processes of homonuclear and heteronuclear diatomic molecules. *J. Mex. Chem. Soc.* **2008**, *52*, 19–30.

98. López-Rosa, S. Information Theoretic Measures of Atomic and Molecular Systems. Ph.D. Thesis, University of Granada, Granada, Spain, 2010.

99. López-Rosa, S.; Esquivel, R.O.; Angulo, J.C.; Antolín, J.; Dehesa, J.S.; Flores-Gallegos, N. Fisher information study in position and momentum spaces for elementary chemical reactions. *J. Chem. Theory Comput.* **2010**, *6*, 145–154. [CrossRef] [PubMed]

100. Esquivel, R.O.; Liu, S.; Angulo, J.C.; Dehesa, J.S.; Antolín, J.; Molina-Espíritu, M. Fisher information a steric effect: Study of internal rotation barrier in ethane. *J. Phys. Chem. A* **2011**, *115*, 4406–4415. [CrossRef] [PubMed]

101. Lieb, E. *Topics in Quantum Entropy and Entanglement*; Three lectures at Priceton Condensed Matter Summer School: Princeton, NJ, USA, 30–31 July 2014.

102. Davydov, A.S. *Quantum Mechanics*; Pergamon Press: Oxford, UK, 1965.

103. Cohen-Tannoudji Diu, B.; Laloë, F. *Quantum Mechanics*; Wiley: New York, NY, USA, 1977.

104. Bader, R.F.W. *Atoms in Molecules*; Oxford University Press: New York, NY, USA, 1990.

Article

Atomic-Scale Understanding of Structure and Properties of Complex Pyrophosphate Crystals by First-Principles Calculations

Redouane Khaoulaf [1,2], Puja Adhikari [3], Mohamed Harcharras [4], Khalid Brouzi [5], Hamid Ez-Zahraouy [2] and Wai-Yim Ching [3,*]

[1] Laboratory of Spectroscopy, Molecular Modeling, Materials, Nanomaterials, Water and Environment, (LS3MN2E-CERNE2D), Faculty of Sciences, Mohammed V University, Av Ibn Battouta, B.P 1014, Rabat 10000, Morocco; rkhaoulaf2000@yahoo.fr

[2] Laboratory of Condensed Matter and Interdisciplinary Sciences (LAMCSCI), Faculty of Sciences, University Mohammed V, Rabat B.P 1014, Morocco; ezahamid@gmail.com

[3] Department of Physics and Astronomy, University of Missouri, Kansas City, MO 64110, USA; paz67@mail.umkc.edu

[4] Laboratory of Materials Engineering and Environment, Université Ibn Tofail, Kenitra B.P 242, Morocco; mharch2009@gmail.com

[5] Energie Matériaux et Développement Durable, EMDD-CERNE2D, Mohammed V University, Rabat B.P 1014, Morocco; khbrouzi@hotmail.com

* Correspondence: Chingw@umkc.edu; Tel.: +1-816-235-2503

Received: 11 December 2018; Accepted: 11 February 2019; Published: 27 February 2019

Featured Application: Pyrophosphate crystals with low effective mass leading to high electron mobility could be used in radiation detectors. In addition, with a large range of absorption in the infrared, visible, and ultraviolet spectral regions, they could be used in the optical field. The prediction of novel mechanical properties of pyrophosphate may lead to new applications.

Abstract: The electronic structure and mechanical and optical properties of five pyrophosphate crystals with very complex structures are studied by first principles density functional theory calculations. The results show the complex interplay of the minor differences in specific local structures and compositions can result in large differences in reactivity and interaction that are rare in other classes of inorganic crystals. These are discussed by dividing the pyrophosphate crystals into three structural units. $H_2P_2O_7$ is the most important and dominating unit in pyrophosphates. The other two are the influential cationic group with metals and water molecules. The strongest P-O bond in P_2O_5 is the strongest bond for crystal cohesion, but O-H and N-H bonds also play an important part. Different type of bonding between O and H atoms such as O-H, hydrogen bonding, and bridging bonds are present. Metallic cations such as Mg, Zn, and Cu form octahedral bonding with O. The water molecule provides the unique H···O bonds, and metallic elements can influence the structure and bonding to a certain extent. The two Cu-containing phosphates show the presence of narrow metallic bands near the valence band edge. All this complex bonding affects their physical properties, indicating that fundamental understanding remains an open question.

Keywords: Pyrophosphate; electronic structure; mechanical properties; optical properties; first-principles calculations

---

1. Introduction

In comparison with oxides or nitrides, the structure and properties of phosphates are much more complex and less well studied. However, they are critically important and explored in multiple disciplines such as physics, chemistry, biology, and materials science. In the elementary classification, there are three types of phosphates—monophosphates, condensed phosphates, and oxyphosphates [1]. Phosphoric anion consists of PO_4 as a basic unit and is present in any phosphoric anion with P-O-P bonds in the condensed phase. One of the condensed phosphates is the pyrophosphate, also known as diphosphate or dipolyphosphate. This is a subset of family of crystals that originates from pyrophosphoric acid $P_2O_5 \cdot 2H_2O$ ($H_4P_2O_7$) and contains the pentoxide group (P_2O_5). Pyrophosphates exist in both crystalline as well as non-crystalline glassy forms. Phosphate glasses are quite different from the silicate-based optical glasses due to low dispersion and high refractive indices. Alkaline earth phosphate glasses are particularly important due to its high transparency in the ultraviolet region [2].

Phosphates have a wide range of scientific and technological applications. Phosphate glasses are used as host materials for rare-earth ions in solid state laser as low temperature sealing glasses [2], and in lithium batteries [3]. They have also been used as source in near-infrared lasers [4]. Phosphate phosphor can be used in field of lighting due to effective excitation in near-ultraviolet range [5], Other uses of phosphate glasses include nuclear power production [6] and piezoelectric material for pressure sensor applications as demonstrate in Ga phosphate crystal [7].

In biomolecular science, phosphates always play a tremendous role. For example, adenosine triphosphate (ATP) and creatine phosphate are the main reservoirs of biochemical energy. Phosphates and especially pyrophosphates are the usual "leaving group" in metabolic reaction [8] i.e., pyrophosphate leaves the reactant, making it more reactive, and this characteristic is used in inhibiting infections [9]. Another example is the use of phosphates in materials, especially calcium orthophosphate, for osteoporosis patients in hip fracture and joint replacement [10] in the human body, which demands special mechanical properties such as high fracture toughness and yield strength. Phosphate-based glasses are increasingly used as the main biomaterials for hard and soft tissue engineering that require a specific biological response [11]. It should also be noted that we actually have a track record in studying the structure and properties of bio-related phosphates including hydroxyapatite (HAp) [12–14] and tricalcium phosphate (TCP) [15].

In addition to biomedical applications, phosphates are also known for applications in several other areas such as energy science, sensors, catalysis, and many more. One of the major applications of phosphates is in agriculture and the development of green energy. They are particularly important in a country like Morocco with a long history of development and natural applications. Indeed, in Morocco, phosphates are exceptionally rich, both in terms of the content of their P_2O_5 containing ore and the enormity of their reserves that form the valuable sedimentary series. According to the report published in January 2018 by the United States Geological Survey (USGS), the world phosphate rock reserves are estimated at about 70 billion tons, of which 50 billion tons are in Morocco, which represents three quarters of the world's reserves.

Phosphates are very important for the economy and industrial development in a country like Morocco. The OCP group (Cherifian Phosphates Office) is a Moroccan company created in 1920 that specializes in the extraction, recovery, and marketing of phosphate (P_2O_5) and its derived products. It makes Morocco the world's leading market in the production of phosphate products. The treatment of phosphate rocks gives the following marketed products: phosphoric acid (H_3PO_4), single superphosphate (SSP), monoammonium phosphate (MAP), diammonium phosphate (DAP), fertilizer (NPK), and triple superphosphate (TSP). The strategy used by the OCP group for the distribution of phosphate (P_2O_5) was based on the three sectors, fertilizers (85% by volume of P_2O_5), animal nutrition (8%) as well as a wide variety of industrial uses (7%), including laundry detergents and food products such as sodas, etc. Another use is that the purified phosphoric acid is used as salts for the food industry such as yeast, cheese, preservation of meat and fish, treatment of drinking water, etc. Other industrial uses include metal processing, textiles, cements etc.

Given the complex structures of phosphates and their ubiquitous presence, a fundamental understanding of their structures and properties is a subject of paramount importance. However, such studies are lagging far behind other inorganic materials such as silicates. High-level quantum mechanical calculations always play a role in such cases in providing the information on electronic structure, molecular interaction, and reactivity. They have implications on physical properties including vibrational, optical, and mechanical properties. Such studies on phosphate crystals can facilitate identifying the potential usage of these complex materials. Several such investigations have appeared in the literature in recent years. Tang et al. [16] investigated the behavior of phosphate species using ab initio molecular dynamics. Application of pyrophosphate for water oxidation catalysts was investigated by Kim et al. [17] using both experimental and computational methods. Other density functional theory (DFT)-based calculations for different kinds of pyrophosphates has been done by Witko et al. [18], Zhang et al. [19], and Xiang et al. [20].

In this work, we present the results of calculations of five pyrophosphate crystals including the one that was published a year ago on $K_2Mg(H_2P_2O_7)_2 \cdot 2H_2O$ [21]. The only difference in the structural components of these five crystals are the alkali metal K or NH_4 and metallic elements such as Mg, Zn, or Cu. They have the same pyrophosphate group $H_2P_2O_7$ and same number of H_2O molecules. As expected, the electronic structure and bonding in these five crystals are very similar, but to our great surprise, they show some considerable variations. The presence of narrow metallic bands at the top of valence band (VB) in the two Cu-containing pyrophosphates is identified. Also, the mechanical properties with the parameter gauging brittleness and ductility can differ by as much as 50%. Since the mechanical properties of all materials ultimately depends on the strengths of interatomic bonding, what could be the atomistic origin of such unusual behavior for the pyrophosphate crystals? With the detailed and accurate calculations of the electronic structure and mechanical and optical properties, we offer some tangible insights that defies the conventional wisdom. In the next two sections, we briefly describe the crystal structures and computational methods used in the calculation. The main results are presented and discussed in Section 4. The paper ends with a summary and some conclusions.

2. Crystal Structures

The five pyrophosphates in this paper are $K_2Mg(H_2P_2O_7)_2 \cdot 2H_2O$, $(NH_4)_2Zn(H_2P_2O_7)_2 \cdot 2H_2O$, $K_2Cu(H_2P_2O_7)_2 \cdot 2H_2O$, $4[K_2Cu(H_2P_2O_7)_2 \cdot 2H_2O]$, and $4[K_2Zn(H_2P_2O_7)_2 \cdot 2H_2O]$. These crystals were synthesized using the wet method. This method is generally applied by the strategy followed by our scientific team (RK, KB, and MH) to obtain multicomponent compounds such as acidic pyrophosphates studied in this work. This method is based on the preparation of a solution of the various precursors, which are generally sodium or potassium pyrophosphates $(Na_4P_2O_7(H_2O)10, K_4P_2O_7)$, NH_4Cl, $MgCl_2$ $(H_2O)_6$, $ZnCl_2(H_2O)_4$, and $CuCl_2(H_2O)_2$. We then resort to the solubilization of all these precursors by the addition of a strong acid of the HCl type (pH of solute must be controlled), sometimes with slow stirring. Afterward, slow evaporation at room temperature for all the solute leads to the formation of single crystals. These pyrophosphates have well-characterized crystal structure as well as vibrational analysis, Raman, and infrared spectra reported such as $K_2Mg(H_2P_2O_7)_2 \cdot 2H_2O$ by Harcharras et al. [22], $(NH_4)_2Zn(H_2P_2O_7)_2 \cdot 2H_2O$ by Capitelli et al. [23], and $4[K_2Zn(H_2P_2O_7)_2 \cdot 2H_2O]$ by Khaoulaf et al. [24]. $K_2Cu(H_2P_2O_7)_2 \cdot 2H_2O$ and $4[K_2Cu (H_2P_2O_7)_2 \cdot 2H_2O]$ are same crystal structure with different space group. i.e., *P*-1 and *Pnma* respectively. Their Raman and infrared vibrational spectroscopic analysis is done by Khaoulaf et al. [25].

The structure of these five crystals after optimization are shown in Figure 1. As already pointed out in ref [21], crystal containing light H atoms have difficulties in accurately locating the positions of H atoms because of the weak intensity signal in the measurement. This results in the unrealistically short O-H bonds. This deficiency has been rectified by accurate ab initio geometry optimization implemented in the Vienna Ab initio Simulation Package (VASP) to be explained in section below. The optimized crystal parameters for all five crystals are shown in Table 1. In the later discussion on the structural characteristics and their physical properties of these five crystals, we will focus on the difference and interplay between the three structural units. The first unit is the cationic group consisting of metallic elements. The only difference in the crystal composition is in the first structural unit K or NH_4 and metallic element Mg, Zn, or Cu. The composition from the other two structural units, pyrophosphate group $H_2P_2O_7$ and water are the same. To save space, we will name these five crystals as C1 to C5 with difference only in the first two components of unit 1: C1 for $K_2Mg(H_2P_2O_7)_2\cdot2H_2O$, C2 for $(NH_4)_2Zn(H_2P_2O_7)_2\cdot2H_2O$, C3 for $K_2Cu(H_2P_2O_7)_2\cdot2H_2O$, C4 for $4[K_2Cu(H_2P_2O_7)_2\cdot2H_2O]$, and C5 for $4[K_2Zn(H_2P_2O_7)_2\cdot2H_2O]$. It should be pointed out that C3 and C4 have the same chemical composition and formula unit but with different crystal symmetry and volumes. The C1, C2, and C3 crystals have the triclinic structure whereas C4 and C5 crystals have orthorhombic structure and are called composites.

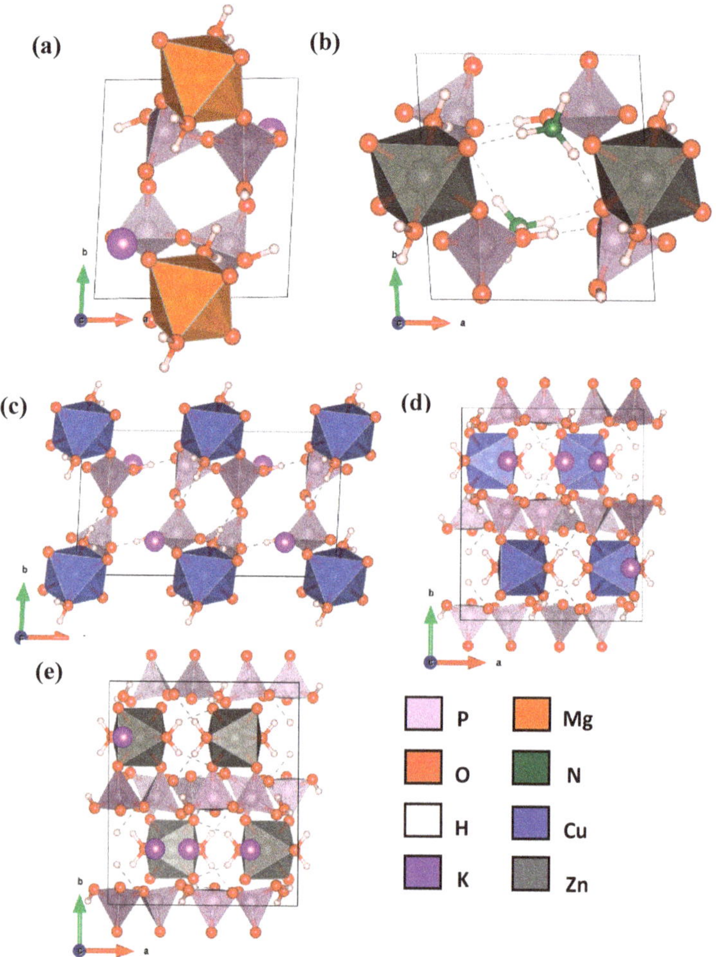

Figure 1. Polyhedra figures for the five pyrophosphate crystals viewed along the c-direction. The labels (**a**–**e**) are $K_2Mg(H_2P_2O_7)_2\cdot2H_2O$, $(NH_4)_2Zn(H_2P_2O_7)_2\cdot2H_2O$, $K_2Cu(H_2P_2O_7)_2\cdot2H_2O$, $4[K_2Cu(H_2P_2O_7)_2\cdot2H_2O]$, and $4[K_2Zn(H_2P_2O_7)_2\cdot2H_2O]$ respectively.

Table 1. Lattice parameters of the five crystals.

	Space Group	a(Å), b(Å), c(Å), α, β, γ	Volume (Å^3)	No. of Atoms
C1	P-1(Ci)	6.954, 7.503, 7.589, 81.166°, 75.522°, 84.257°	378.05	31
C2	P-1(Ci)	7.178, 7.424, 7.808, 81.007°, 71.428°, 90.952°	388.58	39
C3	P-1(Ci)	7.101, 7.430, 7.609, 78.761°, 71.657°, 83.958°	373.38	31
C4	$Pnma$(D$_{2h}^{16}$)	9.757, 11.134, 13.728, 90.000°, 90.000°, 90.000°	1491.35	124
C5	$Pnma$(D$_{2h}^{16}$)	9.770, 11.166, 13.746, 90.000°, 90.000°, 90.000°	1499.60	124

3. Computational Methods

We used two computational packages in this study: VASP [26] and Orthogonal Linear Combination of Atomic Orbitals (OLCAO) [27]. Both are based on the density functional theory. For structural optimization and elastic properties calculations using VASP, we used the PAW-PBE potential [28] with generalized gradient approximation (GGA) for the exchange correlation potential. We used a relatively high energy cutoff of 600 eV. The electronic and ionic force convergence criteria are set at 10^{-9} eV and 10^{-7} eV/Å respectively. A $6 \times 6 \times 6$ k-point mesh was used for C1, C2, and C3 crystals and a less dense mesh of $4 \times 4 \times 4$ for the two larger crystals C4 and C5. No discernable difference in the relaxed structure was observed using other exchange correlation functions such as hybrid functional PBE0, HSE03, and applying van der Waals correction in the case of $K_2Mg(H_2P_2O_7)_2 \cdot 2H_2O$.

The calculation of mechanical properties using the optimized structure has been described before [29,30]. Specifically, we apply a small strain ϵ ($\pm$ 0.25%) to the crystal. The elastic tensor elements are obtained using the stress (σ_j) vs strain (ϵ_i) response analysis scheme to the relaxed structure and then obtained the elastic coefficients C_{ij} (i, j = 1, 2, 3, 4, 5, 6 (using Hooks law) and compliance coefficient S_{ij} by solving the set of linear equations From the calculated C_{ij} and S_{ij}, other mechanical properties such as bulk modulus (K), shear modulus (G), Poisson's ratio (η), and Young's modulus (E) are obtained using Voight-Reuss-Hill (VRH) polycrystals approximation [31,32].

For electronic structure and interatomic bonding, we use the in-house developed OLCAO with the VASP-relaxed structure as input. A more localized minimal basis (MB) is used for the calculation of effective charge Q_α^* and bond order (BO) values. The BO is the overlap population $\rho_{\alpha\beta}$ between any pair of atoms (α, β) based on Mulliken population analysis [33,34]. Mulliken analysis is only meaningful for comparisons between different materials if the same basis set in the same computational package is used as in the present case. Mulliken analysis is more effective if a localized basis is used, and we always use the minimal basis as implemented in OLCAO method with the same atomic basis for all atomic species [27]. The same approach has been successfully applied to many different materials systems including complex biomolecular systems [35–45].

$$Q_\alpha^* = \sum_i \sum_{m,occ} \sum_{j,\beta} C_{i\alpha}^{*m} C_{j\beta}^m S_{i\alpha,j\beta} \tag{1}$$

$$\rho_{\alpha\beta} = \sum_{m,\,occ} \sum_{i,j} C_{i\alpha}^{*m} C_{j\beta}^m S_{i\alpha,j\beta} \tag{2}$$

In the above equation, $S_{i\alpha,j\beta}$ are the overlap integrals between the ith orbital in α^{th} atom in the jth orbital in β^{th} atom. $C_{j\beta}^m$ is the eigenvector coefficients of the mth occupied band. The BO from equation (2) defines the relative strength of the bond. Total bond order density (TBOD) is obtained by normalizing the total BO (TBO) with cell volume. TBOD is a single metric to access the internal cohesion in the crystal and can be decomposed partial components (PBOD) for any structural units or groups of bonded atomic pairs.

For the calculation of interband optical properties in the form of frequency-dependent complex dielectric function $(\hbar\omega) = \varepsilon_1(\hbar\omega) + i\varepsilon_2(\hbar\omega)$, the imaginary part $\varepsilon_2(\hbar\omega)$ is calculated first according to

$$\varepsilon_2(\hbar\omega) = \frac{e^2}{\pi n \omega^2} \int_{BZ} dk^3 \sum_{nl} |\langle \psi_n(k,r)| - i\hbar\nabla |\psi_l(k,r)\rangle|^2 f_l(k)[1 - f_n(k)]\delta[E_n(k) - E_l(k) - \hbar\omega] \quad (3)$$

where l and n are for the occupied and unoccupied states respectively. $\psi_n(k,r)$ are the ab initio Bloch functions from OLCAO calculation using a full basis (FB) set and a larger k-point sampling. $f_l(k)$ and $f_n(k)$ are the Fermi distribution functions. The real part $\varepsilon_1(\hbar\omega)$ is obtained from the imaginary part $\varepsilon_2(\hbar\omega)$ through Kramers-Kronig transformation [46].

The combination of using VASP and OLCAO packages with two different basis expansions is highly effective in revealing the subtle features in the material properties especially on the atomic scale details. The method is particularly useful for highly complex crystals or non-crystalline materials as demonstrated in some of recent publications [35–45].

4. Results and Discussion

4.1. Electronic Structure

The calculated band structures for the five pyrophosphate crystals are shown in Figure 2. They are all insulators with large direct band gaps at Γ ranging from 3.70 eV to 5.22 eV and listed in Table 3. However, on a closer inspection, we find that in C3 and C4 the states near the VB top are actually metallic bands with both occupied and unoccupied band below and above the Fermi level. This is illustrated in Figure 2c′,d′ next to Figure 2c,d.

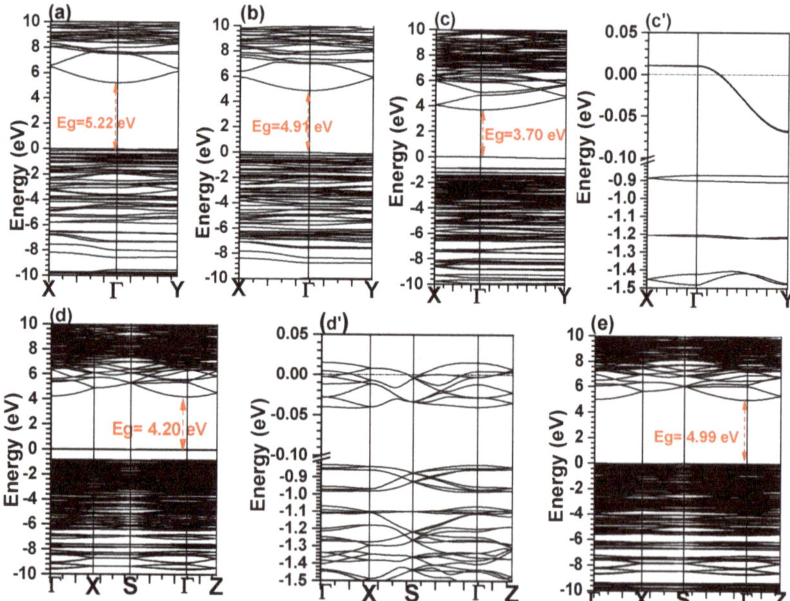

Figure 2. Calculated band structures of five pyrophosphate crystals. (**a**) C1, (**b**) C2, (**c**) C3, (**d**) C4, (**e**) C5, and (**c′**), (**d′**) magnified y-axis from −1.5 eV to 0.05 eV for C3 and C4, respectively, with dashed line showing Fermi level.

The top of valence band (VB) in all five crystals are flat and the bottom of conduction band (CB) have curvatures. The calculated CB effective masses listed in Table 1 are comparable with each other around 0.143 m_e to 0.149 m_e. They can be compared to other wide-gap semiconductors such as AIP (0.13 m_e) [47], GaN (0.19 m_e) [48], ZnSe (0.17 m_e) [49], ZnTe (0.16 m_e) [50], CdS (0.20 m_e) [51], and CdSe (0.13 m_e) [52]. The small effective mass implies these crystals could have large electron mobility in the CB. Therefore, pyrophosphate crystals with wide band gap and high electron mobility could be used in radiation detector [53].

The total density of states (TDOS) for all five crystals with energy range from −25 to 25 eV are shown in Figure 3. The TDOS are further resolved into partial density of states (PDOS) for the four structural units. These give more detailed information on the interaction between different groups and the difference among five crystals. For example, in crystals containing K, a sharp peak at around-10 eV is due to semi-core nature of K-3p orbital. We see a slight double peak for K due to two different types of K based on sites. $(NH_4)_2Zn(H_2P_2O_7)_2 \cdot 2H_2O$ is the only crystal containing NH_4 ligand instead of K. The sharp peak at around −16.38 eV is due to both N and H atoms, and a smaller peak at −6.49 eV is due to N-2p orbital. For C3 and C4, the gap state arises purely from Cu 3d electrons with possible interaction with water and the pyrophosphate group $(H_2P_2O_7)_2$. In Figure 3f, we specifically show the PDOS of Cu in C3 and C4 in the energy range from −8 eV to 8 eV with the zero of energy located at the Fermi level. It should be mentioned that the current calculation is not spin-polarized based on the assumption that the pyrophosphates studied in this paper are all paramagnetic. To get much deeper insight on Cu-containing pyrophosphates, spin-polarized calculation, or DFT +U approach should be attempted.

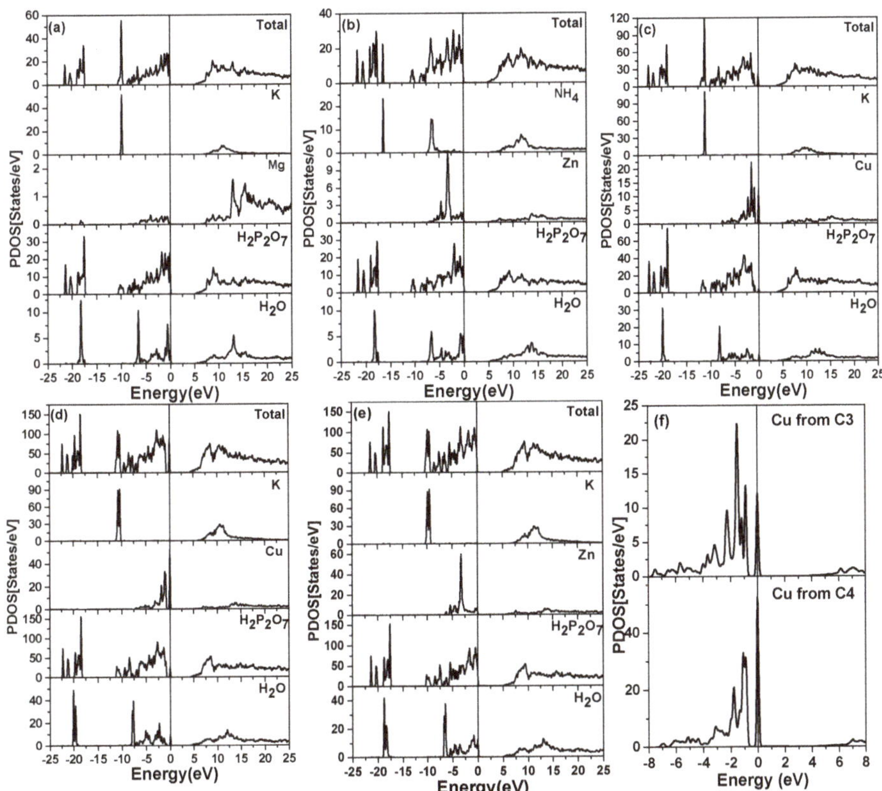

Figure 3. Calculated total density of states (DOS) and partial density of states (PDOS) from four structural unit. (**a**) C1, (**b**) C2, (**c**) C3, (**d**) C4, (**e**) C5, and (**f**) shows PDOS of Cu from C3 and C4.

4.2. Partial Charge Distribution

The partial charge (PC) ΔQ on each atom is defined as the deviation of the effective charge Q_α^* (Equation (1)) from the neutral charge Q^0 on the same atom, or $\Delta Q = Q^0 - Q_\alpha^*$. We have calculated the PC for every atom in all five crystals as shown in Figure 4. As expected P, Mg, K, Cu, and H have positive PC whereas O and N atoms have negative PC and their values are very similar. Minor variations of the PC for P, H, or O only reflect their locations in the structural units of $H_2P_2O_7$ or H_2O. The minor variations can also originate from the different crystal symmetry, triclinic vs orthorhombic. Thus, the PC for each atom in the crystals shown in Figure 4 are not very insightful. On the other hand, grouping the PC for atoms in each of the three structural unit will be quite revealing. They are listed in Table 2. It is noted that there are considerable differences in the PC of the three structural units between the five crystals including the two water molecules ($2 \cdot H_2O$). This indicates that there could be subtle difference in inter molecular interaction and corresponding reactivity.

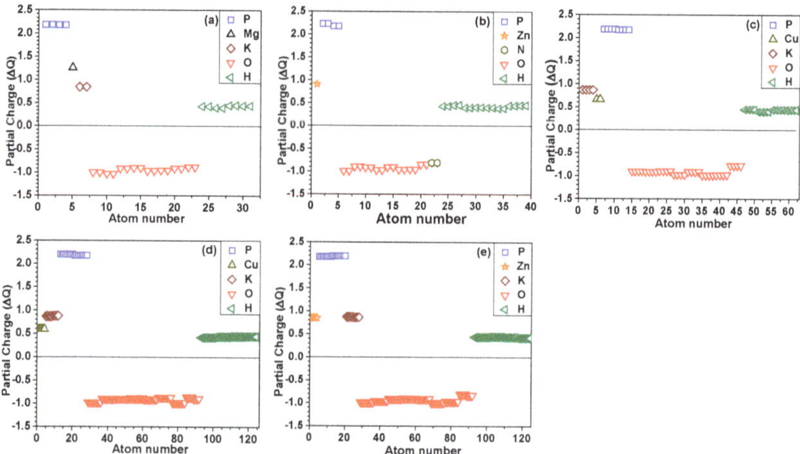

Figure 4. Calculated partial charge distribution in five pyrophosphate crystals. (**a**) C1, (**b**) C2, (**c**) C3, (**d**) C4, (**e**) C5.

Table 2. Partial charge values from three different structural units in five crystals. The first five columns are the elements in the first structural unit.

	K_2	$(NH_4)_2$	Mg	Zn	Cu	$(H_2P_2O_7)_2$	$2 \cdot H_2O$
C1	1.68	-	1.26	-	-	-2.97	0.03
C2	-	1.55	-	0.91	-	-2.57	0.11
C3	1.73	-	-	-	0.67	-2.62	0.22
C4	1.74	-	-	-	0.59	-2.36	0.03
C5	1.73	-	-	0.85	-	-2.71	0.13

4.3. Interatomic Bonding

The best way to describe the interatomic interaction is to show the BO vs BL distribution in each crystal and then use the data to obtain the TBOD and PBOD for each structural unit. The BO vs BL distribution for the five pyrophosphate crystals turns out to be quite similar, except for C3 and C4 containing Cu. We present the distribution for crystal C2 ($(NH_4)_2Zn(H_2P_2O_7)_2 \cdot 2H_2O$) and C4 ($4[K_2Cu(H_2P_2O_7)_2 \cdot 2H_2O$) as examples in Figure 5a,b, respectively. More detailed discussion for C1 has been described in [21]. Essentially, there are seven types of different interatomic bonds for each crystal. The five common bonds in each crystal are covalent bonds (P-O, O-H), hydrogen bond (HB)

(O$\cdots$H), bridging bonds (O-H-O), and negligibly weak (O-H) bonds when far apart. The covalent P-O bonds within the structural unit $H_2P_2O_7$ is the strongest bonds, and they are only slightly different in each crystal, which implies the strong tetrahedral unit PO_4 as is true in all phosphates. Mg-O and Zn-O bonds are stronger in comparison to K-O bonds, and they form octahedral units in the respective crystal. In C2, the unique covalent N-H bonds are very strong since they are part of the intramolecular bonds in NH_4. The BO for Cu-O in C3 and C4 with same chemical composition are quite different (Not shown). In C3, some Cu-O bonds have a larger BL of 2.50 Å with a lower BO of 0.054 e. This shows difference in bonding due to crystal structure. The sum of total bond order values in the crystal when normalized by volume gives the TBOD, a very useful parameter to identify internal cohesion in pyrophosphate crystals, and they are listed in Table 3. It turns out that C2 has the highest cohesion with TBOD of 0.02743 e$_-$/Å^3, much larger than the other four crystals with very similar TBOD. The PBOD from different bond types is shown in Figure 5c in the form of histogram. P-O bonds have highest contribution in all crystals and is responsible for its internal cohesion. Apart from the strong N-H bonds in C2, O-H is second most significant bonds in these crystals. However, all other bonds also play its part in crystal cohesion of pyrophosphates including the O$\cdots$H hydrogen bonds.

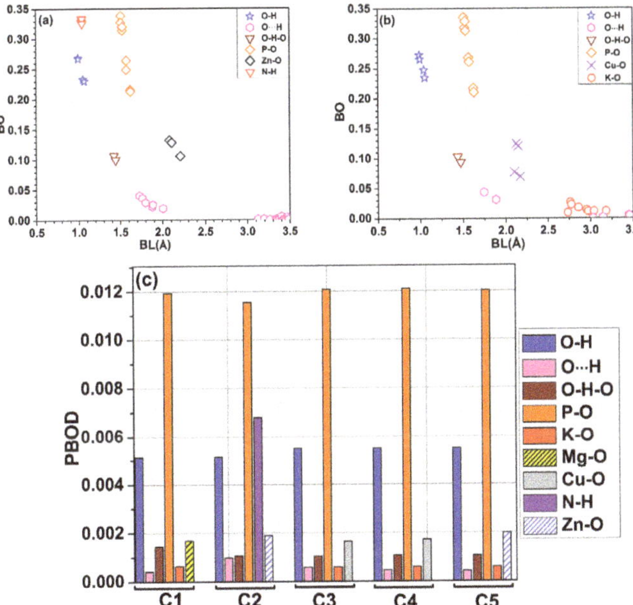

Figure 5. (**a**) Bond order (BO) vs. BL for all pairs of atoms in C2 $(NH_4)_2Zn(H_2P_2O_7)_2\cdot2H_2O$, (**b**) C4 $4[K_2Cu(H_2P_2O_7)_2\cdot2H_2O]$, (**c**) bar graph of contributions from different bonds types in the five pyrophosphate crystals.

Table 3. Calculated physical properties for the five crystals.

	Eg (eV)	m_e^* (m_e)	$\varepsilon1(0)$	n	ω_p (eV)	TBOD
C1	5.22	0.143	2.09	1.44	22.98	0.02125
C2	4.91	0.146	2.11	1.45	20.85	0.02743
C3	3.70	0.146	2.37	1.53	23.14	0.02141
C4	4.20	0.146	290.99	17.06	22.89	0.02139
C5	4.99	0.149	2.15	1.47	22.67	0.02155

4.4. Mechanical Properties

Mechanical properties are essential for any materials with technological applications. However, there is very little information on the mechanical properties for pyrophosphate crystals in contrast to their vibrational properties. To fill this gap, we have calculated the elastic coefficients from the VASP relaxed structure for the five crystals. From the elastic coefficients, the mechanical parameters for these crystals are obtained using the VRH approximation for poly-crystals [17,18]. They are bulk modulus (K), shear modulus (G), Young's modulus (E), and Poisson's ratio (η), and are listed in Table 4. Also included are the Pugh ratio G/K and the universal elastic anisotropy parameter A^U [54]. A^U equal to 0 implies perfect isotropy and is usually much less than 2 and seldom goes beyond 4 for most crystals with different crystalline symmetries [54]. There is a large variations of A^U among the five pyrophosphate crystals with a low value of 0.7182 for C4 and high value of 3.610 for C1, thus C1 is far more elastically anisotropic than C4. Generally speaking, low symmetry triclinic crystal should have higher A^U compared to orthorhombic crystals, however, it is still disconcerting that C5, which is also orthorhombic, has much larger A^U than C4. Each of these mechanical parameters are correlated and all are derived from the elastic coefficients. The higher the bulk modulus, the less compressible is the system. Shear modulus is related to the rigidity of the material and Young's modulus represent stiffness of the material. G/K is a parameter based on Pugh's criteria [55] to estimate brittleness or ductility in pure metals from comparative analysis. i.e., if G/K ratio is high, then it is brittle and ductile if the G/K ratio is low. However, Pugh's ratio may also work in other materials, as it also has relation with Poisson's ratio (η) i.e., higher G/K ratio lower is η and vice versa. All five crystals have low K and G values, with K ranges from 22.88 GPa to 30.28 GPa and G ranges from 13.90 GPa to 15.63 GPa respectively. As a result, G/K ratio ranges from a low value of 0.4591 for C4 to a much higher value of 0.6831 for C1. This is a variation of almost 50% in the Pugh ratio signal the brittle nature for C1 and being ductile for C4. Given the fact that they have almost identical TBOD (C1: 0.02125 and C4: 0.2139), there must be a credible reason for rationalization with the use of Pugh ratio.

Table 4. Calculated Cij and mechanical parameters for the five crystals.

	K(GPa)	G(GPa)	E(GPa)	η	G/K	A^U
C1	22.88	15.63	38.19	0.2218	0.6831	3.6100
C2	30.28	14.84	38.26	0.2894	0.4899	3.1220
C3	28.19	14.66	37.48	0.2784	0.5199	2.7776
C4	30.26	13.90	36.15	0.3009	0.4591	0.7182
C5	24.53	15.58	38.57	0.2379	0.6351	1.9064

In Figure 6, we plot the G vs K in the range for K from 0 to 40 GPa and G from 0 to 20 GPa for the five pyrophosphates C1-C5 together with a selected set of materials within the same range. The slopes of the dashed lines show the G/K ratio which are close to the inverse of the Poisson's ratio for the pyrophosphates. They can be divided into two groups with G/K = 0.611 (C1, C5) and 0.457 (C2, C3, C4). One is ductile and the other is brittle, according to Pugh ratio criterion of greater or larger than G/K around 0.5 [55]. We are not aware of experimental measurement for these pyrophosphate crystals, so the results presented can be considered as theoretical predictions.

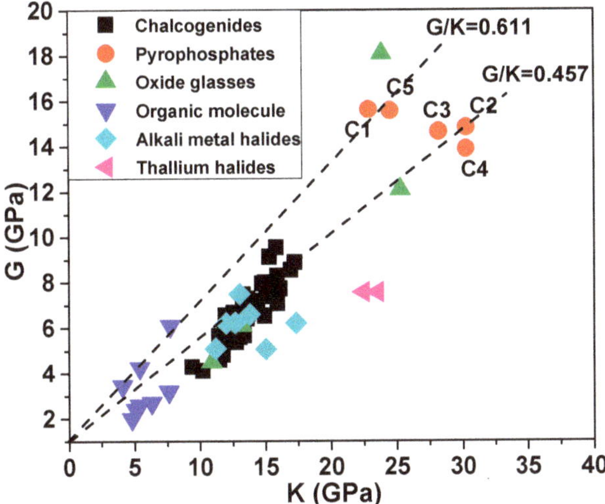

Figure 6. Shear modulus G vs bulk modulus K for the 5 pyrophosphates and other selected crystals such as chalcogenides [56], oxide glasses [57], organic molecules [58], alkali metal halides [59,60], thallium halides [60] within the same range of G and K. The slope of the dashed lines gives the G/K values 0.611 and 0.457.

4.5. Optical Properties

The optical properties for the five pyrophosphate crystals are calculated in the form of complex dielectric function based on the one-electron theory of interband optical transition. The calculated real (ε_1) and imaginary (ε_2) parts of frequency-dependent dielectric function are shown in black and red color respectively in Figure 7 (center panel (a) to (e)). Optical absorption spectra in all five crystals above the absorption threshold of 5.0 eV show five peaks—A, B, C, D, and E—roughly in the ultraviolet region. This feature is related to the similarity in the TDOS of Figure 3 dominated by the PDOS of the $H_2P_2O_7$ unit. However, for the absorption spectra for the Cu containing crystals C3 and C4 in Figure 7c,d, additional absorptions occur in the 0.0 eV to 5.0 eV region, which are in the infrared, visible, ultraviolet spectral region (see the far left column in Figure 7). This is due to the presence of unoccupied part of the metallic band discussed in Section 4.1. This made the optical properties in C4 resemble a metallic material with huge peaks at energy close to 0.0 eV. On the other hand, the optical properties for C3 are quite reasonable, despite the same chemical composition and formula with C4. The optical absorption (ε_2) in C3 occurs at a higher photon energy than in C4. After Kramers-Kronig transformation of ε_2 to ε_1, ε_1 at zero is only slightly larger than those in C1, C2, and C5 but are still reasonable. A plausible explanation on the difference between C3 and C4 is that they have different crystal symmetry (space group P-1(Ci) and $Pnma(D_{2h}^{16})$, respectively) despite these two Cu-containing pyrophosphates have the same chemical formula. This is one of the very rare examples that the crystal symmetry can play a critical role in the optical properties in the infrared frequency region.

The refractive index for the five pyrophosphate crystals can be obtained from the square root of zero-frequency limit of real part of the dielectric function ($\varepsilon_1(0)$). They are listed in Table 3 and are in the range of 1.44 to 1.53. The value of n = 17.06 for C4 is questionable due to large absorption in the infrared region. These refractive index values can be compared with 60% glucose solution in water (1.44) [61], BaF2 (1.47) [62], CaF2 (1.433) [62], SrF2 (1.44) [62], CsF (1.48) [62], RbBr (1.55) [62], and RbCl (1.49) [62] etc.

In the right panel of Figure 7a'–e', we display the calculated energy loss function (ELF) for the five crystals whose optical absorption spectra are shown in the center panel. The peak of ELF is known as plasma frequency (ω_p), ω_p is the frequency of collective excitation of electrons in solid at high energy and can be easily measured experimentally. The ω_p for all five crystals are listed in Table 3. They range from 20.85 eV in C2 to 23.14 eV in C3.

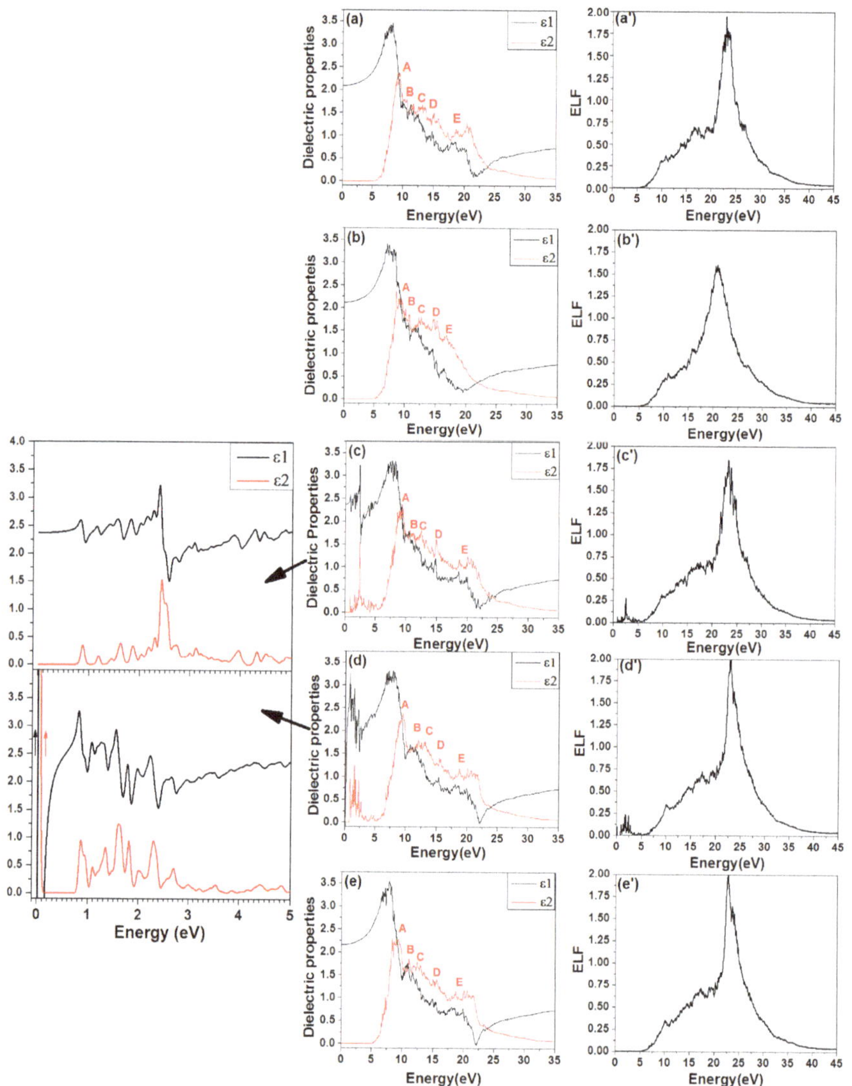

Figure 7. Optical properties for the five pyrophosphate crystals: Center panel: real (ε_1) and imaginary (ε_2) parts of the dielectric function. Far left panel: expanded illustration of ε_1 and ε_2 from 0.0 to 5 eV. Right panel, the electron energy loss function (ELF). (**a**,**a'**) for C1; (**b**,**b'**) for C2; (**c**,**c'**) for C3; (**d**,**d'**) for C4; and (**e**,**e'**) for C5.

5. Summary and Conclusions

We have presented detailed results on the quantum mechanical calculation on electronic structure, partial charge, interatomic bonding, mechanical and optical properties of five pyrophosphates with very complex structures. To our knowledge, these are the first time such calculations have been attempted. This enable us to make detailed comparisons on the structure and properties among them.

The fundamental understanding of pyrophosphate crystals can open the door for many applications. The general conclusions of this study can be succinctly summarized as follows:

1. The electronic structure and bonding in these crystals appear to be quite similar, but there are subtle differences in bonding in these crystals due to differences in their compositions in different metallic elements and crystal geometry.

2. The presence of very narrow metallic state of Cu in the crystals can induce empty states above the Fermi level, which make its optical properties more complicated and interesting.

3. NH_4 is a unique group of atoms to replace the metallic ion. It is the only organic group in the five crystals and has the largest TBOD.

4. Crystal symmetry plays an important role, triclinic vs orthorhombic. This implies short-range intermolecular action is more critical than longer-range crystal symmetry.

5. $H_2P_2O_7$ is the most important and dominating unit in pyrophosphates. The water molecule provides the unique $H\cdots O$ bonds, and metallic elements can influence the structure bonding and reactivity to a certain extent.

6. The mechanical properties of these five crystals vary the most, but detailed correlation to structure and electronic properties remains clear.

The work presented represent the first step in understanding the structure and properties of pyrophosphates. Much work remains to be done, especially in the direction of having different transition metal elements replacing Cu or Zn and invoke spin-polarized calculations for magnetic properties.

Author Contributions: W.-Y.C. and R.K. initiated the project. P.A. and W.-Y.C. did the calculations. R.K., M.H., and K.B. developed the synthesis method of phosphate materials including acidic pyrophosphates. All the authors participated in the discussion and interpretation of the results. W.-Y.C., P.A., and R.K. wrote the paper. All authors edited and proof-read the final manuscript.

Funding: This research received no external funding.

Acknowledgments: This research used the resources of the National Energy Research Scientific Computing Center supported by the DOE under Contract No. DE-AC03-76SF00098 and the Research Computing Support Services (RCSS) of the University of Missouri System. PA is supported by a research grant from the School of Graduate Studies at UMKC.

Conflicts of Interest: The author declares no conflict of interest.

References

1. Durif, A. *Crystal Chemistry of Condensed Phosphates*; Springer Science & Business Media: New York, NY, USA, 1995; Volume 408, ISBN 978-1-4757-9896-8.
2. Brow, R.K. The structure of simple phosphate glasses. *J. Non-Crystal. Solids* **2000**, *263*, 1–28. [CrossRef]
3. Martin, S.W. Ionic conduction in phosphate glasses. *J. Am. Ceram. Soc.* **1991**, *74*, 1767–1784. [CrossRef]
4. Ratnakaram, Y.; Babu, S.; Bharat, L.K.; Nayak, C. Fluorescence characteristics of Nd3+ doped multicomponent fluoro-phosphate glasses for potential solid-state laser applications. *J. Lumin.* **2016**, *175*, 57–66. [CrossRef]
5. Yang, F.; Liu, Y.; Tian, X.; Dong, G.; Yu, Q. Luminescence properties of phosphate phosphor Ba3Y (PO4) 3: Sm3+. *J. Solid State Chem.* **2015**, *225*, 19–23. [CrossRef]
6. Wang, P.; Lu, M.; Gao, F.; Guo, H.; Xu, Y.; Hou, C.; Zhou, Z.; Peng, B. Luminescence in the fluoride-containing phosphate-based glasses: A possible origin of their high resistance to nanosecond pulse laser-induced damage. *Sci. Rep.* **2015**, *5*, 8593. [CrossRef] [PubMed]
7. Krempl, P.; Schleinzer, G.; Wallno, W. Gallium phosphate, GaPO4: A new piezoelectric crystal material for high-temperature sensorics. *Sens. Actuators A Phys.* **1997**, *61*, 361–363. [CrossRef]
8. Westheimer, F.H. Why nature chose phosphates. *Science* **1987**, *235*, 1173–1178. [CrossRef] [PubMed]
9. Andreeva, O.; Efimtseva, E.; Padyukova, N.S.; Kochetkov, S.; Mikhailov, S.; Dixon, H.; Karpeisky, M.Y. Interaction of HIV-1 reverse transcriptase and T7 RNA polymerase with phosphonate analogs of NTP and inorganic pyrophosphate. *Mol. Biol.* **2001**, *35*, 717–729. [CrossRef]
10. Habraken, W.; Habibovic, P.; Epple, M.; Bohner, M. Calcium phosphates in biomedical applications: Materials for the future? *Mater. Today* **2016**, *19*, 69–87. [CrossRef]

11. Knowles, J.C. Phosphate based glasses for biomedical applications. *J. Mater. Chem.* **2003**, *13*, 2395–2401. [CrossRef]

12. Rulis, P.; Ouyang, L.; Ching, W. Electronic structure and bonding in calcium apatite crystals: Hydroxyapatite, fluorapatite, chlorapatite, and bromapatite. *Phys. Rev. B* **2004**, *70*, 155104. [CrossRef]

13. Aryal, S.; Matsunaga, K.; Ching, W.-Y. Ab initio simulation of elastic and mechanical properties of Zn-and Mg-doped hydroxyapatite (HAP). *J. Mech. Behav. Biomed. Mater.* **2015**, *47*, 135–146. [CrossRef] [PubMed]

14. Rulis, P.; Yao, H.; Ouyang, L.; Ching, W. Electronic structure, bonding, charge distribution, and X-ray absorption spectra of the (001) surfaces of fluorapatite and hydroxyapatite from first principles. *Phys. Rev. B* **2007**, *76*, 245410. [CrossRef]

15. Liang, L.; Rulis, P.; Ching, W. Mechanical properties, electronic structure and bonding of α-and β-tricalcium phosphates with surface characterization. *Acta Biomate.* **2010**, *6*, 3763–3771. [CrossRef] [PubMed]

16. Tang, E. *Computational Studies of Phosphate Clusters and Bioglasses*; UCL (University College London): London, UK, 2011.

17. Kim, H.; Park, J.; Park, I.; Jin, K.; Jerng, S.E.; Kim, S.H.; Nam, K.T.; Kang, K. Coordination tuning of cobalt phosphates towards efficient water oxidation catalyst. *Nat. Commun.* **2015**, *6*, 8253. [CrossRef] [PubMed]

18. Witko, M.; Tokarz, R.; Haber, J.; Hermann, K. Electronic structure of vanadyl pyrophosphate: Cluster model studies. *J. Mol. Catal. A Chem.* **2001**, *166*, 59–72. [CrossRef]

19. Zhang, Y.; Cheng, W.; Wu, D.; Zhang, H.; Chen, D.; Gong, Y.; Kan, Z. Crystal and band structures, bonding, and optical properties of solid compounds of alkaline indium (III) pyrophosphates MInP2O7 (M= Na, K, Rb, Cs). *Chem. Mater.* **2004**, *16*, 4150–4159. [CrossRef]

20. Xiang, H.; Feng, Z.; Zhou, Y. Ab initio computations of electronic, mechanical, lattice dynamical and thermal properties of ZrP2O7. *J. Eur. Ceram. Soc.* **2014**, *34*, 1809–1818. [CrossRef]

21. Adhikari, P.; Khaoulaf, R.; Ez-Zahraouy, H.; Ching, W.-Y. Complex interplay of interatomic bonding in a multi-component pyrophosphate crystal: $K_2Mg (H_2P_2O_7)_2 \cdot 2H_2O$. *R. Soc. Open Sci.* **2017**, *4*, 170982. [CrossRef] [PubMed]

22. Harcharras, M.; Capitelli, F.; Ennaciri, A.; Brouzi, K.; Moliterni, A.; Mattei, G.; Bertolasi, V. Synthesis, X-ray crystal structure and vibrational spectroscopy of the acidic pyrophosphate $KMg_{0.5}H_2P_2O_7 \cdot H_2O$. *J. Solid State Chem.* **2003**, *176*, 27–32. [CrossRef]

23. Capitelli, F.; Khaoulaf, R.; Harcharras, M.; Ennaciri, A.; Habyby, S.H.; Valentini, V.; Mattei, G.; Bertolasi, V. Crystal structure and vibrational spectroscopy of the new acidic diphosphate $(NH_4)_2Zn (H_2P_2O_7)_2 \cdot 2H_2O$. *Z. Kristallogr.-Crystal. Mater.* **2005**, *220*, 25–30. [CrossRef]

24. Khaoulaf, R.; Ezzaafrani, M.; Ennaciri, A.; Harcharras, M.; Capitelli, F. Vibrational Study of Dipotassium Zinc Bis (Dihydrogendiphosphate) Dihydrate, $K_2ZN(H_2P_2O_7)_2 \cdot 2H_2O$. *Phosphorus Sulfur Silicon Relat. Elem.* **2012**, *187*, 1367–1376. [CrossRef]

25. Khaoulaf, R.; Ennaciri, A.; Ezzaafrani, M.; Capitelli, F. Structure and Vibrational Spectra of a New Acidic Diphosphate $K_2Cu(H_2P_2O_7)_2 \cdot 2H_2O$. *Phosphorus Sulfur Silicon Relat. Elem.* **2013**, *188*, 1038–1052. [CrossRef]

26. Kresse, G.; Furthmüller, J. Efficiency of ab-initio total energy calculations for metals and semiconductors using a plane-wave basis set. *Comput. Mater. Sci.* **1996**, *6*, 15–50. [CrossRef]

27. Ching, W.-Y.; Rulis, P. *Electronic Structure Methods for Complex Materials: The Orthogonalized Linear Combination of Atomic Orbitals*; Oxford University Press: Oxford, UK, 2012; ISBN 9780199575800.

28. Kresse, G.; Joubert, D. From ultrasoft pseudopotentials to the projector augmented-wave method. *Phys. Rev. B* **1999**, *59*, 1758. [CrossRef]

29. Nielsen, O.; Martin, R.M. First-principles calculation of stress. *Phys. Rev. Lett.* **1983**, *50*, 697. [CrossRef]

30. Yao, H.; Ouyang, L.; Ching, W.Y. Ab initio calculation of elastic constants of ceramic crystals. *J. Am. Ceram. Soc.* **2007**, *90*, 3194–3204. [CrossRef]

31. Reuss, A. Berechnung der fließgrenze von mischkristallen auf grund der plastizitätsbedingung für einkristalle. *ZAMM-J. Appl. Math. Mech./Z. Angew. Math. Mech.* **1929**, *9*, 49–58. [CrossRef]

32. Hill, R. The elastic behaviour of a crystalline aggregate. *Proc. Phys. Soc. Sect. A* **1952**, *65*, 349. [CrossRef]

33. Mulliken, R.S. Electronic population analysis on LCAO–MO molecular wave functions. I. *J. Chem. Phys.* **1955**, *23*, 1833–1840. [CrossRef]

34. Mulliken, R. Electronic population analysis on LCAO–MO molecular wave functions. II. Overlap populations, bond orders, and covalent bond energies. *J. Chem. Phys.* **1955**, *23*, 1841–1846. [CrossRef]

35. Poudel, L.; Steinmetz, N.F.; French, R.H.; Parsegian, V.A.; Podgornik, R.; Ching, W.-Y. Implication of the solvent effect, metal ions and topology in the electronic structure and hydrogen bonding of human telomeric G-quadruplex DNA. *Phys. Chem. Chem. Phys.* **2016**, *18*, 21573–21585. [CrossRef] [PubMed]

36. Adhikari, P.; Xiong, M.; Li, N.; Zhao, X.; Rulis, P.; Ching, W.-Y. Structure and electronic properties of a continuous random network model of an amorphous Zeolitic Imidazolate Framework (a-ZIF). *J. Phys. Chem. C* **2016**, *120*, 15362–15368. [CrossRef]

37. Adhikari, P.; Dharmawardhana, C.C.; Ching, W.Y. Structure and properties of hydrogrossular mineral series. *J. Am. Ceram. Soc.* **2017**, *100*, 4317–4330. [CrossRef]

38. Poudel, L.; Twarock, R.; Steinmetz, N.F.; Podgornik, R.; Ching, W.-Y. Impact of Hydrogen Bonding in the Binding Site between Capsid Protein and MS2 Bacteriophage ssRNA. *J. Phys. Chem. B* **2017**, *121*, 6321–6330. [CrossRef] [PubMed]

39. Baral, K.; Li, A.; Ching, W.-Y. Ab Initio Modeling of Structure and Properties of Single and Mixed Alkali Silicate Glasses. *J. Phys. Chem. A* **2017**, *121*, 7697–7708. [CrossRef] [PubMed]

40. Poudel, L.; Tamerler, C.; Misra, A.; Ching, W.-Y. Atomic-Scale Quantification of Interfacial Binding between Peptides and Inorganic Crystals: The Case of Calcium Carbonate Binding Peptide on Aragonite. *J. Phys. Chem. C* **2017**, *121*, 28354–28363. [CrossRef]

41. Ching, W.Y.; Yoshiya, M.; Adhikari, P.; Rulis, P.; Ikuhara, Y.; Tanaka, I. First-principles study in an inter-granular glassy film model of silicon nitride. *J. Am. Ceram. Soc.* **2018**, *101*, 2673–2688. [CrossRef]

42. San, S.; Li, N.; Tao, Y.; Zhang, W.; Ching, W.Y. Understanding the atomic and electronic origin of mechanical property in thaumasite and ettringite mineral crystals. *J. Am. Ceram. Soc.* **2018**, *101*, 5177–5187. [CrossRef]

43. Dharmawardhana, C.; Misra, A.; Ching, W.-Y. Theoretical investigation of C-(A)-SH (I) cement hydrates. *Constr. Build. Mater.* **2018**, *184*, 536–548. [CrossRef]

44. Baral, K.; Li, A.; Ching, W.Y. Understanding the atomistic origin of hydration effects in single and mixed bulk alkali-silicate glasses. *J. Am. Ceram. Soc.* **2018**, *102*, 207–221. [CrossRef]

45. Adhikari, P.; Li, N.; Rulis, P.; Ching, W.-Y. Deformation behavior of amorphous zeolitic imidazolate framework-from supersoft material to complex organometallic alloy. *Phys. Chem. Chem. Phys.* **2018**. [CrossRef] [PubMed]

46. Martin, P.C. Sum rules, Kramers-Kronig relations, and transport coefficients in charged systems. *Phys. Rev.* **1967**, *161*, 143. [CrossRef]

47. Braunstein, R.; Kane, E. The valence band structure of the III–V compounds. *J. Phys. Chem. Solids* **1962**, *23*, 1423–1431. [CrossRef]

48. Kosicki, B.; Powell, R.; Burgiel, J. Optical Absorption and Vacuum-Ultraviolet Reflectance of GaN Thin Films. *Phys. Rev. Lett.* **1970**, *24*, 1421. [CrossRef]

49. Marple, D. Electron effective mass in ZnSe. *J. Appl. Phys.* **1964**, *35*, 1879–1882. [CrossRef]

50. Riccius, H.; Turner, R. Electroabsorption of zinc telluride films. *J. Phys. Chem. Solids* **1968**, *29*, 15–18. [CrossRef]

51. Hopfield, J.; Thomas, D. Fine structure and magneto-optic effects in the exciton spectrum of cadmium sulfide. *Phys. Rev.* **1961**, *122*, 35. [CrossRef]

52. Wheeler, R.; Dimmock, J. Exciton structure and Zeeman effects in cadmium selenide. *Phys. Rev.* **1962**, *125*, 1805. [CrossRef]

53. Knoll, G.F. *Radiation Detection and Measurement*; John Wiley & Sons: Hoboken, NJ, USA, 2010; ISBN 978-0470131489.

54. Ranganathan, S.I.; Ostoja-Starzewski, M. Universal elastic anisotropy index. *Phys. Rev. Lett.* **2008**, *101*, 055504. [CrossRef] [PubMed]

55. Pugh, S. XCII. Relations between the elastic moduli and the plastic properties of polycrystalline pure metals. *Lond. Edinb. Dublin Philos. Mag. J. Sci.* **1954**, *45*, 823–843. [CrossRef]

56. Sreeram, A.; Varshneya, A.; Swiler, D. Molar volume and elastic properties of multicomponent chalcogenide glasses. *J. Non-Crystal. Solids* **1991**, *128*, 294–309. [CrossRef]

57. Bridge, B.; Patel, N.; Waters, D. On the elastic constants and structure of the pure inorganic oxide glasses. *Phys. Status Solidi (a)* **1983**, *77*, 655–668. [CrossRef]

58. Roberts, R.; Rowe, R.; York, P. The relationship between Young's modulus of elasticity of organic solids and their molecular structure. *Powder Technol.* **1991**, *65*, 139–146. [CrossRef]

59. Anderson, O.L.; Nafe, J.E. The bulk modulus-volume relationship for oxide compounds and related geophysical problems. *J. Geophys. Res.* **1965**, *70*, 3951–3963. [CrossRef]
60. *The Crystran Handbook of Infra-Red and Ultra-Violet Optical Materials*; Crystan LTD Broom Road Business Park: Poole, UK, 2014; p. 112.
61. *CRC Handbook of Physics and Chemistry*, 8th ed.; The Chemical Rubber Company: Cleveland, OH, USA, 2001; ISBN 978-0-8493-0482-8.
62. Weber, M.J. *Handbook of Optical Materials*; CRC Press: Boca Raton, FL, USA, 2002.

Review

Structure, Properties, and Reactivity of Porphyrins on Surfaces and Nanostructures with Periodic DFT Calculations

Bhaskar Chilukuri *, Ursula Mazur and K. W. Hipps

Department of Chemistry, Washington State University, Pullman, WA 99163-4630, USA;
umazur@wsu.edu (U.M.); hipps@wsu.edu (K.W.H.)
* Correspondence: bhaskar.chilukuri@wsu.edu

Received: 30 December 2019; Accepted: 17 January 2020; Published: 21 January 2020

Featured Application: **Choosing appropriate DFT methodology for periodic modeling of porphyrin containing systems.**

Abstract: Porphyrins are fascinating molecules with applications spanning various scientific fields. In this review we present the use of periodic density functional theory (PDFT) calculations to study the structure, electronic properties, and reactivity of porphyrins on ordered two dimensional surfaces and in the formation of nanostructures. The focus of the review is to describe the application of PDFT calculations for bridging the gaps in experimental studies on porphyrin nanostructures and self-assembly on 2D surfaces. A survey of different DFT functionals used to study the porphyrin-based system as well as their advantages and disadvantages in studying these systems is presented.

Keywords: porphyrins, density functional theory; DFT; surfaces; self-assembly; scanning tunneling microscopy; dispersion; nanostructures; solid state; condensed phase

---

1. Introduction

Porphyrins are tetrapyrrolic macrocycles with π-conjugated electronic system that are ubiquitous in nature and have numerous biological and technological applications. For example, porphyrins are found in our body as prosthetic groups in hemeproteins. In plants, porphyrins are important components of chlorophyll which is a pigment playing an essential role in photosynthesis. Porphyrins also have numerous biomedical applications including photoimmunotherapy, photo diagnosis [1], biosensors [2], cancer therapy, etc. Porphyrins also play an important role in organic synthesis of dendrimers [3], metal-organic frameworks (MOFs) [4], biomimetic reactions [5], and as photo-catalysts [6] in numerous oxidation/reduction reactions. Finally, porphyrins act as important components in various technological applications like solar cells [7], chemical sensors [8], optoelectronics [9], spintronics [10], field effect transistors (FETs) [11–14], and in nanotechnology like single molecule junctions [15], nanowires, nanomotors [10], etc.

All the applications listed above are possible only due to the characteristic, yet tunable chemical structure and properties of porphyrins [16]. The backbone of each porphyrin molecule is the porphine group which constitutes four pyrrole groups linked with methine (-CH-) bridges, Figure 1a. Each porphine group has 22 π electrons forming a conjugated system. Due to their large pi-conjugation, porphyrins have strong absorption in the UV and visible regions forming colored compounds. In addition, the large π-electron system is responsible for many properties of porphyrins including optical [17,18], electronic [19], mechanical [20], and chemical [21,22] properties. Additionally, porphyrins from many coordinate covalent complexes with transition metals (metalloporphyrins) and some non-metals at the center of the porphyrin core. In addition, the peripheral substituents at α, β, and meso positions (Figure 1b) can be modified to yield tunable molecular and crystal properties such as solubility, reactivity, conductivity, and photophysics. Metalloporphyrins tend to react with ligands to form numerous (porphyrin)metal-ligand complexes [23] that are also useful for a variety of applications.

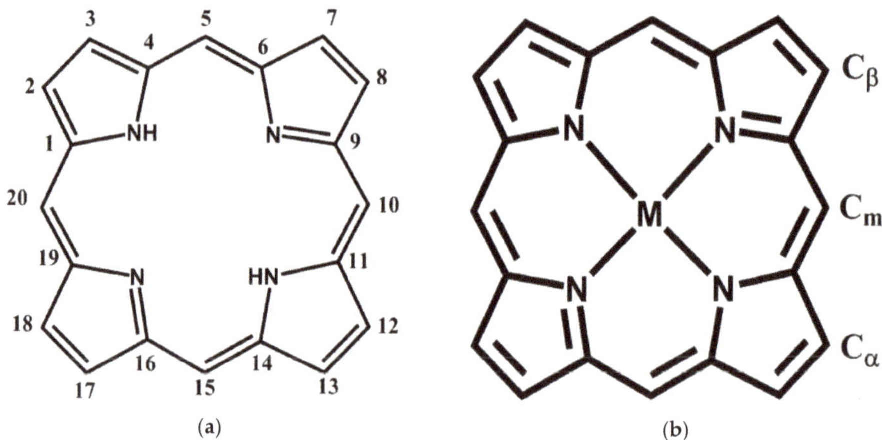

Figure 1. Chemical structures of (**a**) freebase porphine/porphyrin, (**b**) metalloporphyrin. Three unique peripheral substituent positions, alpha (C_α) at 2, 7, 12, 18; beta (C_β) at 3, 8, 13, 18; and meso (C_m) at 5, 10, 15, 20 positions are shown the figure.

A broad scan of the literature of theoretical modeling of porphyrins showed that a variety of computational methods were used to study porphyrins and their derivatives. A review by Shubina [24] listed these computational methods, which ranged from linear combination of atomic orbitals (LCAO), molecular mechanics, semi-empirical methods, through self-consistent field method (SCF) in the earlier literature, to the modern day methods which include molecular dynamics (MD), density functional theory (DFT) [25], Moller-Plesset perturbation theory (MPn), configuration interaction (CI), coupled cluster (CC), and CASSCF/CASPT2 [26] methods. Among the many computational methods used, recent literature is flooded with DFT calculations of molecular porphyrin and its derivatives primarily to understand their frontier orbital configuration, electron occupancy [27], charge transfer, and excited state properties [25]. The DFT functionals used to study the porphyrin complexes include all the rungs of the "Jacob's Ladder" [28] with variable approximations which include local density approximation (LDA), generalized gradient approximations (GGA), meta-GGAs, hybrid, and hybrid-meta GGA functionals [29,30].

While molecular DFT calculations of porphyrin complexes are prevalent, periodic DFT calculations of porphyrins are relatively limited. Periodic DFT (PDFT) calculations [31] refers to the use of density functional theory to describe the electronic structure of periodic systems. PDFT simulations are performed on lattice structures, surfaces, interfaces, and molecules with a defined unit cell in real and reciprocal spaces (k-space). The reciprocal space is obtained from the Brillouin zone of the unit cell. A grid of k-points is used to sample the Brillouin zone by using Bloch's theorem applied to the Kohn-Sham wavefunctions. Optimization and single point calculations with PDFT are performed using various self-consistent field (SCF) iteration schemes. While Gaussian basis sets can be used to perform PDFT calculations, plane wave pseudopotential basis sets are computationally less intensive and are used in many PDFT codes [32].

In this review we present a collection of DFT simulations of periodic systems containing porphyrins. While we tried to include many PDFT studies of porphyrins in this review, these are not necessarily a complete collection in the literature. Omission of a particular citation is not a reflection on the quality of that work.

Periodic DFT calculations of porphyrins can be broadly classified into two categories. First, simulation of porphyrin crystals, supramolecular compounds, and nanostructures like nanorods, nanowires, and nanosheets. Typical interests for these simulations involve understanding the intermolecular or packing interactions of porphyrins and determining their electronic properties, specifically band structure and density of states. The second category involves porphyrin interactions with solid supports like metals, oxides, carbon, and silicon surfaces. PDFT simulations on surfaces typically involve a molecule or monolayer of porphyrins or metalloporphyrins adsorbed on solid supports to study the adsorption configuration, binding energies, molecule-surface interactions and reactions, electronic structure using density of states (DOS) and band structure, chemical reactivity, interfacial charge distribution, and magnetic properties. Some of the ab initio DFT codes used for porphyrin PDFT simulations include VASP [33–35], Quantum Espresso [36], CP2K [37], CASTEP [38], Dmol [39], SIESTA [40], CPMD [41], etc. Omission of any code is not a reflection of the quality of the software for PDFT simulations. However, we note that codes involving plane-wave basis sets are more popular than software that are based on Gaussian type orbital (GTO) and natural atomic orbital (NAO) basis sets.

2. PDFT Simulations of Porphyrins in Nanostructures

Due to their rich chemistry, porphyrins can be synthesized though chemical bonding or through self-assembly into multiple structural forms like crystals, needles, wires, rods, sheets, plates, supramolecular frameworks, tubes, spheres, etc. PDFT simulations of these porphyrin structures primarily involves studying the stacking and intermolecular interactions inside a given geometry and their relation to its material properties. PDFT simulations of porphyrin nanostructures in the bulk were performed either from an experimental crystal structure or a built model based on experimental data and molecular structure. Most of these simulations aim to determine the electronic band structure and density of states. In the following sections we have classified the PDFT simulations on the porphyrin nanostructures based on their shape in the periodic structure.

2.1. Porphyrin Nanostructures Using Crystal Geometries

There are many known single crystal structures of porphyrins and metalloporphyrins in the literature but PDFT simulations on lattice structures of porphyrins are seldom found. Single crystal structures from x-ray crystallography of porphyrin nanostructures are even rare. Adinehnia et al. [42] determined the first single crystal structure of ionic porphyrin nanorods involving meso-tetra(N-methyl-4-pyridyl)porphyrin (TMPyP) and meso-tetra(4-sulfonatophenyl)porphyrin (TSPP). Using various spectroscopic, diffraction, and imaging techniques, they reported the structure-function relationship of TMPyP:TSPP nanorods. The crystal structure was correlated to the morphology and photoconductive behavior of the nanorods, and PDFT calculations on the crystal structure showed that the π-π stacking of TMPyP:TSPP is responsible for their conductivity.

Figure 2 shows the crystal structure and the corresponding band structure of TMPyP:TSPP nanorods. The band structure was determined using optB88-van der Waals (vdW) GGA [43] functional with projector augmented wave (PAW) [35,44] Perdew-Burke-Ernzerhof (PBE) [45] pseudopotentials. The authors note that the band structure obtained with PDFT underestimates the band gap (0.90 eV) and extended Huckel tight binding (EHTB) was used to determine the band gap (1.3 eV) which matches closer to experiment. Hybrid DFT functionals such as Heyd-Scuseria-Ernzerhof (HSE) [46,47] have shown improved band gap prediction over GGA and LDA functionals. Although HSE can improve band gap prediction, geometric optimizations showed little change in contrast to GGA and LDA functionals. On the other hand, vdW-DF [43,48] functionals were more reliable for geometric optimizations in systems with considerable dispersion interactions.

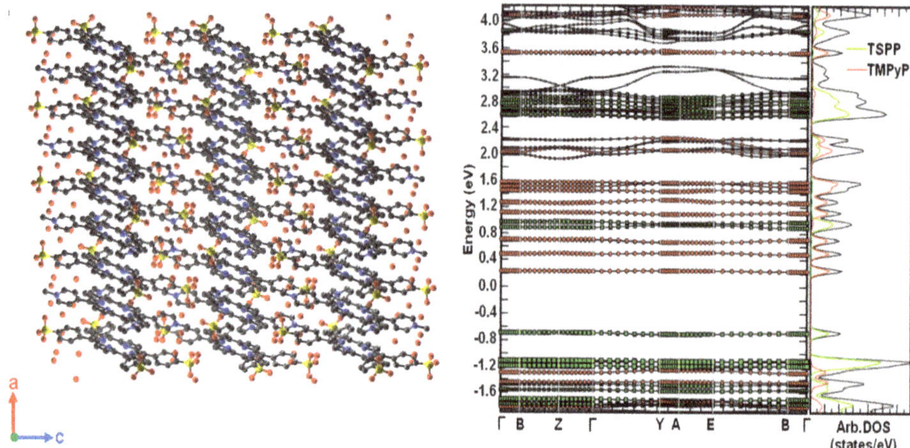

Figure 2. On the left, crystal structure of meso-tetra(N-methyl-4-pyridyl)porphyrin (TMPyP):meso-tetra(4-sulfonatophenyl)porphyrin (TSPP) nanorods in direction normal to the crystallographic b axis, showing the alternating cationic and anionic porphyrin tectons within the columns. Color codes: blue, N; gray, C; yellow, S; red, O. On the right is the projected density of states and band structure for the TMPyP:TSPP crystal computed from periodic density functional theory (PDFT). The Fermi level (E_f) is set as zero. The high symmetry points the Brillouin zone are as follows, G = (0,0,0), Z = (0,0,0.5), Y = (0,0.5,0), X = (0.5,0,0), A = (−0.5,0,0.5), E = (−0.5,0.5,0.5), B = (0,0,0.5). Reproduced from reference [42] published by The Royal Society of Chemistry.

In the case of TMPyP:TSPP, DFT calculations have been useful to predict the appropriate partial density of states (pDOS) that showed that the top of the valence band is populated by the contributions from TSPP and the bottom of the conduction band is populated by the TMPyP with no orbital hybridization in the vicinity of the bandgap. This prediction was consistent with the experimental data from UV-visible, diffuse reflectance and photoconductivity action spectra. This shows that PDFT calculations played a critical role in elucidating the photoconductive mechanism in TMPyP:TSPP nanorods. In the same study the authors reported that decreasing the porphyrin stacking distance would not necessarily change the band gap but would increase the dispersion in the band structure which would improve charge mobility in the nanorods.

The work on TMPyP:TSPP nanorods has been expanded by Borders et al. [49], by selective metalation of TMPyP and TSPP porphyrin cores with Ni and Cu transition metals. A single crystal structure of H_2TMPyP:NiTSPP nanorods was determined and it was shown that metalation of ionic porphyrins led to exhibition of dark conductivity at moderately high temperatures and that conductivity increases upon photoexcitation. Additionally, the photoresponse of the H_2TMPyP:CuTSPP substituted crystals is significantly higher than that of the CuTMPyP:H_2TSPP and the Ni-substituted crystals. To understand the reasons behind this discrepancy in the conductive behavior with different metalation, PDFT calculations were performed on H_2TMPyP:(M)TSPP and (M)TMPyP:H_2TSPP systems, where M = Ni, Cu. The crystal structure of H_2TMPyP:NiTSPP was used to create the lattice structure of other metalated binary ionic porphyrins by changing the core substitution of the H_2TMPyP:NiTSPP crystals and then optimizing the structure. The band structure of each H_2TMPyP:(M)TSPP and (M)TMPyP:H_2TSPP systems are shown in Figure 3. It was reported that adding a metal to the freebase porphyrins reduces the band gap of the corresponding nanorods. Additionally, it was shown that Ni and Cu metalation causes distinct changes in the frontier bands of porphyrin nanostructures.

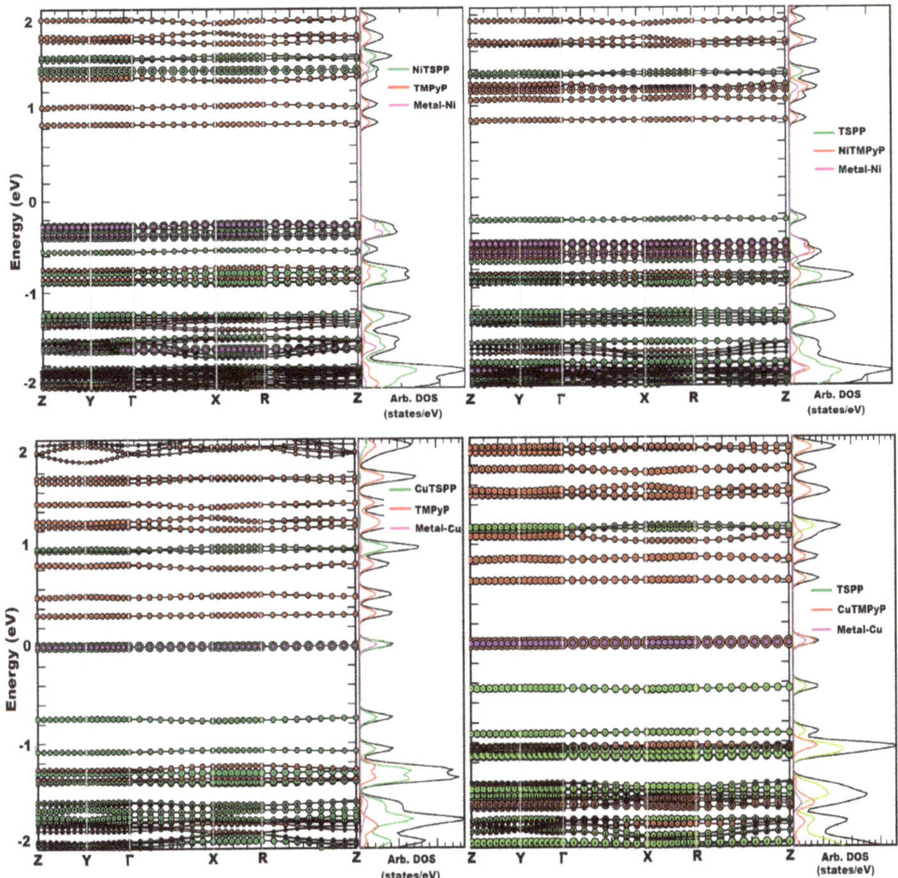

Figure 3. Projected density of states and the band structure for the H_2TMPyP:NiTSPP, NiTMPyP:H_2TSPP, H_2TMPyP:CuTSPP, and CuTMPyP:H_2TSPP crystals computed from DFT. The Fermi level (Ef) is set at zero. The high symmetry points of the Brillouin zone are as follows, G = (0,0,0), Z = (0,0,0.5), Y = (0,0.5,0), X = (0.5,0,0), and R = (0.5,0.5,0.5). Reproduced from reference [49] published by The Royal Society of Chemistry.

As mentioned earlier, PDFT studies of x-ray crystal structures of porphyrins are rare. Some of the other studies include determination of surface free energy of different crystal faces of FeTPPCl (TPP = tetra-phenyl porphyrin) and FeTPPOH·H_2O nanocrystals by Tian et al. [50]. They used PDFT calculated surface energies of {001}, {100}, {110}, {011} crystal faces to understand and tune the growth and shape of the FeTPP based nanocrystals. The calculations were performed using GGA-PBE [45] functional on predetermined crystal structures of FeTPPCl and FeTPPOH·H2O from the Cambridge Crystallographic Data Centre (CCDC) database. Krasnov et al. [51] performed a PDFT study of porphyrin:fullerene supramolecular compounds using models constructed from x-ray crystal structures determined by Boyd et al. [52]. The PBE functional [45] with Grimme DFT-D2 [53] dispersion interaction correction was used to optimize the models of porphyrin:fullerene compounds. The band structure and absorbance spectra of various optimized structures were determined using HSE [47] functional and DFPT [54] method, respectively. The HSE functional is a hybrid functional used for improved bang gap prediction.

2.2. Porphyrins in Organic Frameworks

Porphyrins are used to develop many supramolecular frameworks like metal organic (MOF), covalent organic (COF), surface metal organic (SURMOF) frameworks with applications for gas storage/separation, catalysis, drug delivery, photovoltaics, etc. Hence, PDFT calculations of porphyrin organic frameworks have gained importance for understanding their electronic structure to tune their applications. Hamad et al. [55], studied the electronic structure of porphyrin-based MOFs (PMOF) with porphyrins connected through phenyl-carboxyl ligands and AlOH species to assess their suitability for the photocatalysis of fuel production reactions using sunlight. They used the rhombohedral primitive cell obtained from the orthorhombic crystal structure of Al-PMOF [56] and replaced the porphyrin core with either hydrogens or various 3d transition metals and performed PDFT calculations of each model. The calculations were performed with GGA-PBE [45] functional starting with optimization of each lattice structure, followed by single point calculations for determination of DOS and band structure with HSE06 functional [46,47]. A typical crystal structure of the PMOF unit cell is presented in Figure 4a. From PDFT calculations, they reported that the bandgaps for PMOFs are in a favorable range (2.0–2.6 eV) for efficient adsorption of solar light. Furthermore, it was shown that the MOFs' band edges align with the redox potentials (Figure 4b) for water splitting and carbon dioxide reduction with reactions that can occur at neutral pH. This study was followed up with another PDFT study by Aziz et al. [4] by modifying the octahedral Aluminum metal center of the lattice structure with Fe^{+3} metal to determine the changes in the band structure and electronic properties of PMOFs. It was reported that adding Fe at the porphyrin core slightly raises the valence band edge, while Fe at the octahedral node significantly lowers the conduction band edge. So, iron can be used as a good dopant for band structure alignment in porphyrin-based MOFs.

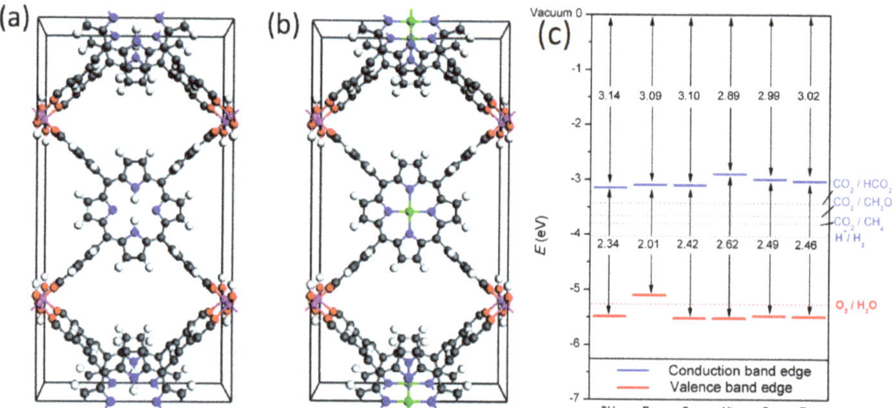

Figure 4. Perspective view of the porphyrin-based metal organic framework (MOF) investigated in this study in (**a**) the protonated case, and (**b**) the metal-substituted case. Color code: gray = carbon, white = hydrogen, red = oxygen, blue = nitrogen, magenta = aluminum, green = transition metal. (c) Bandgaps and band edge positions of MOFs with respect to the vacuum level, as calculated with the HSE06 functional. Energy levels corresponding to redox potentials of water splitting and carbon dioxide reduction reactions producing methane, methanol, and formic acid at pH = 7 are also shown with dotted lines. Reproduced from reference [55] published by The Royal Society of Chemistry.

PDFT calculations of porphyrin based SURMOFs has been reported by Liu et al. [57,58], who studied the photophysical properties of Zn(II)porphyrin-based SURMOFs. Using PDFT band structure with PBE functional [45] its was shown that a small dispersion of occupied and unoccupied bands in the Γ-Z direction [57], which is the porphyrin stacking direction, leads to the formation of a small indirect band gap. In a follow up study [58], the effect of introducing an electron-donating diphenylamine (DPA) into Zn SURMOFs was studied and PDFT simulations showed that DPA causes a shift in the charge localization pattern in the valence band minimum. This charge shift was attributed to the DPA groups which causes a shift of the optical absorption spectrum and the improved photocurrent generation in Zn SURMOFs. PDFT simulations are also used to study the stability of the porphyrin-COFs upon gas adsorption. Ghosh et al. used GGA-PW91 [59] and LDA-PZ [60] functionals to study hydrogen storage in H_2P-COF. The structural stability of COF upon introducing pyridine molecules to bridge the interlayer gaps in porphyrin COFs is studied.

2.3. Porphyrins as Nano Wires, Sheets, Tubes, and Ladders

As mentioned earlier, porphyrins can form many structural shapes due to their mechanical flexibility and rich chemistry. Hence, PDFT simulations were used to study the unique electronic structures and their applications that are possible because of the multidimensionality of porphyrins. Figure 5 shows typical shapes—various porphyrin nanowires, nanotubes, and nanosheets. Posligua et al. [61] studied the band structures of porphyrin nano sheets and tubes formed through covalent linkers. They used screened hybrid density functional theory simulations and Wannier function interpolation to obtain accurate band structures. The structural optimizations were performed with PBE-D2 [45,53] functional and single point calculations for band structure were obtained using the HSE06 [46,47] functional. It was reported that the electronic properties exhibit strong variations with the number of linking carbon atoms (C0 = no carbon atoms, C2 = two carbon atoms, C4 = four carbon atoms). For example, all C0 nanostructures exhibit gapless or metallic band structures, whereas band gaps open for the C2 or C4 structures. PDFT simulations showed that it is possible to design porphyrin nanostructures with tailored electronic properties such as specific band gap values and band structures by varying the type of the linkage used between each porphyrin units and the type of self-assembled formations (linear chains, nanosheets, nanotubes, and nanorings). A previous study on porphyrin nanotubes formed with acetyl linkers was done by Allec et al. [62] who reported large oscillations in bandgaps of porphyrin nanotubes with increase in their size. The simulations were performed with a variety of periodic DFT functionals which show similar oscillation trend in the band gaps irrespective of the functional. Additionally, the authors report that the bandgap is a direct-bandgap which can be observed with photoelectron spectroscopic experiments.

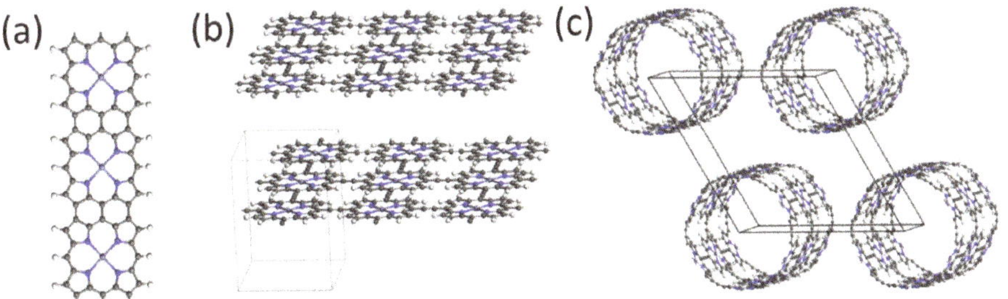

Figure 5. Typical structures of porphyrin (**a**) nanowire, (**b**) nanosheets, and (**c**) nanotubes. Reprinted with permission from reference [61] Copyright 2018 American Chemical Society.

Various porphyrin nanosheet structures with variable porphyrin core metals and substituents were also studied extensively with PDFT calculations. Using the GGA-PBE functional [45], Singh et al. [63] predicted the stable formation of a 2D ferromagnetic half-metal based on vanadium polyporphyrin (PP). The stability of the 2D metal was determined by comparing the Curie temperature (Tc) and phonon dispersion to other known 2D structures like manganese phthalocyanine (MnPc) and CrPP. The authors also note that the predicted 2D VPP is quite suitable for use in flexible spintronic devices. A similar study with Fe, Co, Li, Zn, and H_2PPs was done by Zhu et al. [64] who reported that H_2, Li, and Zn PPs behave as direct bandgap semiconductors while Fe, and Co PP behave as half-metals.

They also report that 2D PP systems behave like n-type semiconducting materials with strong electron mobilities, which were obtained using PDFT combined with Boltzmann transport method with relaxation time approximation. An interesting PDFT study of porphyrin nanosheets and nanowires fused at meso-meso-, β-β-, and β-β-linked array (referred to as SA) and a directly β-fused array (referred to as SB), was performed by Yamaguchi [65]. In this study it was reported that the bandgaps of nanosheets is slightly lower than the nanowires and more importantly it was found that SA type linked arrays have significantly lower bandgaps than SB type. Another example of fused polyporphyrins (PP) with pyrazine linkage was performed by Kumar et al. [66] with transition metals Cr, Mn, Fe, Co, Ni, Cu, and Zn at the porphyrin cores of the nanosheet. Using PDFT calculations with GGA-PBE [45] and GGA+U [67,68] approaches, it was shown that metal PPs have excellent thermal stability with the MnPP system having a ferromagnetic character and half-metallic behavior.

Instead of 2D polyporphyrin (PP) nanosheets, nanowires of 1D PPs were also studied using PDFT. Some 1D porphyrins are also referred to as tape porphyrins in the literature. Gao et. al. [69] studied the electronic structure of metal-polyporphyrin (MPP) and metal-polyhexaphyrin (MPHP) tapes using GGA-PBE functionals. While different MPHP and MPHP (M = Co, Ni, Cu, Zn, and Ru) tapes were studied for their conductive (metals/half-metals/semiconductors) behavior, it was reported that doubly linked CoPHP, NiPHP, and double, triple-linked RuPP has half-metallic nature with potential applications for spintronic devices. Zheng et al. studied 1D PP nanowires linked with acetyl linkers with various transition metals (Cr, Mn, Co, Ni, Cu, and Zn) in the porphyrin cores using GGA-PBE [45] and GGA+U [67,68] functionals. Of all the PPs, ZnPP and NiPP nanowires are nonmagnetic while the rest are magnetic with magnetic moments like their corresponding monomer structures. Among all the metal-PP nanowires with acetyl linkers, only MnPP nanowires exhibit half-metallic behavior.

3. PDFT Simulations of Porphyrins on Surfaces

Porphyrins have extended pi-electronic structures which makes them excellent candidates for adsorption on surfaces. They can also bind covalently to substrates through selective meso, α, β, peripheral substituents. PDFT simulations of porphyrins on surfaces are aimed to understand the adsorption configurations, surface reactivity, binding energetics, charge distribution, and transport at surfaces and interfaces and the corresponding magnetic and electronic properties (DOS and band structure) when porphyrin bind to substrates. In this section we present a survey of various PDFT simulations of porphyrins on surfaces based on the type of porphyrin, type of substrate, and application of interest.

3.1. Conformational Studies of Meso-Substituted Porphyrins on Substrates

Meso-substituted porphyrins have multiple structural configurations due to the flexibility of the porphyrin molecule [70]. When porphyrins adsorb on atomically flat conductive surfaces, scanning tunneling microscopy (STM) is used as the preferred technique [71] to understand their surface structures. The adsorption configuration of meso-substituted porphyrins cannot be easily obtained from STM alone due to the limits of STM resolution and structural flexibility of meso-substituted porphyrins. Figure 6 shows some typical confirmations of meso-substituted porphyrins on the surface. Thus, PDFT simulations are complimentary to experimental studies of conformation of porphyrins on substrates. One of the earlier PDFT studies of meso-substituted porphyrins on surfaces was performed by Zotti et al. [72] who studied the adsorption of freebase tetrapyridyl porphyrin (TPyP) and FeTPyP on Ag(111). The PDFT simulations were performed with GGA-PW91 [59] functional in conjunction with STM experiments. It was reported that TPyP adsorbs in a flat geometry at 5.6 Å from the surface. The dihedral angle of the pyridyl rings is found to be 70° with adlayer structure primarily directed by lateral intermolecular interactions. Another PDFT study of MnTPyP on Cu(111) [73] with GGA-PBE [45] showed that MnTPyP adsorbs in a saddle shape due to the rotation and inclination of the pyridyl groups towards Cu adatoms, which stabilize the metal-organic chains.

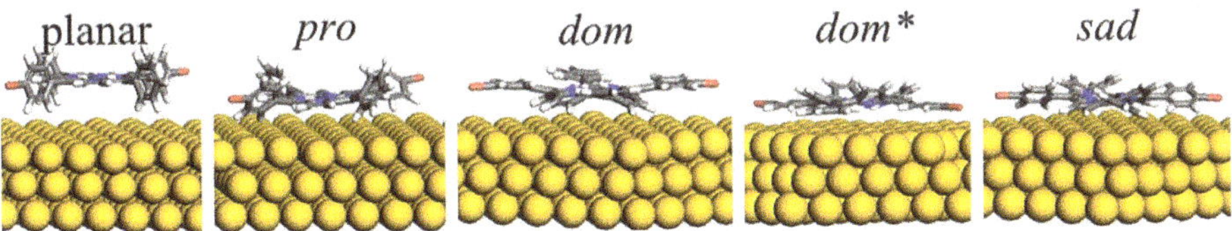

Figure 6. Typical confirmations of meso-substituted tetraphenyl porphyrins on a surface. Reprinted with permission from reference [74] Copyright 2017 American Chemical Society.

Metalated and non-metalated meso-substituted phenyl porphyrins are some of the extensively studied porphyrins on surfaces [71]. Like the TPyP porphyrins, tetra-phenyl porphyrins (TPP) have high degree of structural flexibility. Rojas et al. [75] studied the adsorption of freebase TPPs on Ag(111) and Cu(111) metal substrates using GGA-HCTH functional [76,77]. They reported that TPPs form a 2D network on Ag(111), driven by attractive intermolecular interactions, small migration barrier, and minimal charge transfer. In contrast 2H-TPP/Cu(111) has significant charge transfer, resulting in repulsive forces between the molecules that prevent molecular adlayer network formation. A similar result was observed by Lepper et al. [78] who reported an inverted TPP on Cu(111) surface due to coordination of the two iminic nitrogen atoms to the Cu(111) surface via their lone pairs and thus significant charge transfer.

Extensive PDFT studies of adsorption of 3d transition metal (TM) TPP (TM = Co [79–82], Ni [83]) molecule on Ag and Cu substrates were done by various research groups. Both LDA and GGA functionals were used to study the TM-TPP/substrate system. CoTPP on Ag(111) and Cu(111) surfaces exhibited two adsorption properties: first, an asymmetric saddle deformation of CoTPP with an enhanced tilting of the upwards bent pyrroles and second, a single adsorption site where the Co center occupies a bridge position and one molecular axis aligned with the [1-10] substrate direction [79]. On Cu(110) [81], CoTPP molecules adsorb at the short-bridge site with substantial chemical interaction between the molecular core and the surface causing the porphyrin macrocycle to accommodate close to the surface in a flat geometry, which induces considerable tilting distortions in the phenyl groups. NiTPP [83] also has an asymmetric saddle deformation on Cu(111) with observed chemical shifts of Ni $2p_{3/2}$ caused by Ni 3d orbital interaction with Cu(111) substrate.

Due to their structural flexibility, meso-substituted porphyrin molecules interact with surface adatoms on metallic substrates. Hötger et al. [84] studied the surface transmetalation of central metals in TPP and TPyP molecules on Au(111) surface. They reported that Fe^{+2} cation of FeTPP can be replaced by Co in a redox transmetalation-like reaction on Au(111) surface. Likewise, Cu can be replaced by Co. The reverse reaction does not occur, i.e., Fe does not replace Co in the porphyrin. The mechanism and energetics for the surface transmetalation reaction was determined using PDFT calculations with GGA-PBE [45] functional and DFT-D3 [53,85] van der Waals (vdW) corrections. They also report that while identical transmetalation in TPyP molecules were observed, they are not prevalent as in TPP molecules. The reason for this is attributed to peripheral pyridyl groups offering additional coordination sites for the metals, thus suppressing the metal exchange. Moreno-López et al. [86], used PDFT studies with GGA-PBE [45] with DFT-D3 [53,85] and vdW-DF [43,48] functionals to understand the adsorption and coupling of Cl_4TPP molecules on Cu(111). Using DFT, they reported two coupling reaction pathways: direct dechlorination and Cu adatom-mediated Ullmann coupling. The latter is barrierless, whereas the former faces a barrier of about 0.9 eV for inverted Cl_4TPP on Cu(111). Adatoms of Au(111) also interact with H_2TPP [87] forming different surface electronic structures. These observations were confirmed by simulated STM images from PDFT calculations of H_2TPP on Au(111) in various configurations.

Tetraphenyl porphyrins on non-metallic substrates have also been studied using PDFT simulations. Bassiouk et al. [88], studied the self-assembly of H_2TPP, CoTPP, and NiTPP molecules on HOPG (highly ordered pyrolytic graphite) surface using STM. It was reported that these TPP molecules only adsorb at surface defects due to weak pi-pi or vdW interactions between TPPs and HOPG. PDFT calculations were performed to understand the TPP adsorption at the step edges and defects of HOPG. Simulations with PW91 [59] LDA functional showed that the electronic structure is modified significantly at the surface defects and edges of HOPG causing the adsorption and nucleation TPP molecules. TPP molecules on SiC(110) substrate were studied by Catellani et al. [89] using PDFT with GGA-PBE [45]. The sensitization of SiC(110) substrate based on the adsorbed components of a TPP like pyrrole group, phenyl group, and the whole TPP molecule was studied. It was reported that none of these molecules changes the polarity of the SiC(110) substrate even with dispersion interactions. El Garah et al. [90], and Boukari et al. [91], studied the adsorption if Cu-5,10,15,20-tetrakis(3,5-ditert-butyl-phenyl) porphyrin (Cu-TBPP) on Si(111) and boron-defect Si(111)-B surfaces respectively using PDFT simulation and in both cases the CuTBPP confirmations on the substrate were determined.

All the PDFT studies of meso-substituted porphyrins listed above involve aromatic substituents in the meso positions. PDFT studies of meso-substituted porphyrins with tetra-alkyl groups were also reported in the literature. Chin et al. [92] studied the adsorption of tetranonadecyl ($C_{19}H_{39}$) porphyrin on HOPG using STM and DFT. While the structural optimization was carried with molecular DFT, the authors used to determine the STM structure with PDFT simulations with GGA-PW91 [59] functional. Due to long alkyl chains in the molecule, ONIOM [93] method with quantum mechanical simulations on the porphyrin macrocycle and molecular mechanics simulations on alkyl substituents was used to determine the optimized geometries. The STM simulation of the optimized structure matched with STM experiments. In a later study, Reimers et al. [94], studied the adsorption of terta-alkyl porphyrins with alkyl chain lengths (C_nH_{2n+1} with n = 6–28) on HOPG using multiple computational methods. It was reported that molecular QM/MM calculations and PDFT calculations with PBE-D3 method predicted similar properties for the chain-length dependence of monolayer formation and polymorphism.

3.2. Conformational Studies of Non-Meso-Substituted Porphyrins on Substrates

Porphyrins without the meso substituents lack the structural flexibility of meso-substituted porphyrins [70]. Octaethylporphyrins (OEP) are some of the commonly studied non-meso-substituted porphyrins on substrates [71] using STM. Fanetti et al. [95] studied the adsorption of CoOEP on Ag(110) surface and reported that CoOEP molecule bind to Ag(110) with a tilt angle of 15° with respect to the substrate due to strong hybridization of the adlayer with the Ag substrate which the authors confirm by PDFT simulations using GGA-PBE [45] functional. Kim et al. studied the adsorption of PtOEP molecule on bare Au(111) and on NaCl/Au(111) surfaces using STM and PDFT with GGA-PBE [45] functional and reported that the top of the valence band has a downward shift in NaCl/Au(111) substrate relative to Au(111). In either of these studies, the adsorption energy of OEP molecules on these substrates has not been reported.

A first comprehensive study of OEP on substrates using PDFT simulations was reported by Chilukuri et al. [96]. It was reported that using standard GGA [45] and LDA [97] functionals would significantly underestimate the adsorption energies of porphyrins on substrates compared to calculations using van der Waals corrected DFT methods like vdW-DF [43,48] or DFT-D3 [53,85]. The binding energies of CoOEP on Au(111) with GGA functional was reported to be −0.31 eV, while the value is −4.34 eV with vdW-DF functional. On HOPG the binding energies are −1.18 and −2.42 eV with LDA and vdW-DF functionals respectively. These results indicate that traditional LDA and GGA functionals significantly underestimate dispersion energies with metallic substrates but to a lesser degree with carbon supports. The authors also report the interfacial charge redistribution, DOS and work function changes upon adsorption of CoOEP on substrates. They also reported the first PDFT simulated bias dependent STM images (Figure 7) of OEP molecules on Au(111) and HOPG which matches with experimental observations. The advantage of using dispersion DFT methods on OEPs is further corroborated by Tada et al. [98], who reported that $Rh^{III}(OEP)(Cl)$ molecule would not have bound to the basal plane of HOPG if not for dispersion corrected DFT functional.

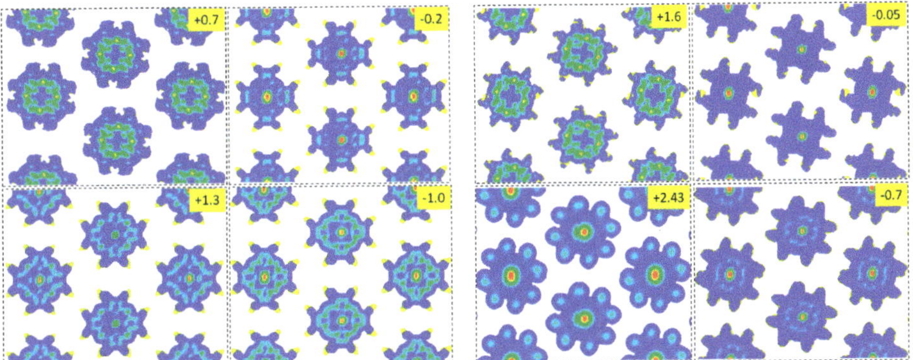

Figure 7. PDFT simulated bias-dependent STM images of Cobalt Octaethylporphyrins (OEP) on Au(111) and HOPG surfaces. Reproduced from reference [96] published by the PCCP Owner Societies.

Iron based haem(b) porphyrin has tetraethyl and tetramethyl substituents in the non-meso positions. Sena et al. [99], studied the adsorption of haem(b) porphyrin on Si(111):H substrate using GGA-PW91 [59] functional and determined the STM images using Tersoff-Hamann approach [100]. The binding energy was estimated to be only 0.42 eV which is likely significantly underestimated because they did not include dispersion interactions in their functional.

Non-substituted porphins (P) are also studied extensively using PDFT simulations. Hanke et al. [101], used PDFT simulations with GGA-PW91 [59] and vdW-DF [43,48] functionals to determine the surface configurations of H_2P molecule on Cu(110) surface and the respective adatom interactions were reported. A similar study was performed by Dyer et al. [102], using vdW-DF method who reported that H_2P is chemisorbed to the surface, caused by electron donation into down-shifted and nearly degenerate unoccupied porphine π-orbitals accompanied with electron back-donation from molecular π-orbitals. Miller et al. [103], reported the electronic and spin structure of FeP on Pt(111) surface using multiple vdW-DF [43,48] and functionals with added Hubbard U term. They report that vdW-DF-optPBE and vdW-DF-optB88 functionals found the same binding site to be the most stable and yielded binding energies that were within ~20% of each other, whereas vdW-DF-revPBE functional were substantially weaker. One of the earlier PDFT studies of porphine adsorption was done by Leung et al. [104] using MnP and PdP porphins on Au(111). DFT+U [68] technique with PDFT simulations using LDA [97] functionals were used to determine the face-on and side-on interactions of porphins on Au(111) substrate. Buimaga-Iarinca et al. studied the effect of translation on binding energy for transition-metal (V, Cr, Mn, Fe, Co, Ni) porphins adsorbed on Ag(111) surface using vdW-DF-cx [105]. They concluded that the bridge positions of Ag(111) are favorable for all transition metal porphins.

PDFT calculations of not porphyrins but porphyrin-based molecules are also studied in the literature. For example, Zhang et al. studied the adsorption of Ni(Salophen) molecule on Au(111) surface using vdW-DFT functionals and determined the adsorption energy to be 2.74 eV which is about 2/3 of the adsorption energy of similar porphyrin [96,106] molecules on Au(111). Gurdal et al. studied the adsorption of pyrphyrin molecules on Au(111) [107] and Ag(111) [108] surfaces using various GGA and vdW-GGA functionals. They reported the effect of surface herringbone reconstruction of Au(111) surface on the adsorption dynamics of Co(Pyrphyrin) molecules. It was reported that the dominant contribution to the adsorption energy are dispersion forces, followed by the interaction of the cyano groups with the metal. The monolayer formation and geometrical configuration of the assembly are mainly driven by the molecule/molecule interactions.

Surface adsorption of porphyrins as part of a multicomponent mixture were also studied using PDFT simulations. Jahanbekam et al. [109], reported the competitive adsorption of CoOEP and coronene molecules as a function of concentration at the 1-phenyloctane/Au(111) solution solid interface using

STM. In this work, it was reported that CoOEP prefers to bind to the substrate unto a molar ratio of more than 55:1 coronene:CoOEP. At 55:1 ratio, only coronene molecules are seen on the Au(111) surface with STM. Using PDFT simulations of coronene, CoOEP, and coronene:CoOEP models, that authors reported that the strong preference of CoOEP binding to Au(111) is that CoOEP has about ~1.8 times larger binding energy than coronene on Au(111). vdW-DF functionals with PAW pseudopotential basis sets were used to determine the adsorption energetics. Additionally, they report that the 1-1 coronene:CoOEP (A in Figure 8) structure is stable between 22:1 to 45:1 molar ratios because the ethyl groups of the CoOEP molecule trap the coronenes on to the Au(111) surface until further changes in the molar ratios. Additionally, the authors used PDFT calculations to determine the interfacial charge distribution (B in Figure 8) and potential energy surfaces for adsorption of CoOEP and coronene molecules on Au(111) surface and used these energies to determine the vibrational frequencies for molecular desorption. It is important to note that coronene molecules exhibit cooperativity and coverage dependency [110] when desorbing from Au(111) substrates.

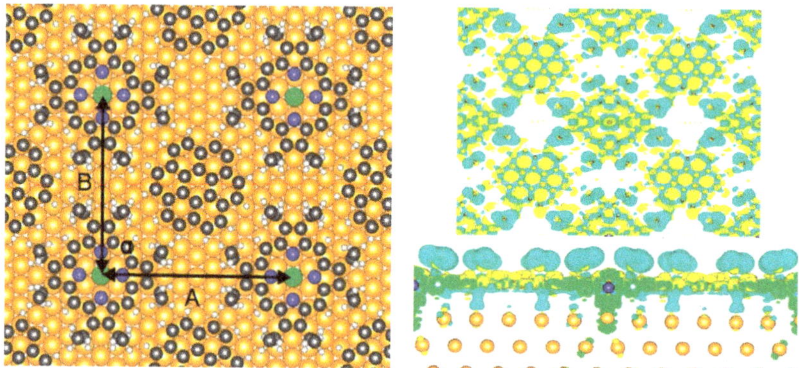

Figure 8. Model of coronene:CoOEP 1:1 structure (**A**) with corresponding surface charge distribution (**B**) obtained from PDFT simulations. Reprinted with permission from reference [109] Copyright 2015 American Chemical Society.

3.3. Porphyrins on Single Layer Substrates

Porphyrins can be used for functionalizing and tuning material properties of various single layered substrates like graphene, carbon nanotubes (CNT) and boron nitride (BN) nanotubes. PDFT simulations were used to study the binding and functional changes caused by porphyrins on such substrates. Touzeau et al. [111], studied the adsorption of Metal (Fe, Zn, Mn, Ti) TPP and Zn tetraalkyl porphyrins on graphene surface using PDFT with GGA-PBE [45] functional. The goal was to understand the effect of the peripheral substituents and metal-centers of porphyrins on the functionalization of graphene. PDFT simulations revealed that graphene functionalization with porphyrin-like molecule is suitable for band-gap opening in graphene. They showed porphyrin adsorption on graphene is controlled by the size of the atomic radii, the occupation of the metal 3d orbitals and the host porphyrin structure. Zeng et al. [112], used PDFT simulations to study the spin filter characteristics metal (M = Cr, Mn, Fe, Co) porphyrins functionalized to edges of graphene. Using LDA functional and non-equilibrium Green's functions (NEGF) [113], they determined that Mn-porphyrin bound to graphene exhibits an extremely high spin polarization coefficient in a parallel magnetic configuration which plays a significant role in making a high-performance spin filter.

Functionalization of single walled carbon nanotubes (SWCNT) with various metalloporphyrin (M = Co, Ni, Cu, Zn) molecules using PDFT simulations was reported Zhao et al. [114]. The authors used semiconducting (10,0) and metallic (6,6) SWNTs for functionalization studies using the GGA-PBE [45] functional. DFT calculations indicate that porphyrins can be used to separate conducting vs. semiconducting in SWCNTs. This is due to hybridization and charge transfer between porphyrins and CNTs. Additionally, metalloporphyrins were found to retain unpaired electrons during functionalization which makes the porphyrin-CNT system a good candidate for optical and spintronic devices. Correa et al. [115], determined the optical response from freebase, Zn-porphyrins/CNT and phthalocyanine/CNT systems using PDFT simulations with vdW-DF [43,48] functionals. They propose that CNT-porphyrins and CNT-phthalocyanines have variable absorption in the visible region, thereby causing increased conversion energy efficiency in an optical device made with both macrocycles. Another interesting study on functionalization of SWCNTs with porphyrins is conducted by Ruiz-Tagle et al. [116], by comparing the effect of physisorption and chemisorption of FeP on metallic and semiconducting SWCNTs. For physisorption studies, LDA [97] functional with vdW-DF [43,48] formalism was used while only LDA [97] functional was used for chemisorption study. The results showed that non-covalent functionalization caused the least change in the electronic and optical properties of SWCNTs. On the other hand, covalent functionalization with metallic SWCNTs would have better electrocatalytic properties than with semiconducting SWCNTs. Porphyrins were also used for functionalizing boron nitride nanotubes (BNNT) similar to SWCNTs. Zhao and Ding [117] performed a PDFT study of BNNT functionalized with metalloporphyrins (M = Fe, Co, Ni, Cu, Zn) using GGA-PBE [45] functional. The authors found that metalloporphyrins energetically prefer to bind the metal with the binding energies ranging from 0.17 to 0.91 eV.

3.4. Porphyrins on Oxide Supports

Porphyrins are widely used as dyes in dye sensitized solar cells where the solar energy trapped by porphyrins is transferred into conductive oxide substrates like TiO_2, ZnO, etc. In this section, we present a collection of PDFT studies involving interactions of porphyrins with oxide supports used in photovoltaic devices and in catalysis.

Gomez et al. [118], used PDFT simulations with GGA-PW91 [59] functionals to study the surface interactions and charge transfer of [COOH-TPP-Zn(II)] porphyrin on TiO_2(110) surface. Using PDFT simulations they identified the stable binding site for the anchoring group (-COO^-) to the TiO_2 surface that facilitates electron injection. Using frontier orbitals and DOS from DFT simulations, the authors report that Zn(II) porphyrin is capable of electron injection into TiO_2, as has been shown from experiments. A similar study was performed by Lin et al. [119], and they used PDFT simulations to study the optical and charge transfer properties of Zn porphyrin adsorbed on the TiO_2 surface. If the TiO_2(110) surface is hydroxylated, Lovat et al. [120], showed that the iminic nitrogens of free base porphyrins (OEP, TPP, and tetra-butyl TPP) capture the hydrogen atoms from the TiO_2(110) surface. They used x-ray photoelectron spectroscopy (XPS) experiments combined with PDFT simulations using GGA-PBE [45] functional and DFT-D corrections, and showed the favorable energetics and mechanism for hydrogen capture by porphyrins.

In-situ metalation of TPP using Ni atoms adsorbed on TiO_2(110) surface was studied by Wang et al. [121] using STM. PDFT simulations were done with the DFT-FIREBALL [122] method. STM images and currents were simulated using Keldysh-Green function formalism. The surface electronic structure from STM experiments matched with the simulated STM images. PDFT simulations [123] are also used to determine the band gap of porphyrin MOFs adsorbed on TiO_2(110) surface, and it was found that HSE06 functional [47] predicted the bandgap and TiO_2 structure matching the experimental data. Xie et al. [124], studied the effect of asymmetric modification of meso-substituents of TPPs on TiO_2(110) using PDFT simulations. Experiments have shown that asymmetric modifications can improve the light-harvesting properties and enhance the electron distribution, but the surface adsorbed structure was unknown. In this work, the authors used molecular and PDFT simulations with GGA-PW91 [59] functional to optimize the adsorption geometries of various asymmetric meso-substituted porphyrins on TiO_2 and determine their electronic properties. While TiO_2 is the extensively used oxide support for many PDFT simulations involving porphyrin binding, supports like ZnO [125] and SiO_2 [126] are also studied. In the case of ZnO, the support was used as an alternative for TiO_2 in dye sensitized solarcells and PDFT calculations were used to understand the charge transfer characteristics. SiO_2-porphyrin [126] studies are used to determine the binding strength for porphyrins to SiO_2 support for application as a trapping agent in petroleum industry.

3.5. PDFT Simulations of Substrate Bound Porphyrin Reactions and Catalysis

Porphyrins on solid substrates can act as active sites for various catalysis reactions. PDFT simulations are typically used to determine the reaction mechanisms of catalytic reactions on surfaces. Quinn et al. [127], used Cu porphyrin to functionalize the graphene surface for methane catalytic reaction. Using PDFT simulations with GGA-PBE [45] functional, it was determined that the porphyrin Cu metal center acts as an active site for the direct oxidation of methane to methanol. The PDFT simulations elucidate the step-by-step reaction mechanism and energetics to understand the catalytic reaction with CuP functionalized graphene at the atomic level. The reaction coordinate (Figure 9) and corresponding geometries of the oxidation reaction was obtained using climbing image nudged elastic band method [128,129] and charge analysis using Bader charges [130].

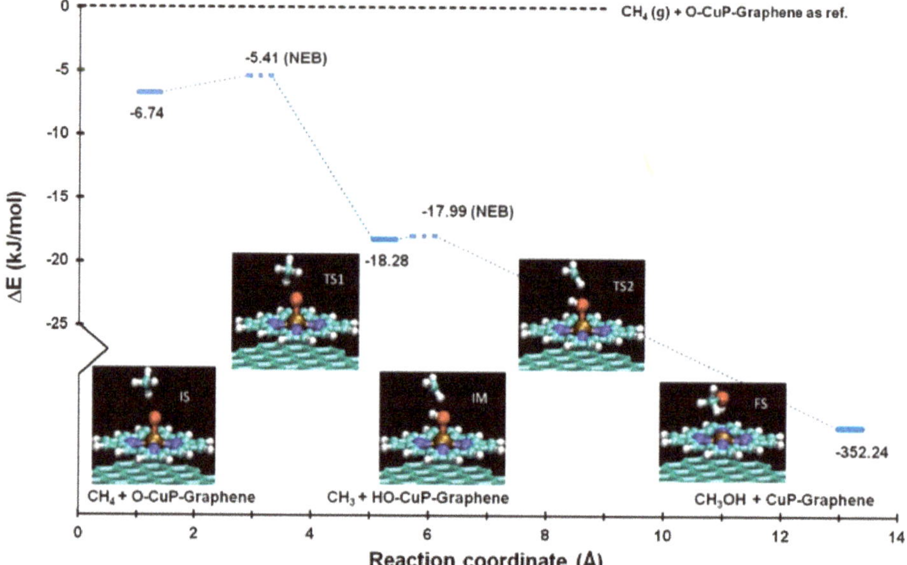

Figure 9. Reaction energy profile for the reaction of CH_4 with pre-adsorbed O atom on CuP functionalized graphene surface. The numbers indicate the energies relative to the reference system of gas phase CH_4 and O bound to the metal center of CuP functionalized graphene surface. The transition states are shown as the maxima on the energy landscape. The x-axis corresponds to the reaction coordinate along the reaction pathway. Reprinted with permission from reference [127] Copyright 2017 Elsevier.

On surface reactions of organic molecules with porphyrins supported on surfaces and monitored by STM experiments has gained significant interest since the pioneering work by Grill et al. [131]. PDFT simulations were used to understand the on-surface reaction mechanisms, energetics and dynamics of porphyrin. Shi et al. [132] studied the Heck reaction of alkene attached aryl bromides to TPP porphyrins on Au(111) surface. The surface structure of TPP changes upon the attachment of the aryl group. The reactions was found to be catalyzed by Pd attached to Au(111) surface. PDFT simulations using GGA-PBE [45] functional with DFT-D3 correction was used to determine the molecular surface configurations during debromination and coupling of Hack reaction on Au(111). Shu et al. [133], studied the on-surface reactions of aryl-chloride and Cu(111) attached porphyrins. PDFT simulations with GGA-PBE [45] functional and DFT-D3 correction was used to determine the surface reaction dynamics and energetics of dehalogenation, cross coupling and cyclodehydrogenation reactions on Cu(111) surface.

3.6. Magnetic Couplings in Substrate Bound Porphyrins

Porphyrins adsorbed on ferromagnetic substrates can act as excellent candidates for spintronic devices. The ferromagnetic substrates typically involve Co or Ni films deposited on metallic substrates. Mn, Fe, Co, Ni, and Cu porphyrins on Co and Ni substrates were studied to understand the magnetic coupling between metalloporphyrins and metallic surfaces. PDFT simulations are especially helpful to understand the magnetic properties of porphyrin/substrate systems. Wende et al. [134] studied the adsorption of FeOEP on Ni and Co films bound to Cu(100) surface using X-ray absorption spectroscopy (XAS) and X-ray magnetic circular dichroism (XMCD). The experiments were combined with PDFT simulations of corresponding porphyrin/substrate systems using GGA-PW91 [59] functional and DFT+U approach [68]. Binding energetics of ClFeOEP and FeOEP on Co and Ni substrates were determined, and it was noted that the porphyrin loses Cl^- upon adsorption. PDFT simulations provided a deeper understanding of Fe-substrate exchange coupling from DOS, charge analysis and magnetization densities. A similar study with free base and Fe porphyrins on Co substrate was carried out by Oppeneer et al. [135] also using GGA-PW91 [59] functional and DFT+U approach [68]. The PDFT simulations were used to understand the origins of the substrate induced magnetic ordering of metalloporphyrins. It was demonstrated that FeOEP ferromagnetically exchange couples, while ClFeOEP antiferromagnetic couples with the substrate. The same research group also studied the magnetic coupling of FeOEP on $c(2 \times 2)$ oxygen reconstructed Co(100) surface [136] with the same PDFT methodology as their earlier studies and showed that FeOEP couples antiferromagnetically with oxygen reconstructed Co(100) surface.

Magnetic coupling with Mn porphyrins on Co substrates was studied by Ali et al. [137] using GGA-PW91 [59] functional and DFT+U approach [68]. The authors reported that Mn porphyrins can adsorb or chemisorb on the Co substrate with MnP-Co binding distances at 3.5 Å and 2.1 Å, respectively. This variable surface binding caused distinct magnetic exchange interactions between porphyrin and substrate, but it was found that Mn magnetic switching occurs at both binding distances. Chylarecka et al. [138] showed that ClMnTPP porphyrin involves in indirect magnetic coupling with Co substrate using PDFT and, STM, XAS, and XMCD studies. GGA-PBE [45] functional with Gaussian type orbitals and with DFT+U approach were used to determine the surface DOS and magnetic properties of ClMnTPP on Co substrate. It was found that if the chloride ion of the MnTPPCl molecule orients away (Co-Mn-Cl) from the Co surface, a weak ferromagnetic molecule-substrate coupling is observed. PDFT simulations with DFT+U approach were also used to understand the magnetic coupling interactions of Co porphyrins on Ni [139] and graphene [140] substrates.

3.7. Porphyrin Molecular Junctions

Porphyrins and substituted porphyrins were widely studied in single molecule junctions [14]. These junctions typically involve a molecule covalently linking two electrodes. PDFT studies were used to understand the binding and charge transport properties of the molecule and electrodes in the junction. Lamoen et al. [141] performed one of the early studies of the covalently bound Pd porphyrin to gold electrode. They studied the side on (hydrogens of one pyrrole) interactions of Pd porphyrin on Au(111) using the LDA functional. Although this initial model is not a full Pd porphyrin molecular junction (because only one gold electrode was used), the simulation was helpful to gain atomic level understanding of rectifying behavior and charging effects associated with molecular conduction via single-electron tunneling. Another study with the same Pd porphyrin in a one sided junction with Al(111) electrode was performed by Picozzi et al. [142], with both LDA-PW [59] and

GGA-PBE [45] functionals. The energy level alignments, binding energy, and DOS were calculated, and it was reported that PdP/Al(111) interaction is weak yet rectifying.

PDFT study of a complete junction involving a porphyrin sandwiched between two gold electrodes was conducted by Long et al. [143]. A nonequilibrium Green's functions (NEGF) [113] approach with PDFT was used to determine the electron transport properties of the porphyrin molecular junction. The porphyrin molecule is linked to the gold electrodes through two (5,15)-meso-diphenyl substituents. In addition, the role of electron donating and withdrawing substituent (at the 10-meso position) on the electron transport properties was determined. The authors report a negative differential resistance behavior from PDFT calculations which is a critical property for many molecular electronics applications. In another PDFT study, An et al. [144] studied high efficiency switching in porphyrin ethyne benzene/gold molecular junction using GGA-PBE functional. The electron transport properties from PDFT simulations indicated that ethyne-bridged phenyl porphyrin molecules are good candidates for molecular switching devices. The rotation of the phenyl substituent allows for high and low currents due to change in orbital overlaps upon rotation. Additionally, placing amino and nitro substituents caused high ON/OFF current ratios with larger substituent effect when entire porphyrin molecule is in a co-planar geometry than in perpendicular configuration.

Liu et al. [145], developed an optimally-tuned range-separated hybrid (OT-RSH) functional [146] using DFT+Σ method [147,148] and NEGF [113] approach for accurate description of molecular conduction in junctions. The effect of central metal (M = 2H, Ni, Co, Cu) of porphyrins in the molecular conductance between two gold electrodes was studied using PDFT simulations. It was reported that changing the central metal can change the conductance by nearly a factor of 2. Cho et al. [149], studied magnetic and charge transport properties of a metal (M = H2, Cr, Mn Fe, Co) porphyrin array (PA$_\infty$) connected through thiol groups to two gold electrodes. PDFT simulations with GGA-PBE functional and NGEF formalism was used to determine the conductance and band structure of metal porphyrin junction. It was reported that CrPA$_\infty$ exhibits half-metallic behavior originating from the high spin state of Cr which causes spin asymmetry of the conduction band in CrPA$_\infty$. Additionally, it was reported that spin-filtering ability occurs with an array size of 2, Cr-PA$_2$. Sedghi et al. [150] studied the effect of the length of the substituent side chain in oligo porphyrin molecular wires on the long-range electron tunneling and conductance properties in molecular junction. Three oligo-porphyrins with oligomer length of 1, 2, 3 were sandwiched between gold electrodes and, PDFT simulations and scattering theory were used to study the molecular conductance. It was reported that phase-coherent tunneling occurs through the whole molecular junction.

Graphene and carbon nanotubes (CNT) were also used as electrodes in porphyrin molecular junctions. Suárez et al. [151] studied the low voltage transport response of porphyrin molecular wires bridging two graphene sheets via physisorption. They used the vdW-DF [43,48] functional to study the Breit-Wigner molecular resonances as a function of translation of graphene sheets and porphyrin wires. It was reported that the conductance values are dependent upon the sampling of k-points during simulation. Li et al. used PDFT simulations to study the molecular conductance of porphyrin bridged CNTs. Maximally localized Wannier functions (MLWF) in conjunction with NEGF formalism was used to determine the conductance and quantum interference in the transport properties of porphyrin/CNT junctions. Using molecular conductance data, they reported that tape porphyrins can act as molecular size memory units with many-valued logic.

3.8. Ligand-Porphyrin Reactions on Surfaces

Central metals in metalloporphyrins can react with many axial ligands [152] forming porphyrin-ligand complexes. These ligands can be mono or bi-axial leading to pentavalent and hexavalent complexes. When metalloporphyrins bind to solid surfaces, the surface may act like an axial ligand in the fifth coordination site [153–155]. In this section we present a collection of PDFT studies on coordination of ligands to metalloporphyrins that are adsorbed to surfaces. Coordination of gaseous molecules like CO [156], NO [155], O$_2$ [157], etc. to porphyrins on surfaces were studied by various experimental techniques. PDFT simulations were used to understand the binding mechanism of these ligands to substrate bound porphyrins.

Nandi et al. [106], used PDFT simulations with vdW-DF [43,48] functional to understand the binding mechanism of imidazole (Im) ligand to Ni-octaethylporphyrin (NiOEP) bound on HOPG surface. Using STM, solution-spectroscopy, and molecular DFT calculations the authors reported that Im ligand does not bind to NiOEP in solution or in gas-phase but does bind when NiOEP is on the HOPG surface. The reactivity of imidazole toward NiOEP adsorbed on HOPG is attributed to charge donation from the graphite stabilizing the Im-Ni bond (Figure 10). This charge transfer pathway is supported by molecular and periodic modeling calculations which indicate that the Im ligand behaves as a π-acceptor. DFT calculations also show that the nickel ion in the Im-NiOEP/HOPG complex is in a singlet ground state. This is surprising because the gas phase Im-NiOEP complex is found to be stable in a triplet ground state. Integrated charge transfer data (Figure 10) from PDFT also showed that HOPG donates the charge to Imidazole ligand via the NiOEP macrocycle, which indicates that the porphyrin molecule only acts as a charge mediator.

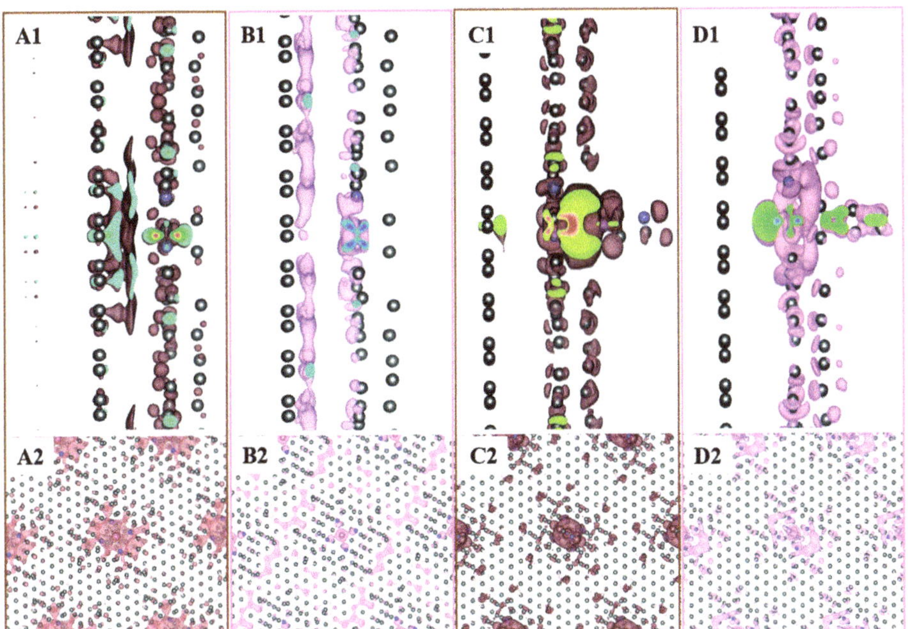

Figure 10. From left to right, charge density difference mappings for positive (colored in brown) and negative (colored in pink) charges for NiOEP/HOPG (A and B) and Im-NiOEP/HOPG (C and D) systems respectively. The images in the top row (**A1–D1**) represent side-view (along the a-axis) and the bottom row (**A2–D2**) represent top-view (along c-axis). Element colors are carbon—gray, nitrogen—blue, nickel—yellow (not visible). Hydrogens are masked for clarity. In the cross-section (**A1–D1**, top row) the rainbow colors (blue to red) indicate charge with blue being highly negative and red being highly positive. Reproduced from reference [106] published by the PCCP Owner Societies.

Binding of gaseous molecules to substrate bound porphyrins using PDFT were carried with tetraphenyl-porphyrins and tape porphyrins. Hieringer et al. [155], reported the 'surface trans effect' of NO axial coordination to Co-porphyrin on Ag(111). GGA-PBE [45] functional with dispersion corrections was used to model the NO/CoP/Ag(111) surface and the corresponding structural and energetics were determined. They reported that competition effects, like the trans effect, play a central role and lead to a mutual interference of the two axial ligands, NO and Ag, and their bonds to the metal center. Wäckerlin et al. [158], used PDFT simulations with DFT+U approach [68] and showed that surface magnetization of porphyrin/ferromagnetic surfaces can be tuned via the choice of axial ligands. NO (S = 1/2) and NH_3 (S = 0) ligands were used to coordinate with FeTPP and MnTPP porphyrins on Ni and Co ferromagnetic surfaces. PDFT simulations revealed that they reported that the structural trans effect on the surface rules the molecular spin state, as well as the sign and strength of the exchange interaction with the substrate. In another study, Janet et al. [159] compared DFT+U approach and semi-local DFT simulations with GGA-PBE [45] functional using O_2 coordination with CoTPP/Au(111) interface. They reported that semi-local DFT simulations can optimize a structure but DFT+U approach is better for charge and spin predictions in the system. It was also reported that O_2 binding to CoTPP was over stabilized by GGA, while DFT+U predicted reliable energetics especially with spin active systems. Ghosh et al. [160], used DFT+U approach to determine the spin states of CO, NO and O_2 bound to Mn porphyrin on Au(111). The PDFT simulations were used to demonstrate reversible spin-switching of ligand bound porphyrin/Au(111) system by conformations changing of porphyrin structure on the substrate. Ligand binding on porphyrin nanowires were also studied using PDFT simulations. Binding of NO molecule to metal tape-porphyrins [161–163] with PDFT calculations reveal molecular structure of metal tape-porphyrins has negligible change upon ligand binding but considerable change in the electronic structure was observed. Additionally, a significant band gap reduction has been observed upon NO ligand binding.

4. Summary

Periodic density functional theory (PDFT) calculations have been indispensable to bridge the gaps between observable properties at the condensed phase and the electronic structure of the periodic system. They led to fundamental understanding and tuning of the solid-state behavior of many functional materials and interfaces. In this review, a collection of PDFT simulations of porphyrins in nanostructures and on surfaces were presented. Porphyrins are important compounds used for many have numerous biological and technological applications. While many reviews of porphyrins and their derivatives are available in the literature, a review of periodic porphyrin structures has never been reported to our knowledge. We organized the review based on applications of PDFT simulations to understand specific structural, conformational, adsorption, electronic, magnetic, charge transfer properties, and reactivity of porphyrins on surfaces. The typical properties calculated using PDFT simulations include optimized geometries, binding energetics, density of states, band structure, spin switching and magnetization, charge transport, STM images (local electronic structure), reaction intermediates in catalytic reactions, etc. Hence, this review should be of great interest for the porphyrin research community and to the broader audience performing PDFT simulations.

In our survey of periodic simulations on porphyrins with DFT, plane wave pseudopotential basis sets were the predominant choice rather than Gaussian type orbital basis sets. Most of the early simulations were performed with the bottom two rungs of the "Jacob's ladder" which are the GGA and LDA type functionals. The drawbacks of using these functionals include underestimation of electronic bandgaps and weak dispersion interactions, especially with systems involving organic molecules like porphyrins. LDA overestimates, while GGA underestimates the binding energies in condensed phase systems like transition metal, metal oxides, carbon, and silicon crystals. Both functionals underestimate interactions involving organic molecules like porphyrins.

Recently many PDFT studies were performed with dispersion corrected DFT functionals or with empirical dispersion corrections. Inclusion of vdW interactions greatly improved the calculated binding energies of porphyrins leading to data that better matched experiments. It was reported that dispersion interactions are more important for porphyrins on metallic substrates than on non-metal substrates. Within the last decade hybrid functionals like B3LYP and HSE were used for PDFT simulations of porphyrins. These functionals improved the band gap and geometric optimizations of porphyrin nanostructures and interfaces. Our review of the literature also found that DFT+U is the method of choice for calculations involving magnetically coupled and spin active porphyrin systems. With the improvement of DFT functionals and computing capability more and more expensive calculations were being performed for a fundamental understanding of porphyrin behavior in periodic systems.

Like many quantum mechanical calculations, PDFT calculations have limitations with respect to the size of the modeled system and accuracy of the energies obtained from DFT calculations. Additionally, choosing the right DFT functional for modeling a heterogeneous periodic system like porphyrins is challenging and no single PDFT functional is deemed appropriate for all applications. Based on the problem of interest like electronic structure, binding energies, excited state properties, etc., a variety of PDFT functionals were used to study bulkier systems like porphyrins and phthalocyanines. Unlike molecular DFT functionals used for porphyrin based systems [24], the number of PDFT functionals are limited especially for performing time-dependent or excited state systems. In addition, simulations of larger periodic systems [110] with PDFT is computationally expensive, while ab initio molecular dynamics (MD) simulations of larger systems are practically impossible.

Funding: This material is based upon work supported by the National Science Foundation under Grant No. (CHE-1800070).

Conflicts of Interest: The authors declare no conflict of interest.

References

1. Huang, H.; Song, W.; Rieffel, J.; Lovell, J.F. Emerging applications of porphyrins in photomedicine. *Front. Phys.* **2015**, *3*, 3. [CrossRef] [PubMed]
2. Imran, M.; Ramzan, M.; Qureshi, A.K.; Khan, M.A.; Tariq, M. Emerging Applications of Porphyrins and Metalloporphyrins in Biomedicine and Diagnostic Magnetic Resonance Imaging. *Biosensors* **2018**, *8*, 95. [CrossRef] [PubMed]
3. Li, W.; Aida, T. Dendrimer Porphyrins and Phthalocyanines. *Chem. Rev.* **2009**, *109*, 6047–6076. [CrossRef] [PubMed]
4. Aziz, A.; Ruiz-Salvador, A.; Hernández, N.C.; Calero, S.; Hamad, S.; Grau-Crespo, R. Porphyrin-based metal-organic frameworks for solar fuel synthesis photocatalysis: Band gap tuning via iron substitutions. *J. Mater. Chem. A* **2017**, *5*, 11894–11904. [CrossRef]
5. Maldotti, A.; Amadelli, R.; Bartocci, C.; Carassiti, V.; Polo, E.; Varani, G. Photochemistry of Iron-porphyrin complexes. Biomimetics and catalysis. *Coord. Chem. Rev.* **1993**, *125*, 143–154. [CrossRef]
6. Barona-Castaño, J.C.; Carmona-Vargas, C.C.; Brocksom, T.J.; de Oliveira, K.T. Porphyrins as Catalysts in Scalable Organic Reactions. *Molecules* **2016**, *21*, 310. [CrossRef]
7. Li, L.; Diau, E.W. Porphyrin-sensitized solar cells. *Chem. Soc. Rev.* **2012**, *42*, 291–304. [CrossRef]
8. Paolesse, R.; Nardis, S.; Monti, D.; Stefanelli, M.; Di Natale, C. Porphyrinoids for Chemical Sensor Applications. *Chem. Rev.* **2017**, *117*, 2517–2583. [CrossRef]
9. Jurow, M.; Schuckman, A.E.; Batteas, J.D.; Drain, C.M. Porphyrins as Molecular Electronic Components of Functional Devices. *Coord. Chem. Rev.* **2010**, *254*, 2297–2310. [CrossRef]
10. Lopes, D.M.; Araújo-Chaves, J.C.; Menezes, L.R.; Nantes-Cardoso, I.L. Technological Applications of Porphyrins and Related Compounds: Spintronics and Micro-/Nanomotors. *Solid State Phys.* **2019**. [CrossRef]
11. Hoang, M.H.; Choi, D.H.; Lee, S.J. Organic field-effect transistors based on semiconducting porphyrin single crystals. *Synth. Met.* **2012**, *162*, 419–425. [CrossRef]
12. Hoang, M.H.; Ngo, T.T.; Nguyen, D.N. Effect of molecular packing of zinc(II) porphyrins on the performance of field-effect transistors. *Adv. Nat. Sci. Nanosci. Nanotechnol.* **2014**, *5*, 045012. [CrossRef]
13. Che, C.; Xiang, H.; Chui, S.S.; Xu, Z.; Roy, V.A.L.; Yan, J.J.; Fu, W.; Lai, P.T.; Williams, I.D. A High-Performance Organic Field-Effect Transistor Based on Platinum(II) Porphyrin: Peripheral Substituents on Porphyrin Ligand Significantly Affect Film Structure and Charge Mobility. *Chem. Asian J.* **2008**, *3*, 1092–1103. [CrossRef] [PubMed]
14. Seol, M.; Choi, S.; Kim, C.; Moon, D.; Choi, Y. Porphyrin–Silicon Hybrid Field-Effect Transistor with Individually Addressable Top-gate Structure. *ACS Nano* **2012**, *6*, 183–189. [CrossRef] [PubMed]
15. El Abbassi, M.; Zwick, P.; Rates, A.; Stefani, D.; Prescimone, A.; Mayor, M.; Van Der Zant, H.S.J.; Dulić, D. Unravelling the conductance path through single-porphyrin junctions. *Chem. Sci.* **2019**, *10*, 8299–8305. [CrossRef] [PubMed]

16. Feixas, F.; Solà, M.; Swart, M. Chemical bonding and aromaticity in metalloporphyrins. *Can. J. Chem.* **2009**, *87*, 1063–1073. [CrossRef]

17. Senge, M.O.; Fazekas, M.; Notaras, E.G.A.; Blau, W.J.; Zawadzka, M.; Locos, O.B.; Ni Mhuircheartaigh, E.M. Nonlinear Optical Properties of Porphyrins. *Adv. Mater.* **2007**, *19*, 2737–2774. [CrossRef]

18. Tran-Thi, T. Assemblies of phthalocyanines with porphyrins and porphyrazines: Ground and excited state optical properties. *Coord. Chem. Rev.* **1997**, *160*, 53–91. [CrossRef]

19. Saito, S.; Osuka, A. Expanded Porphyrins: Intriguing Structures, Electronic Properties, and Reactivities. *Angew. Chem. Int. Ed.* **2011**, *50*, 4342–4373. [CrossRef]

20. Adinehnia, M.; Eskelsen, J.R.; Hipps, K.W.; Mazur, U. Mechanical behavior of crystalline ionic porphyrins. *J. Porphyr. Phthalocyanines* **2019**, *23*, 154–165. [CrossRef]

21. Phillips, J.N. Chapter II—Physico-chemical Properties of Porphyrins. In *Comprehensive Biochemistry*; Florkin, M., Stotz, E.H., Eds.; Elsevier: Amsterdam, The Netherlands, 1963; Volume 9, pp. 34–72.

22. Jensen, K.P.; Ryde, U. Comparison of the Chemical Properties of Iron and Cobalt Porphyrins and Corrins. *ChemBioChem* **2003**, *4*, 413–424. [CrossRef] [PubMed]

23. Brothers, P.J.; Collman, J.P. The organometallic chemistry of transition-metal porphyrin complexes. *Acc. Chem. Res.* **1986**, *19*, 209–215. [CrossRef]

24. Shubina, T.E. Computational Studies on Properties, Formation, and Complexation of M(II)-Porphyrins. In *Advances in Inorganic Chemistry*; van Eldik, R., Harvey, J., Eds.; Academic Press: New York, NY, USA, 2010; Volume 62, pp. 261–299.

25. Baerends, E.; Ricciardi, G.; Rosa, A.; Van Gisbergen, S. A DFT/TDDFT interpretation of the ground and excited states of porphyrin and porphyrazine complexes. *Coord. Chem. Rev.* **2002**, *230*, 5–27. [CrossRef]

26. Kepenekian, M.; Calborean, A.; Vetere, V.; Le Guennic, B.; Robert, V.; Maldivi, P. Toward Reliable DFT Investigations of Mn-Porphyrins through CASPT2/DFT Comparison. *J. Chem. Theory Comput.* **2011**, *7*, 3532–3539. [CrossRef] [PubMed]

27. Aydin, M. Geometric and Electronic Properties of Porphyrin and its Derivatives. In *Applications of Molecular Spectroscopy to Current Research in the Chemical and Biological Sciences*; IntechOpen: Rijeka, Croatia, 2016; p. 10.

28. Rappoport, D.; Crawford, N.R.M.; Furche, F.; Burke, K. Which functional should I choose? In *Computational Inorganicand Bioinorganic Chemistry*; Solomon, E.I., Scott, R.A., King, B.R., Eds.; Wiley John & Sons, Inc.: Hoboken, NJ, USA, 2009.

29. Liao, M.-S.; Lu, Y.; Scheiner, S. Performance assessment of density-functional methods for study of charge-transfer complexes. *J. Comput. Chem.* **2003**, *24*, 623–631. [CrossRef] [PubMed]

30. De Visser, S.P.; Stillman, M.J. Challenging Density Functional Theory Calculations with Hemes and Porphyrins. *Int. J. Mol. Sci.* **2016**, *17*, 519. [CrossRef]

31. Kratzer, P.; Neugebauer, J. The Basics of Electronic Structure Theory for Periodic Systems. *Front. Chem.* **2019**, *7*, 106. [CrossRef]

32. Tosoni, S.; Tuma, C.; Sauer, J.; Civalleri, B.; Ugliengo, P. A comparison between plane wave and Gaussian-type orbital basis sets for hydrogen bonded systems: Formic acid as a test case. *J. Chem. Phys.* **2007**, *127*, 154102. [CrossRef]

33. Kresse, G.; Hafner, J. Ab initio molecular dynamics for liquid metals. *Phys. Rev. B* **1993**, *47*, 558–561. [CrossRef]

34. Kresse, G.; Furthmüller, J. Efficiency of ab-initio total energy calculations for metals and semiconductors using a plane-wave basis set. *Comput. Mater. Sci.* **1996**, *6*, 15–50. [CrossRef]

35. Kresse, G.; Joubert, D. From ultrasoft pseudopotentials to the projector augmented-wave method. *Phys. Rev. B* **1999**, *59*, 1758–1775. [CrossRef]

36. Giannozzi, P.; Baroni, S.; Bonini, N.; Calandra, M.; Car, R.; Cavazzoni, C.; Ceresoli, D.; Chiarotti, G.L.; Cococcioni, M.; Dabo, I.; et al. QUANTUM ESPRESSO: A modular and open-source software project for quantum simulations of materials. *J. Physics: Condens. Matter* **2009**, *21*, 395502. [CrossRef] [PubMed]

37. Hutter, J.; Iannuzzi, M.; Schiffmann, F.; Vande Vondele, J. cp2k: Atomistic simulations of condensed matter systems. *Wires Comput. Mol. Sci.* **2014**, *4*, 15–25. [CrossRef]

38. Clark, S.J.; Segall, M.D.; Pickard, C.J.; Hasnip, P.J.; Probert, M.I.; Keith, R.; Payne, M.C. First principles methods using CASTEP. *Z. Für Krist. Cryst. Mater.* **2009**, *220*, 567. [CrossRef]

39. Delley, B. DMol, a standard tool for density functional calculations: Review and advances. In *Theoretical and Computational Chemistry*; Seminario, J.M., Politzer, P., Eds.; Elsevier: Amsterdam, The Netherlands, 1995; Volume 2, pp. 221–254.

40. Soler, J.M.; Artacho, E.; Gale, J.D.; Garcia, A.; Junquera, J.; Ordejón, P.; Sánchez-Portal, D. The SIESTA method for ab initio order- N materials simulation. *J. Phys. Condens. Matter* **2002**, *14*, 2745–2779. [CrossRef]

41. Hutter, J.; Marcella, I. CPMD: Car-Parrinello molecular dynamics. *Z. Für Krist. Cryst. Mater.* **2009**, *220*, 549. [CrossRef]

42. Adinehnia, M.; Borders, B.; Ruf, M.; Chilukuri, B.; Hipps, K.W.; Mazur, U. Comprehensive structure–function correlation of photoactive ionic π-conjugated supermolecular assemblies: An experimental and computational study. *J. Mater. Chem. C* **2016**, *4*, 10223–10239. [CrossRef]

43. Klimes, J.; Bowler, D.R.; Michaelides, A. Chemical accuracy for the van der Waals density functional. *J. Phys. Condens. Matter* **2009**, *22*, 022201. [CrossRef]

44. Blöchl, P.E. Projector augmented-wave method. *Phys. Rev. B* **1994**, *50*, 17953. [CrossRef]

45. Perdew, J.P.; Burke, K.; Ernzerhof, M. Generalized Gradient Approximation Made Simple. *Phys. Rev. Lett.* **1996**, *77*, 3865–3868. [CrossRef]

46. Heyd, J.; Scuseria, G.E.; Ernzerhof, M. Hybrid functionals based on a screened Coulomb potential. *J. Chem. Phys.* **2003**, *118*, 8207. [CrossRef]

47. Brothers, E.N.; Izmaylov, A.F.; Normand, J.O.; Barone, V.; Scuseria, G.E. Accurate solid-state band gaps via screened hybrid electronic structure calculations. *J. Chem. Phys.* **2008**, *129*, 11102. [CrossRef] [PubMed]

48. Klimes, J.; Bowler, D.R.; Michaelides, A. Van der Waals density functionals applied to solids. *Phys. Rev. B* **2011**, *83*. [CrossRef]

49. Borders, B.; Adinehnia, M.; Chilukuri, B.; Ruf, M.; Hipps, K.W.; Mazur, U. Tuning the optoelectronic characteristics of ionic organic crystalline assemblies. *J. Mater. Chem. C* **2018**, *6*, 4041–4056. [CrossRef]

50. Tian, X.; Lin, C.; Zhong, Z.; Li, X.; Xu, X.; Liu, J.; Kang, L.; Chai, G.; Yao, J. Effect of Axial Coordination of Iron Porphyrin on Their Nanostructures and Photocatalytic Performance. *Cryst. Growth Des.* **2019**, *19*, 3279–3287. [CrossRef]

51. Krasnov, P.O.; Kuzubov, A.A.; Kholtobina, A.S.; Kovaleva, E.; Kuzubova, M.V. Optical charge transfer transitions in supramolecular fullerene and porphyrin compounds. *J. Struct. Chem.* **2016**, *57*, 642–648. [CrossRef]

52. Boyd, P.D.W.; Hodgson, M.C.; Rickard, C.E.F.; Oliver, A.G.; Chaker, L.; Brothers, P.J.; Bolskar, R.D.; Tham, F.S.; Reed, C.A. Selective Supramolecular Porphyrin/Fullerene Interactions1. *J. Am. Chem. Soc.* **1999**, *121*, 10487–10495. [CrossRef]

53. Grimme, S. Semiempirical GGA-type density functional constructed with a long-range dispersion correction. *J. Comput. Chem.* **2006**, *27*, 1787–1799. [CrossRef]

54. Gajdoš, M.; Hümmer, K.; Kresse, G.; Furthmüller, J.; Bechstedt, F. Linear optical properties in the projector-augmented wave methodology. *Phys. Rev. B* **2006**, *73*, 045112. [CrossRef]

55. Hamad, S.; Hernandez, N.C.; Aziz, A.; Ruiz-Salvador, A.; Calero, S.; Grau-Crespo, R. Electronic structure of porphyrin-based metal–organic frameworks and their suitability for solar fuel production photocatalysis. *J. Mater. Chem A* **2015**, *3*, 23458–23465. [CrossRef]

56. Fateeva, A.; Chater, P.A.; Ireland, C.P.; Tahir, A.A.; Khimyak, Y.Z.; Wiper, P.V.; Darwent, J.R.; Rosseinsky, M.J. A Water-Stable Porphyrin-Based Metal-Organic Framework Active for Visible-Light Photocatalysis. *Angew. Chem. Int. Ed.* **2012**, *51*, 7440–7444. [CrossRef] [PubMed]

57. Liu, J.; Zhou, W.; Liu, J.; Howard, I.; Kilibarda, G.; Schlabach, S.; Coupry, D.; Addicoat, M.; Yoneda, S.; Tsutsui, Y.; et al. Photoinduced Charge-Carrier Generation in Epitaxial MOF Thin Films: High Efficiency as a Result of an Indirect Electronic Band Gap? *Angew. Chem. Int. Ed.* **2015**, *54*, 7441–7445. [CrossRef] [PubMed]

58. Liu, J.; Zhou, W.; Liu, J.; Fujimori, Y.; Higashino, T.; Imahori, H.; Jiang, X.; Zhao, J.; Sakurai, T.; Hattori, Y.; et al. A new class of epitaxial porphyrin metal-organic framework thin films with extremely high photocarrier generation efficiency: Promising materials for all-solid-state solar cells. *J. Mater. Chem. A* **2016**, *4*, 12739–12747. [CrossRef]

59. Perdew, J.P.; Wang, Y. Accurate and simple analytic representation of the electron-gas correlation energy. *Phys. Rev. B* **1992**, *45*, 13244–13249. [CrossRef] [PubMed]

60. Perdew, J.P.; Zunger, A. Self-interaction correction to density-functional approximations for many-electron systems. *Phys. Rev. B* **1981**, *23*, 5048–5079. [CrossRef]

61. Posligua, V.; Aziz, A.; Haver, R.; Peeks, M.D.; Anderson, H.L.; Grau-Crespo, R. Band Structures of Periodic Porphyrin Nanostructures. *J. Phys. Chem. C* **2018**, *122*, 23790–23798. [CrossRef]

62. Allec, S.I.; Ilawe, N.V.; Wong, B.M. Unusual Bandgap Oscillations in Template-Directed Ï€-Conjugated Porphyrin Nanotubes. *J. Phys. Chem. Lett.* **2016**, *7*, 2362–2367. [CrossRef]

63. Singh, H.K.; Kumar, P.; Waghmare, U.V. Theoretical Prediction of a Stable 2D Crystal of Vanadium Porphyrin: A Half-Metallic Ferromagnet. *J. Phys. Chem. C* **2015**, *119*, 25657–25662. [CrossRef]

64. Zhu, B.; Zhang, X.; Zeng, B.; Li, M.; Long, M. First-principles predictions on charge mobility and half-metallicity in two dimensional metal coordination polyporphyrin sheets. *Org. Electron.* **2017**, *49*, 45. [CrossRef]

65. Yamaguchi, Y. Theoretical study of two-dimensionally fused zinc porphyrins: DFT calculations. *Int. J. Quantum Chem.* **2009**, *109*, 1584–1597. [CrossRef]

66. Kumar, S.; Choudhuri, I.; Pathak, B. An atomically thin ferromagnetic half-metallic pyrazine-fused Mn-porphyrin sheet: A slow spin relaxation system. *J. Mater. Chem. C* **2016**, *4*, 9069–9077. [CrossRef]

67. Dudarev, S.L.; Botton, G.A.; Savrasov, S.Y.; Humphreys, C.J.; Sutton, A.P. Electron-energy-loss spectra and the structural stability of nickel oxide: An LSDA+U study. *Phys. Rev. B* **1998**, *57*, 1505–1509. [CrossRef]

68. Anisimov, V.I.; Aryasetiawan, F.; Lichtenstein, A.I. First-principles calculations of the electronic structure and spectra of strongly correlated systems: Dynamical mean-field theory. *J. Phys. Condens. Matter* **1997**, *9*, 767–808. [CrossRef]

69. Gao, G.; Kang, H.S. Engineering of the electronic structures of metal-porphyrin tapes and metal-hexaphyrin tapes: A first-principles study. *Chem. Phys.* **2010**, *369*, 66. [CrossRef]

70. Medforth, C.J.; Senge, M.O.; Smith, K.M.; Sparks, L.D.; Shelnutt, J.A. Nonplanar distortion modes for highly substituted porphyrins. *J. Am. Chem. Soc.* **1992**, *114*, 9859–9869. [CrossRef]

71. Otsuki, J. STM studies on porphyrins. *Coord. Chem. Rev.* **2010**, *254*, 2311–2341. [CrossRef]

72. Zotti, L.A.; Teobaldi, G.; Hofer, W.A.; Auwärter, W.; Weber-Bargioni, A.; Barth, J.V. Ab-initio calculations and STM observations on tetrapyridyl and Fe(II)-tetrapyridyl-porphyrin molecules on Ag (1 1 1). *Surf. Sci.* **2007**, *601*, 2409–2414. [CrossRef]

73. Chen, X.; Lei, S.; Lotze, C.; Czekelius, C.; Paulus, B.; Franke, K.J. Conformational adaptation and manipulation of manganese tetra(4-pyridyl)porphyrin molecules on Cu(111). *J. Chem. Phys.* **2017**, *146*, 092316. [CrossRef]

74. Zhang, Q.; Zheng, X.; Kuang, G.; Wang, W.; Zhu, L.; Pang, R.; Shi, X.; Shang, X.; Huang, X.; Liu, P.N.; et al. Single-Molecule Investigations of Conformation Adaptation of Porphyrins on Surfaces. *J. Phys. Chem. Lett.* **2017**, *8*, 1241–1247. [CrossRef]

75. Rojas, G.; Chen, X.; Bravo, C.; Kim, J.-H.; Kim, J.-S.; Xiao, J.; Dowben, P.A.; Gao, Y.; Zeng, X.C.; Choe, W.; et al. Self-Assembly and Properties of Nonmetalated Tetraphenyl-Porphyrin on Metal Substrates. *J. Phys. Chem. C* **2010**, *114*, 9408–9415. [CrossRef]

76. Hamprecht, F.A.; Cohen, A.J.; Tozer, D.J.; Handy, N.C. Development and assessment of new exchange-correlation functionals. *J. Chem. Phys.* **1998**, *109*, 6264–6271. [CrossRef]

77. Boese, A.D.; Doltsinis, N.L.; Handy, N.C.; Sprik, M. New generalized gradient approximation functionals. *J. Chem. Phys.* **2000**, *112*, 1670–1678. [CrossRef]

78. Lepper, M.; Köbl, J.; Schmitt, T.; Gurrath, M.; Siervo, A.d.; Schneider, M.A.; Steinrück, H.; Meyer, B.; Marbach, H.; Hieringer, W. "Inverted" porphyrins: A distorted adsorption geometry of free-base porphyrins on Cu (111). *Chem. Commun.* **2017**, *53*, 8207–8210. [CrossRef] [PubMed]

79. Houwaart, T.; Le Bahers, T.; Sautet, P.; Auwarter, W.; Seufert, K.; Barth, J.V.; Bocquet, M. Scrutinizing individual CoTPP molecule adsorbed on coinage metal surfaces from the interplay of STM experiment and theory. *Surf. Sci.* **2015**, *635*, 108–114. [CrossRef]

80. Weber-Bargioni, A.; Auwärter, W.; Klappenberger, F.; Reichert, J.; Lefrançois, S.; Strunskus, T.; Wöll, C.; Schiffrin, A.; Pennec, Y.; Barth, J.V. Visualizing the Frontier Orbitals of a Conformationally Adapted Metalloporphyrin. *ChemPhysChem* **2008**, *9*, 89–94. [CrossRef]

81. Donovan, P.; Robin, A.; Dyer, M.S.; Persson, M.; Raval, R. Unexpected Deformations Induced by Surface Interaction and Chiral Self-Assembly of CoII-Tetraphenylporphyrin (Co-TPP) Adsorbed on Cu (110): A Combined STM and Periodic DFT Study. *Chem. A Eur. J.* **2010**, *16*, 11641–11652. [CrossRef]

82. Auwärter, W.; Seufert, K.; Klappenberger, F.; Reichert, J.; Weber-Bargioni, A.; Verdini, A.; Cvetko, D.; Dell'Angela, M.; Floreano, L.; Cossaro, A.; et al. Site-specific electronic and geometric interface structure of Co-tetraphenyl-porphyrin layers on Ag(111). *Phys. Rev. B* **2010**, *81*, 245403. [CrossRef]

83. Fatayer, S.; Veiga, R.G.A.; Prieto, M.J.; Perim, E.; Landers, R.; Miwa, R.H.; De Siervo, A. Self-assembly of NiTPP on Cu(111): A transition from disordered 1D wires to 2D chiral domains. *Phys. Chem. Chem. Phys.* **2015**, *17*, 18344–18352. [CrossRef]

84. Hötger, D.; Abufager, P.F.; Morchutt, C.; Alexa, P.; Grumelli, D.E.; Dreiser, J.; Stepanow, S.; Gambardella, P.; Busnengo, H.F.; Etzkorn, M.; et al. On-surface transmetalation of metalloporphyrins. *Nanoscale* **2018**, *10*, 21116–21122. [CrossRef]

85. Grimme, S.; Antony, J.; Ehrlich, S.; Krieg, H. A consistent and accurate ab initio parametrization of density functional dispersion correction (DFT-D) for the 94 elements H-Pu. *J. Chem. Phys.* **2010**, *132*, 154104. [CrossRef]

86. Moreno-López, J.C.; Mowbray, D.J.; Paz, A.P.; Ferreira, R.C.D.C.; Dos Santos, A.C.; Ayala, P.; De Siervo, A. Roles of Precursor Conformation and Adatoms in Ullmann Coupling: An Inverted Porphyrin on Cu(111). *Chem. Mater.* **2019**, *31*, 3009–3017. [CrossRef]

87. Mielke, J.; Hanke, F.; Peters, M.V.; Hecht, S.; Persson, M.; Grill, L. Adatoms underneath Single Porphyrin Molecules on Au(111). *J. Am. Chem. Soc.* **2015**, *137*, 1844–1849. [CrossRef] [PubMed]

88. Bassiouk, M.; Alvarez-Zauco, E.; Basiuk, V.A. Theoretical analysis of the effect of surface defects on porphyrin adsorption and self-assembly on graphite. *J. Comput. Nanosci.* **2012**, *9*, 532–540. [CrossRef]

89. Catellani, A.; Calzolari, A. Functionalization of SiC(110) Surfaces via Porphyrin Adsorption: Ab Initio Results. *J. Phys. Chem. C* **2012**, *116*, 886–892. [CrossRef]

90. El Garah, M.; Makoudi, Y.; Palmino, F.; Duverger, E.; Sonnet, P.; Chaput, L.; Gourdon, A.; Cherioux, F. STM and DFT Investigations of Isolated Porphyrin on a Silicon-Based Semiconductor at Room Temperature. *ChemPhysChem* **2009**, *10*, 3190–3193. [CrossRef]

91. Boukari, K.; Sonnet, P.; Duverger, E. DFT-D Studies of Single Porphyrin Molecule on Doped Boron Silicon Surfaces. *ChemPhysChem* **2012**, *13*, 3945–3951. [CrossRef]

92. Chin, Y.; Panduwinata, D.; Sintic, M.; Sum, T.J.; Hush, N.S.; Crossley, M.J.; Reimers, J.R. Atomic-Resolution Kinked Structure of an Alkylporphyrin on Highly Ordered Pyrolytic Graphite. *J. Phys. Chem. Lett.* **2011**, *2*, 62–66. [CrossRef]

93. Dapprich, S.; Komáromi, I.; Byun, K.S.; Morokuma, K.; Frisch, M.J. A new ONIOM implementation in Gaussian98. Part I. The calculation of energies, gradients, vibrational frequencies and electric field derivatives1Dedicated to Professor Keiji Morokuma in celebration of his 65th birthday.1. *J. Mol. Struct.* **1999**, *461*, 1–21. [CrossRef]

94. Reimers, J.R.; Panduwinata, D.; Visser, J.; Chin, Y.; Tang, C.; Goerigk, L.; Ford, M.J.; Baker, M.; Sum, T.J.; Coenen, M.J.J.; et al. From Chaos to Order: Chain-Length Dependence of the Free Energy of Formation of Meso-tetraalkylporphyrin Self-Assembled Monolayer Polymorphs. *J. Phys. Chem. C* **2016**, *120*, 1739–1748. [CrossRef]

95. Fanetti, M.; Calzolari, A.; Vilmercati, P.; Castellarin-Cudia, C.; Borghetti, P.; Di Santo, G.; Floreano, L.; Verdini, A.; Cossaro, A.; Vobornik, I.; et al. Structure and Molecule–Substrate Interaction in a Co-octaethyl Porphyrin Monolayer on the Ag(110) Surface. *J. Phys. Chem. C* **2011**, *115*, 11560–11568. [CrossRef]

96. Chilukuri, B.; Mazur, U.; Hipps, K.W. Effect of dispersion on surface interactions of cobalt(II) octaethylporphyrin monolayer on Au(111) and HOPG(0001) substrates: A comparative first principles study. *Phys. Chem. Chem. Phys.* **2014**, *16*, 14096–14107. [CrossRef] [PubMed]

97. Kohn, W.; Sham, L.J. Self-Consistent Equations Including Exchange and Correlation Effects. *Phys. Rev.* **1965**, *140*, A1133–A1138. [CrossRef]

98. Tada, K.; Maeda, Y.; Ozaki, H.; Tanaka, S.; Yamazaki, S. Theoretical investigation on the interaction between RhIII octaethylporphyrin and a graphite basal surface: A comparison study of DFT, DFT-D, and AFM. *Phys. Chem. Chem. Phys.* **2018**, *20*, 20235–20246. [CrossRef] [PubMed]

99. Sena, A.M.P.; Brazdova, V.; Bowler, D.R. Density functional theory study of the iron-based porphyrin haem(b) on the Si(111):H surface. *Phys. Rev. B* **2009**, *79*, 245404/1–245404/7. [CrossRef]

100. Tersoff, J.; Hamann, D.R. Theory of the scanning tunneling microscope. *Phys. Rev. B* **1985**, *31*, 805–813. [CrossRef] [PubMed]

101. Hanke, F.; Haq, S.; Raval, R.; Persson, M. Heat-to-connect: Surface commensurability directs organometallic one-dimensional self-assembly. *ACS Nano* **2011**, *5*, 9093–9103. [CrossRef]

102. Dyer, M.S.; Robin, A.; Haq, S.; Raval, R.; Persson, M.; Klimeš, J. Understanding the Interaction of the Porphyrin Macrocycle to Reactive Metal Substrates: Structure, Bonding, and Adatom Capture. *ACS Nano* **2011**, *5*, 1831–1838. [CrossRef]

103. Miller, D.P.; Hooper, J.; Simpson, S.; Costa, P.S.; Tyminska, N.; McDonnell, S.M.; Bennett, J.A.; Enders, A.; Zurek, E. Electronic Structure of Iron Porphyrin Adsorbed to the Pt(111) Surface. *J. Phys. Chem. C* **2016**, *120*, 29173–29181. [CrossRef]

104. Leung, K.; Rempe, S.B.; Schultz, P.A.; Sproviero, E.M.; Batista, V.S.; Chandross, M.E.; Medforth, C.J. Density Functional Theory and DFT+U Study of Transition Metal Porphines Adsorbed on Au(111) Surfaces and Effects of Applied Electric Fields. *J. Am. Chem. Soc.* **2006**, *128*, 3659–3668. [CrossRef]

105. Berland, K.; Hyldgaard, P. Exchange functional that tests the robustness of the plasmon description of the van der Waals density functional. *Phys. Rev. B* **2014**, *89*, 035412. [CrossRef]

106. Nandi, G.; Chilukuri, B.; Hipps, K.W.; Mazur, U. Surface directed reversible imidazole ligation to nickel (II) octaethylporphyrin at the solution/solid interface: A single molecule level study. *Phys. Chem. Chem. Phys.* **2016**, *18*, 20819–20829. [CrossRef] [PubMed]

107. Gurdal, Y.; Hutter, J.; Iannuzzi, M. Insight into (Co)Pyrphyrin Adsorption on Au(111): Effects of Herringbone Reconstruction and Dynamics of Metalation. *J. Phys. Chem. C* **2017**, *121*, 11416–11427. [CrossRef]

108. Gurdal, Y. Theoretical investigation of metalated and unmetalated pyrphyrins immobilized on Ag(111) surface. *J. Incl. Phenom. Macrocycl. Chem.* **2019**, *95*, 273–283. [CrossRef]

109. Jahanbekam, A.; Chilukuri, B.; Mazur, U.; Hipps, K.W. Kinetically Trapped Two-Component Self-Assembled Adlayer. *J. Phys. Chem. C* **2015**, *119*, 25364–25376. [CrossRef]

110. Chilukuri, B.; Mazur, U.; Hipps, K.W. Cooperativity and coverage dependent molecular desorption in self-assembled monolayers: Computational case study with coronene on Au (111) and HOPG. *Phys. Chem. Chem. Phys.* **2019**, *21*, 10505–10513. [CrossRef]

111. Touzeau, J.; Barbault, F.; Maurel, F.; Seydou, M. Insights on porphyrin-functionalized graphene: Theoretical study of substituent and metal-center effects on adsorption. *Chem. Phys. Lett.* **2018**, *713*, 172–179. [CrossRef]

112. Zeng, J.; Chen, K. A nearly perfect spin filter and a spin logic gate based on a porphyrin/graphene hybrid material. *Phys. Chem. Chem. Phys.* **2018**, *20*, 3997–4004. [CrossRef]

113. Meir, Y.; Wingreen, N.S. Landauer formula for the current through an interacting electron region. *Phys. Rev. Lett* **1992**, *68*, 2512–2515. [CrossRef]

114. Zhao, J.; Ding, Y. Functionalization of Single-Walled Carbon Nanotubes with Metalloporphyrin Complexes: A Theoretical Study. *J. Phys. Chem. C* **2008**, *112*, 11130–11134. [CrossRef]

115. Correa, J.D.; Orellana, W. Optical response of carbon nanotubes functionalized with (free-base, Zn) porphyrins, and phthalocyanines: A DFT study. *Phys. Rev. B* **2012**, *86*, 125417. [CrossRef]

116. Ruiz-Tagle, I.; Orellana, W. Iron porphyrin attached to single-walled carbon nanotubes: Electronic and dynamical properties from ab-initio calculations. *Phys. Rev. B* **2010**, *82*, 115406. [CrossRef]

117. Zhao, J.; Ding, Y. Theoretical studies of chemical functionalization of the (8,0) boron nitride nanotube with various metalloporphyrin MP (M=Fe, Co, Ni, Cu, and Zn) complexes. *Mater. Chem. Phys.* **2009**, *116*, 21–27. [CrossRef]

118. Gomez, T.; Zarate, X.; Schott, E.; Arratia-Perez, R. Role of the main adsorption modes in the interaction of the dye [COOH–TPP-Zn(II)] on a periodic TiO2 slab exposing a rutile (110) surface in a dye-sensitized solar cell. *RSC Adv.* **2014**, *4*, 9639–9646. [CrossRef]

119. Lin, Y.; Zhu, C.; Jiang, Z.; Zhao, Y.; Wang, Q.; Zhang, R.; Lin, S.H. Enhanced photovoltaic performance of dye-sensitized solar cells by the adsorption of Zn-porphyrin dye molecule on TiO2 surfaces. *J. Alloy. Compd.* **2019**, *794*, 35–44. [CrossRef]

120. Lovat, G.; Forrer, D.; Abadia, M.; Dominguez, M.; Casarin, M.; Rogero, C.; Vittadini, A.; Floreano, L. Hydrogen capture by porphyrins at the TiO2(110) surface. *Phys. Chem. Chem. Phys.* **2015**, *17*, 30119–30124. [CrossRef] [PubMed]

121. Wang, C.; Fan, Q.; Han, Y.; Martínez, J.I.; Martín-Gago, J.A.; Wang, W.; Ju, H.; Gottfried, J.M.; Zhu, J. Metalation of tetraphenylporphyrin with nickel on a TiO2(110)-1 × 2 surface. *Nanoscale* **2015**, *8*, 1123–1132. [CrossRef]

122. Lewis, J.P.; Jelinek, P.; Ortega, J.; Demkov, A.A.; Trabada, D.G.; Haycock, B.; Wang, H.; Adams, G.; Tomfohr, J.K.; Abad, E.; et al. Advances and applications in the FIREBALL ab initio tight-binding molecular-dynamics formalism. *Phys. Status Solidi* **2011**, *248*, 1989–2007. [CrossRef]

123. Spoerke, E.D.; Small, L.J.; Foster, M.E.; Wheeler, J.; Ullman, A.M.; Stavila, V.; Rodriguez, M.; Allendorf, M.D. MOF-Sensitized Solar Cells Enabled by a Pillared Porphyrin Framework. *J. Phys. Chem. C* **2017**, *121*, 4816–4824. [CrossRef]

124. Xie, M.; Bai, F.; Wang, J.; Zheng, Y.; Lin, Z. Theoretical investigations on the unsymmetrical effect of β-link Zn-porphyrin sensitizers on the performance for dye-sensitized solar cells. *Phys. Chem. Chem. Phys.* **2018**, *20*, 3741–3751. [CrossRef]

125. Niskanen, M.; Kuisma, M.; Cramariuc, O.; Golovanov, V.; Hukka, T.I.; Tkachenko, N.; Rantala, T.T. Porphyrin adsorbed on the (1010) surface of the wurtzite structure of ZnO—conformation induced effects on the electron transfer characteristics. *Phys. Chem. Chem. Phys.* **2013**, *15*, 17408–17418. [CrossRef]

126. Torres, A.; Amaya Suarez, J.; Remesal, E.R.; Marquez, A.M.; Fernandez Sanz, J.; Rincon Canibano, C. Adsorption of Prototypical Asphaltenes on Silica: First-Principles DFT Simulations Including Dispersion Corrections. *J. Phys. Chem. B* **2017**. Ahead of Print. [CrossRef] [PubMed]

127. Quinn, T.; Choudhury, P. Direct oxidation of methane to methanol on single-site copper-oxo species of copper porphyrin functionalized graphene: A DFT study. *Mol. Catal.* **2017**, *431*, 9–14. [CrossRef]

128. Henkelman, G.; Uberuaga, B.P.; Jónsson, H. A climbing image nudged elastic band method for finding saddle points and minimum energy paths. *J. Chem. Phys.* **2000**, *113*, 9901–9904. [CrossRef]

129. Sheppard, D.; Terrell, R.; Henkelman, G. Optimization methods for finding minimum energy paths. *J. Chem. Phys.* **2008**, *128*, 134106. [CrossRef]

130. Henkelman, G.; Arnaldsson, A.; Jónsson, H. A fast and robust algorithm for Bader decomposition of charge density. *Comput. Mater. Sci.* **2006**, *36*, 354–360. [CrossRef]

131. Grill, L.; Dyer, M.; Lafferentz, L.; Persson, M.; Peters, M.V.; Hecht, S. Nano-architectures by covalent assembly of molecular building blocks. *Nat. Nanotechnol.* **2007**, *2*, 687–691. [CrossRef]

132. Shi, K.; Shu, C.; Wang, C.; Wu, X.; Tian, H.; Liu, P. On-Surface Heck Reaction of Aryl Bromides with Alkene on Au(111) with Palladium as Catalyst. *Org. Lett.* **2017**, *19*, 2801–2804. [CrossRef]

133. Shu, C.; Xie, Y.; Wang, A.; Shi, K.; Zhang, W.; Li, D.; Liu, P. On-surface reactions of aryl chloride and porphyrin macrocycles via merging two reactive sites into a single precursor. *Chem. Commun.* **2018**, *54*, 12626–12629. [CrossRef]

134. Wende, H.; Bernien, M.; Luo, J.; Sorg, C.; Ponpandian, N.; Kurde, J.; Miguel, J.; Piantek, M.; Xu, X.; Eckhold, P.; et al. Substrate-induced magnetic ordering and switching of iron porphyrin molecules. *Nat. Mater.* **2007**, *6*, 516–520. [CrossRef]

135. Oppeneer, P.M.; Panchmatia, P.M.; Sanyal, B.; Eriksson, O.; Ali, M.E. Nature of the magnetic interaction between Fe-porphyrin molecules and ferromagnetic surfaces. *Prog. Surf. Sci.* **2009**, *84*, 18–29. [CrossRef]

136. Weber, A.P.; Caruso, A.N.; Vescovo, E.; Ali, M.E.; Tarafder, K.; Janjua, S.Z.; Sadowski, J.T.; Oppeneer, P.M. Magnetic coupling of Fe-porphyrin molecules adsorbed on clean and c (2 × 2) oxygen-reconstructed Co(100) investigated by spin-polarized photoemission spectroscopy. *Phys. Rev. B* **2013**, *87*, 184411. [CrossRef]

137. Ali, M.E.; Sanyal, B.; Oppeneer, P.M. Tuning the Magnetic Interaction between Manganese Porphyrins and Ferromagnetic Co Substrate through Dedicated Control of the Adsorption. *J. Phys. Chem. C* **2009**, *113*, 14381–14383. [CrossRef]

138. Chylarecka, D.; Kim, T.K.; Tarafder, K.; Müller, K.; Gödel, K.; Czekaj, I.; Wäckerlin, C.; Cinchetti, M.; Ali, M.E.; Piamonteze, C.; et al. Indirect Magnetic Coupling of Manganese Porphyrin to a Ferromagnetic Cobalt Substrate. *J. Phys. Chem. C* **2011**, *115*, 1295–1301. [CrossRef]

139. Waeckerlin, C.; Maldonado, P.; Arnold, L.; Shchyrba, A.; Girovsky, J.; Nowakowski, J.; Ali, M.E.; Haehlen, T.; Baljozovic, M.; Siewert, D.; et al. Magnetic exchange coupling of a synthetic Co(II)-complex to a ferromagnetic Ni substrate. *Chem. Commun.* **2013**, *49*, 10736–10738. [CrossRef]

140. Hermanns, C.F.; Tarafder, K.; Bernien, M.; Kruger, A.; Chang, Y.; Oppeneer, P.M.; Kuch, W. Magnetic coupling of porphyrin molecules through graphene. *Adv. Mater.* **2013**, *25*, 3473–3477. [CrossRef]

141. Lamoen, N.; Ballone, N.; Parrinello, N. Electronic structure, screening, and charging effects at a metal/organic tunneling junction: A first-principles study. *Phys. Rev. B* **1996**, *54*, 5097–5105. [CrossRef]

142. Picozzi, S.; Pecchia, A.; Gheorghe, M.; Di Carlo, A.; Lugli, P.; Delley, B.; Elstner, M. Organic/metal interfaces: An ab initio study of their structural and electronic properties. *Surf. Sci.* **2004**, *566*, 628–632. [CrossRef]

143. Long, M.; Chen, K.; Wang, L.; Qing, W.; Zou, B.S.; Shuai, Z. Negative differential resistance behaviors in porphyrin molecular junctions modulated with side groups. *Appl. Phys. Lett.* **2008**, *92*, 243303. [CrossRef]

144. An, Y.; Yang, Z.; Ratner, M.A. High-efficiency switching effect in porphyrin-ethyne-benzene conjugates. *J. Chem. Phys.* **2011**, *135*, 044706. [CrossRef]

145. Liu, Z.; Wei, S.; Yoon, H.; Adak, O.; Ponce, I.; Jiang, Y.; Jang, W.; Campos, L.M.; Venkataraman, L.; Neaton, J.B. Control of Single-Molecule Junction Conductance of Porphyrins via a Transition-Metal Center. *Nano Lett.* **2014**, *14*, 5365–5370. [CrossRef]

146. Stein, T.; Eisenberg, H.; Kronik, L.; Baer, R. Fundamental Gaps in Finite Systems from Eigenvalues of a Generalized Kohn-Sham Method. *Phys. Rev. Lett.* **2010**, *105*, 266802. [CrossRef] [PubMed]

147. Quek, S.Y.; Venkataraman, L.; Choi, H.J.; Louie, S.G.; Hybertsen, M.S.; Neaton, J.B. Amine–Gold Linked Single-Molecule Circuits: Experiment and Theory. *Nano Lett.* **2007**, *7*, 3477–3482. [CrossRef] [PubMed]

148. Quek, S.Y.; Choi, H.J.; Louie, S.G.; Neaton, J.B. Length Dependence of Conductance in Aromatic Single-Molecule Junctions. *Nano Lett.* **2009**, *9*, 3949–3953. [CrossRef] [PubMed]

149. Cho, W.J.; Cho, Y.; Min, S.K.; Kim, W.Y.; Kim, K.S. Chromium Porphyrin Arrays as Spintronic Devices. *J. Am. Chem. Soc.* **2011**, *133*, 9364–9369. [CrossRef]

150. Sedghi, G.; García-Suárez, V.M.; Esdaile, L.J.; Anderson, H.L.; Lambert, C.J.; Martín, S.; Bethell, D.; Higgins, S.J.; Elliott, M.; Bennett, N.; et al. Long-range electron tunnelling in oligo-porphyrin molecular wires. *Nat. Nanotechnol.* **2011**, *6*, 517–523. [CrossRef]

151. García-Suárez, V.M.; Ferradás, R.; Carrascal, D.; Ferrer, J. Universality in the low-voltage transport response of molecular wires physisorbed onto graphene electrodes. *Phys. Rev. B* **2013**, *87*. [CrossRef]

152. Sanders, J.K.; Bampos, N.; Clyde-Watson, Z.; Darling, S.L.; Hawley, J.C.; Kim, H.J.; Webb, S.J. Axial Coordination Chemistry of Metalloporphyrins. In *The Porphyrin Handbook*; Kadish, K.M., Smith, K.M., Guilard, R., Eds.; Academic: San Diego, CA, USA, 2000; pp. 1–48.

153. Gottfried, J.M. Surface chemistry of porphyrins and phthalocyanines. *Surf. Sci. Rep.* **2015**, *70*, 259–379. [CrossRef]

154. Gottfried, J.M.; Marbach, H. Surface-Confined Coordination Chemistry with Porphyrins and Phthalocyanines: Aspects of Formation, Electronic Structure, and Reactivity. *Zeitschrift Fur Physikalische Chemie-Int. J. Res. Phys. Chem. Chem. Phys.* **2009**, *223*, 53–74. [CrossRef]

155. Hieringer, W.; Flechtner, K.; Kretschmann, A.; Seufert, K.; Auwärter, W.; Barth, J.V.; Görling, A.; Steinrück, H.; Gottfried, J.M. The surface trans effect: Influence of axial ligands on the surface chemical bonds of adsorbed metalloporphyrins. *J. Am. Chem. Soc.* **2011**, *133*, 6206–6222. [CrossRef]

156. Seufert, K.; Auwärter, W.; Barth, J.V. Discriminative Response of Surface-Confined Metalloporphyrin Molecules to Carbon and Nitrogen Monoxide. *J. Am. Chem. Soc.* **2010**, *132*, 18141–18146. [CrossRef]

157. Friesen, B.A.; Bhattarai, A.; Mazur, U.; Hipps, K.W. Single Molecule Imaging of Oxygenation of Cobalt Octaethylporphyrin at the Solution/Solid Interface: Thermodynamics from Microscopy. *J. Am. Chem. Soc.* **2012**, *134*, 14897–14904. [CrossRef] [PubMed]

158. Wäckerlin, C.; Tarafder, K.; Siewert, D.; Girovsky, J.; Hählen, T.; Iacovita, C.; Kleibert, A.; Nolting, F.; Jung, T.A.; Oppeneer, P.M.; et al. On-surface coordination chemistry of planar molecular spin systems: Novel magnetochemical effects induced by axial ligands. *Chem. Sci.* **2012**, *3*, 3154–3160. [CrossRef]

159. Janet, J.P.; Zhao, Q.; Ioannidis, E.I.; Kulik, H.J. Density functional theory for modelling large molecular adsorbate–surface interactions: A mini-review and worked example. *Mol. Simul.* **2017**, *43*, 327–345. [CrossRef]

160. Ghosh, D.; Parida, P.; Pati, S.K. Spin-state switching of manganese porphyrin by conformational modification. *J. Phys. Chem. C* **2016**, *120*, 3625–3634. [CrossRef]

161. Nguyen, T.Q.; Aspera, S.M.; Nakanishi, H.; Kasai, H. NO adsorption effects on various functional molecular nanowires. *Comput. Mater. Sci.* **2009**, *47*, 111–120. [CrossRef]

162. Nguyen, T.Q.; Escaño, M.C.S.; Shimoji, N.; Nakanishi, H.; Kasai, H. Adsorption of diatomic molecules on iron tape-porphyrin: A comparative study. *Phys. Rev. B* **2008**, *77*, 195307. [CrossRef]

163. Nguyen, T.Q.; Escaño, M.C.S.; Shimoji, N.; Nakanishi, H.; Kasai, H. DFT study on the adsorption of NO on iron tape-porphyrin. *Surf. Interface Anal.* **2008**, *40*, 1082–1084. [CrossRef]

Review

Understanding Electronic Structure and Chemical Reactivity: Quantum-Information Perspective

Roman F. Nalewajski

Department of Theoretical Chemistry, Jagiellonian University, Gronostajowa 2, 30-387 Kraków, Poland;
nalewajs@chemia.uj.edu.pl

Received: 18 February 2019; Accepted: 22 March 2019; Published: 26 March 2019

Abstract: Several applications of quantum mechanics and information theory to chemical reactivity problems are presented with emphasis on equivalence of variational principles for the constrained minima of the system electronic energy and its kinetic energy component, which also determines the overall gradient information. Continuities of molecular probability and current distributions, reflecting the modulus and phase components of molecular wavefunctions, respectively, are summarized. Resultant measures of the entropy/information descriptors of electronic states, combining the classical (probability) and nonclassical (phase/current) contributions, are introduced, and information production in quantum states is shown to be of a nonclassical origin. Importance of resultant information descriptors for distinguishing the bonded (entangled) and nonbonded (disentangled) states of reactants in acid(A)–base(B) systems is stressed and generalized entropy concepts are used to determine the phase equilibria in molecular systems. The grand-canonical principles for the minima of electronic energy and overall gradient information allow one to explore relations between energetic and information criteria of chemical reactivity in open molecules. The populational derivatives of electronic energy and resultant gradient information give identical predictions of electronic flows between reactants. The role of electronic kinetic energy (resultant gradient information) in chemical-bond formation is examined, the virial theorem implications for the Hammond postulate of reactivity theory are explored, and changes of the overall structure information in chemical processes are addressed. The frontier-electron basis of the hard (soft) acids and bases (HSAB) principle is reexamined and covalent/ionic characters of the intra- and inter-reactant communications in donor-acceptor systems are explored. The complementary A–B coordination is compared with its regional HSAB analog, and polarizational/relaxational flows in such reactive systems are explored.

Keywords: chemical reactivity theory; HSAB principle; information theory; quantum mechanics; regional complementarity rule; virial theorem

---

1. Introduction

The quantum mechanics (QM) and information theory (IT) establish a solid basis for both determining the electronic structure of molecules and understanding, in chemical terms, general trends in their chemical behavior. The energy principle of QM has been recently interpreted [1–3] as equivalent variational rule for the overall content of the gradient information in the system electronic wavefunction, proportional to the state average kinetic energy. In the grand-ensemble representation of thermodynamic (mixed) states they both determine the same equilibrium of an externally open molecular system. This equivalence parallels identical predictions resulting from the minimum-energy and maximum-entropy principles in ordinary thermodynamics [4].

The generalized, Fisher-type gradient information in the specified electronic state is proportional to the system average kinetic energy. This allows one to interpret the variational principle for electronic energy as equivalent information rule. The energy and resultant-information/kinetic-energy rules thus represent equivalent sources of reactivity criteria, the populational derivatives of ensemble-average values of electronic energy or overall information, e.g., the system chemical potential (negative electronegativity) and hardness/softness descriptors. The IT transcription of the variational principle for the minimum of electronic energy allows one to interpret the familiar (energetical) criteria of chemical reactivity, the populational derivatives of electronic energy, in terms of the corresponding derivatives of the state-resultant information content. The latter combines the classical (probability) and nonclassical (current) contributions to the state kinetic energy of electrons, generated by the modulus and phase components of molecular wavefunctions, respectively. This proportionality between the state resultant gradient information and its kinetic energy also allows one to use the molecular virial theorem [5] in general reactivity considerations [1–3].

The resultant measures combining the probability and phase/current contributions allow one to distinguish the information content of states generating the same electron density but differing in their phase/current composition. To paraphrase Prigogine [6], the electron density alone reflects only the molecular static structure of "being", missing the dynamic structure of "becoming" contained in the state current distribution. Both these manifestations of electronic "organization" in molecules ultimately contribute to resultant IT descriptors of the structural entropy/information content in generally complex electronic wavefunctions [7–10]. In quantum information theory (QIT) [7], the classical information contribution probes the entropic content of incoherent (disentangled) local "events" while its nonclassical supplement provides the information complement due to their coherence (entanglement).

The classical IT [11–18] has been already successfully applied to interpret the molecular probability distributions, e.g., References [19–22]. Information principles have been explored [1–3,23–28] and density pieces attributed to atoms-in-molecules (AIM) have been approached [22,26–30] providing the information basis of the intuitive (stockholder) division of Hirshfeld [31]. Patterns of chemical bonds have been extracted from electronic communications in molecules [7,19–21,32–42] and entropy/information distributions in molecules have been explored [7,19–21,43,44]. The nonadditive Fisher information [7,19–21,45,46] has been linked to electron localization function (ELF) [47–49] of modern density functional theory (DFT) [50–55]. This analysis has also formulated the contragradience (CG) probe [7,19–21,56] for localizaing chemical bonds and the orbital communication theory (OCT) of the chemical bond has identified the bridge bonds originating from the cascade propagations of information between AIM, which involve intermediate orbitals [7,21,57–62].

In entropic theories of molecular electronic structure, one ultimately requires the quantum (resultant) extensions of the familiar complementary measures of Fisher [11] and Shannon [13], of the information and entropy content in probability distributions, which are appropriate for the complex probability amplitudes (wavefunctions) of molecular QM. The wavefunction phase, or its gradient determining the current density and the associated velocity field, gives rise to nonclassical supplements in resultant measures of an overall entropic content of molecular states [7,63–68]. The information distinction between the bonded (entangled) and nonbonded (disentangled) states of subsystems, e.g., molecular substrates of a chemical reaction, also calls for such generalized information descriptors [69–72]. The extremum principles for the global and local measures of the resultant entropy have been used to determine the phase-equilibrium states of molecular systems, identified by their optimum (local, probability-dependent) "thermodynamic" phase.

Various DFT-based approaches to classical issues in reactivity theory [73–79] use the energy-centered arguments in justifying the observed reaction paths and relative yields of their products. Qualitative considerations on preferences in chemical reactions usually emphasize changes in energies of both reactants and of the whole reactive system, which are induced by displacements (perturbations) in parameters describing the relevant (real or hypothetical) electronic state. In such classical treatments, also covering the linear responses to these primary shifts, one also explores reactivity implications of the electronic equilibrium and stability criteria. For example, in charge sensitivity analysis (CSA) [73,74] the "principal" (energy) derivatives with respect to the system external potential (v), due to the fixed nuclei defining molecular geometry, and its overall number of electrons (N), as well as the associated charge responses of both the whole reactive system and its constituent subsystems, have been used as reactivity criteria. In R = acid(A)$\leftarrow$ base(B) $\equiv$ A–B complexes, consisting of the coordinated electron-acceptor and electron-donor reactants, respectively, such responses can be subsequently combined into the corresponding in situ descriptors characterizing the B$\rightarrow$A charge transfer (CT).

We begin this overview with a summary of the probability and current distributions, the physical attributes reflecting the modulus and phase components of quantum states, and an introduction to the resultant QIT descriptors. The phase equilibria, representing extrema of the overall entropy measures, will be explored and molecular orbital (MO) contributions to the overall gradient-information measure will be examined. Using the molecular virial theorem, the role of electronic kinetic energy, also reflecting the system resultant information, in shaping the electronic structure of molecules, will be examined. This analysis of the theorem implications will cover the bond-formation process and the qualitative Hammond [80] postulate of reactivity theory. The hypothetical stages of chemical reactions invoked in reactivity theory will be explored and the in situ populational derivatives will be applied to determine the optimum amount of CT in donor-acceptor coordinations. Populational derivatives of the resultant gradient information will be advocated as alternative indices of chemical reactivity, related to their energetical analogs. They will be shown to be capable of predicting both the direction and magnitude of electron flows in A–B systems. The frontier-electron (FE) [81–83] framework for describing molecular interactions will be used to reexamine Pearson's [84] hard (soft) acids and bases (HSAB) principle of structural chemistry (see also Reference [85]) and electron communications between reactants will be commented upon. The ionic and covalent interactions between the "frontier" MO will be invoked to fully explain the HSAB stability predictions, and the "complementary" A–B complex will be compared with its regional-HSAB analog. The complementary preference will be explained by examining physical implications of the polarizational and relaxational flows in these alternative reactive complexes. In appendices, the continuity relations for the probability and phase distributions of molecular electronic states resulting from the Schrödinger equation (SE) of QM will be summarized, the dynamics of resultant gradient information will be addressed, the nonclassical origin of the overall gradient-information production will be demonstrated, and the grand-ensemble representation of open molecular systems will be outlined.

2. Physical Attributes of Quantum States and Generalized Information Descriptors

The electronic wavefunctions of molecules are determined by SE of QM. This fundamental equation also determines the dynamics of the modulus (probability) and phase (current) attributes of such elementary quantum states. In a discussion of "productions" of the resultant entropy/information quantities [7,69,72], it is of interest to examine implications of SE for the dynamics of these fundamental physical distributions of quantum states. For simplicity, let us first consider a single electron at time t in state $|\psi(t)\rangle \equiv |\psi(t)\rangle \equiv |\psi\rangle$, described by the (complex) wavefunction in position representation,

$$\psi(\mathbf{r}, t) = \langle \mathbf{r} \,|\, \psi(t)\rangle = R(\mathbf{r}, t)\exp[i\phi(\mathbf{r}, t)] \equiv R(t)\exp[i\phi(t)] \equiv \psi(t), \tag{1}$$

where the real functions $R(\mathbf{r}, t) \equiv R(t)$ and $\phi(\mathbf{r}, t) \equiv \phi(t)$ stand for its modulus and phase parts, respectively. It determines the state probability distribution at the specified time t,

$$p(\mathbf{r}, t) = \langle \psi(t) \,|\, \mathbf{r}\rangle\langle \mathbf{r} \,|\, \psi(t)\rangle = \psi(\mathbf{r}, t)^{*}\psi(\mathbf{r}, t) = R(t)^{2} \equiv p(t), \tag{2}$$

and its current density

$$j(\mathbf{r}, t) = [\hbar/(2mi)]\,[\psi(\mathbf{r}, t)^{*}\,\nabla\psi(\mathbf{r}, t) - \psi(\mathbf{r}, t)\,\nabla\psi(\mathbf{r}, t)^{*}] = (\hbar/m)\,\mathrm{Im}[\psi(\mathbf{r}, t)^{*}\,\nabla\psi(\mathbf{r}, t)]$$
$$= (\hbar/m)\,p(\mathbf{r}, t)\,\nabla\phi(\mathbf{r}, t) \equiv p(t)\,V(t) \equiv j(t). \tag{3}$$

The effective velocity field $V(r, t) = j(t)/p(t) \equiv V(t)$ of the probability "fluid" measures the local current-per-particle and reflects the state phase gradient:

$$V(t) = j(t)/p(t) = (\hbar/m)\,\nabla\phi(t). \tag{4}$$

The wavefunction modulus, the classical amplitude of the particle probability density, and the state phase, or its gradient determining the effective velocity of the probability flux, thus constitute two physical degrees-of-freedom in the full IT treatment of quantum states of a monoelectronic system:

$$\psi \Leftrightarrow (R, \phi) \Leftrightarrow (p, j). \tag{5}$$

One envisages the electron moving in the external potential $v(r)$, due to the "frozen" nuclei of the Born–Oppenheimer (BO) approximation determining the system geometry, described by the electronic Hamiltonian

$$\hat{H}(r) = -(h^2/2m)\nabla^2 + v(r) \equiv \hat{T}(r) + v(r), \tag{6}$$

where $\hat{T}(r)$ denotes its kinetic part. The quantum dynamics of a general electronic state of Equation (1) is generated by SE

$$\partial\psi(t)/\partial t = (ih)^{-1}H\psi(t), \tag{7}$$

which also determines temporal evolutions of the state physical distributions: The (instantaneous) probability density $p(t)$ and (local) phase $\phi(t)$ or its gradient reflecting the velocity field $V(t)$, the current-per-particle of the probability "fluid" [7,69,72]. The relevant continuity relations resulting from SE are summarized in Appendix A.

To simplify the notation for the the specified time $t = t_0$, let us suppress this parameter in the list of state arguments, e.g., $\psi(r, t_0) \equiv \psi(r) = \langle r\,|\,\psi \rangle$, etc. We, again, examine the mono-electron system in (pure) quantum state $|\,\psi \rangle$. The average Fisher's measure [11,12] of the classical gradient information for locality events, called the intrinsic accuracy, which is contained in the molecular probability density $p(r) = R(r)^2$ is reminiscent of von Weizsäcker's [86] inhomogeneity correction to the density functional for electronic kinetic energy:

$$I[p] = \int p(r)\,[\nabla \ln p(r)]^2 dr = \langle \psi\,|\,(\nabla \ln p)^2\,|\,\psi \rangle = \int [\nabla p(r)]^2/p(r)\,dr = 4\int [\nabla R(r)]^2\,dr \equiv I[R]. \tag{8}$$

This local measure characterizes an effective "narrowness" of the particle spatial probability distribution, i.e., a degree of the particle position determinicity. It represents the complementary measure to the global entropy of Shannon [13,14], the position-uncertainty index,

$$S[p] = -\int p(r)\,\ln p(r)\,dr = -\langle \psi\,|\,\ln p\,|\,\psi \rangle = -2\int R(r)^2\,\ln R(r)\,dr \equiv S[R], \tag{9}$$

which reflects the particle position indeterminicity, a "spread" of probability distribution. This classical descriptor also measures the amount of information received when the uncertainty about particle's location is removed by an appropriate experiment: $I_S[p] \equiv S[p]$.

In QM, these classical measures can be supplemented by the associated nonclassical contributions in the corresponding resultant QIT descriptors [7,45,63–67]. The intrinsic accuracy concept then naturally generalizes into the associated overall descriptor, a functional of the quantum state $|\,\psi \rangle$ itself. This generalized Fisher-type measure is defined by the expectation value of the Hermitian operator $\hat{I}(r)$ [7,45] of the overall gradient information,

$$\hat{I}(r) = -4\Delta = (2i\nabla)^2 = (8m/h^2)\hat{T}(r), \tag{10}$$

related to kinetic energy operator $\hat{T}(r)$ of Equation (6). Using integration by parts then gives:

$$I[\psi] = \langle\psi|\hat{\mathrm{I}}|\psi\rangle = -4\int \psi(\mathbf{r})^*\Delta\psi(\mathbf{r})d\mathbf{r} = 4\int |\nabla\psi(\mathbf{r})|^2 d\mathbf{r}$$
$$= I[p] + 4\int p(\mathbf{r})[\nabla\phi(\mathbf{r})]^2 d\mathbf{r} \equiv \int p(\mathbf{r})[I_p(\mathbf{r}) + I_\phi(\mathbf{r})]d\mathbf{r} \equiv I[p] + I[\phi] \equiv I[p,\phi] \tag{11}$$
$$= I[p] + (2m/h)^2\int p(\mathbf{r})^{-1}j(\mathbf{r})^2 d\mathbf{r} \equiv I[p] + I[j] \equiv I[p,j].$$

The classical and nonclassical densities-per-electron of this information functional read:

$$I_p(\mathbf{r}) = [\nabla p(\mathbf{r})/p(\mathbf{r})]^2 \quad \text{and} \quad I_\phi(\mathbf{r}) = 4[\nabla\phi(\mathbf{r})]^2. \tag{12}$$

The quantum-information concept $I[\psi] = I[p, \phi] = I[p, j]$ thus combines the classical (probability) contribution $I[p]$ of Fisher and the nonclassical (phase/current) supplement $I[\phi] = I[j]$. The positive sign of the latter expresses the fact that a nonvanishing current pattern introduces more structural determinicity (order information) about the system, which also implies less state indeterminicity (disorder information). This dimensionless measure is seen to reflect the average kinetic energy $T[\psi] = \langle\psi|\hat{\mathrm{T}}|\psi\rangle$:

$$I[\psi] = (8m/\hbar^2)\,T[\psi] \equiv \sigma\,T[\psi]. \tag{13}$$

One similarly generalizes the entropy (uncertainty) concept of the disorder information in probability density, e.g., the global quantity of Shannon or the gradient descriptor of Fisher, by supplementing the relevant classical measure of the information contained in probability distribution with the corresponding nonclassical complement due to the state (positive) phase or the associated current pattern [7–10]. The resultant Shannon-type global-entropy measure then reads

$$S[\psi] = -\langle\psi|\ln p + 2\phi|\psi\rangle = S[p] - 2\int p(\mathbf{r})\,\phi(\mathbf{r})\,d\mathbf{r} \equiv S[p] + S[\phi] \equiv S[p,\phi]. \tag{14}$$

It includes the (positive) probability information $I_S[p] \equiv S[p]$ and (negative) nonclassical supplement $S[\phi]$ reflecting the state average phase. These entropy contributions also reflect the real and imaginary parts of the associated complex-entropy concept [8], the quantum expectation value of the non-Hermitian entropy operator $S(\mathbf{r}) = -2\ln\psi(\mathbf{r})$,

$$S[p, \phi] = -2\langle\psi|\ln\psi|\psi\rangle \equiv S[p] + i\,S[\phi]. \tag{15}$$

The resultant gradient entropy similarly combines the (positive) Fisher probability information and (negative) phase contribution due to the current density:

$$M[\psi] = \langle\psi|(\nabla\ln p)^2 - (2\nabla\phi)^2|\psi\rangle = I[p] - I[\phi] \equiv M[p] + M[\phi] \equiv M[p,\phi]. \tag{16}$$

The sign of the latter reflects an extra decrease of the state overall structure indeterminicity due to its nonvanishing current pattern.

The extrema of these resultant entropies identify the same optimum, equilibrium-phase solution $\phi_{eq.} \geq 0$ [7,63–67]:

$$\{\delta S[\psi]/\delta\psi^*(\mathbf{r}) = 0 \quad \text{or} \quad \delta M[\psi]/\delta\psi^*(\mathbf{r}) = 0\} \quad \Rightarrow \quad \phi_{eq.}(\mathbf{r}) = -(1/2)\ln p(\mathbf{r}). \tag{17}$$

This local "thermodynamic" phase generates the associated current contribution reflecting the negative probability gradient:

$$j_{eq.}(\mathbf{r}) = (\hbar/m)\,p(\mathbf{r})\,\nabla\phi_{eq.}(\mathbf{r}) = -[\hbar/(2m)]\,\nabla p(\mathbf{r}). \tag{18}$$

The above one-electron development can be straightforwardly generalized into a general case of N-electron system in the specified (pure) quantum state $|\Psi(N)\rangle$, exhibiting the electron density $\rho(\mathbf{r}) = Np(\mathbf{r})$, where $p(\mathbf{r})$ stands for the density probability (shape) factor. The corresponding N-electron information operator then combines terms due to each particle,

$$\hat{I}(N)= \sum_{i=1}^{N} \hat{I}(r_i)= (8m/h^2)\sum_{i=1}^{N} \hat{T}(r_i)\equiv \sigma\hat{T}(N), \tag{19}$$

and determines the state overall gradient information,

$$I(N) = \langle \Psi(N)|\hat{I}(N)|\Psi(N)\rangle = \sigma\langle \Psi(N)|\hat{T}(N)|\Psi(N)\rangle = \sigma T(N), \tag{20}$$

proportional to the expectation value $T(N)$ of the system kinetic-energy operator $\hat{T}(N)$. The relevant separation of the modulus and phase components of such general N-electron states calls for a wavefunction yielding the specified electron density [52]. For example, this goal can be accomplished using the Harriman–Zumbach–Maschke (HZM) [87,88] construction of DFT. It uses N (complex) equidensity orbitals, each generating the molecular probability distribution $p(r)$ and exhibiting the density-dependent spatial phases, $f(r) = f[\rho; r]$, which safeguard the MO orthogonality.

Consider the Slater-determinant describing an electron configuration defined by N (singly) occupied spin MO,

$$\begin{aligned} \psi = \{\psi_s\} = (\psi_1, \psi_2, \ldots, \psi_N), \quad \{n_s = 1\}, \\ \Psi(N) = |\psi_1\psi_2 \ldots \psi_N|. \end{aligned} \tag{21}$$

The kinetic-energy/gradient-information descriptors then combine additive contributions due to each particle:

$$T(N) = \sum_s n_s\langle\psi_s|\hat{T}|\psi_s\rangle \equiv \sum_s n_s T_s = (h^2/8m)\sum_s n_s\langle\psi_s|\hat{I}|\psi_s\rangle \equiv \sigma^{-1}\sum_s n_s I_s. \tag{22}$$

In the analytical (LCAO MO) representation, with MO expressed as linear combinations of the (orthogonalized) atomic orbitals (AO) $\chi = (\chi_1, \chi_2, \ldots, \chi_k, \ldots)$,

$$|\psi\rangle = |\chi\rangle\,\mathbf{C}, \quad \mathbf{C} = \langle\chi|\psi\rangle = \{C_{k,s} = \langle\chi_k|\psi_s\rangle\}, \tag{23}$$

the average gradient information in $\Psi(N)$ for the unit matrix $\mathbf{n} = \{n_s\,\delta_{s,s'}\} = \{\delta_{s,s'}\}$ of MO occupations, then reads

$$I(N) = \sum_s n_s\langle\psi_s|\hat{I}|\psi_s\rangle = \sum_k \sum_l \left\{\sum_s C_{k,s}n_s C_{s,l}^*\right\}\langle\chi_l|\hat{I}|\chi_k\rangle \equiv \sum_k \sum_l \gamma_{k,s}I_{l,k} = \text{tr}(\gamma\mathbf{I}). \tag{24}$$

Here, the AO matrix representation of the *gradient*-information operator,

$$\mathbf{I} = \{I_{k,l} = \langle\chi_k|\hat{I}|\chi_l\rangle = \sigma\langle\chi_k|\hat{T}|\chi_l\rangle = \sigma T_{k,l}\}, \tag{25}$$

and the charge/bond-order (CBO) (density) matrix of LCAO MO theory,

$$\gamma = \mathbf{CnC}^\dagger = \langle\chi|\psi\rangle\mathbf{n}\langle\psi|\chi\rangle \equiv \langle\chi|\hat{P}_\psi|\chi\rangle, \tag{26}$$

is the associated matrix representation of the projection operator onto the occupied MO-subspace,

$$\hat{P}_\psi = N[\sum_s |\psi_s\rangle(n_s/N)\langle\psi_s|] \equiv N[\sum_s |\psi_s\rangle p_s\langle\psi_s|] \equiv N\hat{d}, \tag{27}$$

proportional to the density operator $\hat{d}$ of the configuration MO "ensemble".

This expression for the average overall information in $\Psi(N)$ thus assumes thermodynamic-like form, as trace of the product of the CBO matrix, the AO representation of the (occupation-weighted) MO projector determining the configuration density operator $\hat{d}$, and the corresponding AO matrix of the resultant gradient information related to that of the kinetic energy of electrons. It has been argued elsewhere [7,32–42] that elements of the CBO matrix generate amplitudes of electronic "communications" between AO "events" in the molecule. Therefore, the average gradient information of Equation (24) is seen to represent the communication-weighted (dimentionless) kinetic-energy descriptor.

The SE (7) also determines a temporal evolution of the average resultant descriptor of the gradient-information content in the specified (pure) quantum state (see Equation (11)). The time derivative of this overall information functional $I[\psi]$ is addressed in Appendix B.

3. Probing Formation of the Chemical Bond

The association between the overall gradient information and electronic kinetic energy suggests the use of molecular virial theorem in extracting the physical origins of the chemical bonding [89–92] and for understanding general reactivity rules [1–3,5]. The previous analyses [89–91] have focused on the interplay between the longitudinal (in the bond direction, along "z" coordinate) and transverse (perpendicular to the bond axis, due to coordinates "x" and "y") components of electronic kinetic energy. The former appears as the true driving force of the covalent-bond formation and the accompanying electron-delocalization process at an early approach by the two atoms, while the latter reflects an overall transverse contraction of the electron density in the attractive field of both nuclei.

It is of interest to examine the global production (Equation (A20)) of the resultant gradient measure of the electronic information due to the equilibrium current of Equation (18),

$$\sigma_I^{eq.} = -\sigma_M^{eq.} \propto -\int j_{eq.}(\boldsymbol{r}) \cdot \nabla v(\boldsymbol{r})\, d\boldsymbol{r} \propto \int \nabla p(\boldsymbol{r}) \cdot \nabla v(\boldsymbol{r})\, d\boldsymbol{r}, \tag{28}$$

which accompanies a formation of the covalent chemical bond A–B (Figure 1). Reference to this figure shows that in the axial (bond) section of the molecule $\nabla p(\boldsymbol{r}) \cdot \nabla v(\boldsymbol{r}) < 0$, thus confirming a derease of the longitudinal contribution to the average structure information, i.e., an increase in the axial component of the overall gradient entropy (Equation (A24)) as a result of the chemical bond formation: $\sigma_I^{eq.}(\text{axial}) < 0$ and $\sigma_M^{eq}(\text{axial}) > 0$. This accords with the chemical intuition: electron delocalization in the covalent chemical bond at its equilibrium length $R = R_e$ should produce a higher indeterminicity (disorder, entropy) measure and a lower level of the determinicity (order, information) descriptor, particularly in the axial bond region between the two nuclei.

Since the gradient measure of the state overall gradient information reflects the system kinetic-energy content, one could indeed relate these conclusions to the known profiles of the longitudinal and transverse components of this energy contribution [89–91]. The former contribution effectively lowers the longitudinal inhomogeneity of molecular probability density, $\sigma_I^{eq.}(\text{axial}) < 0$, particularly in the bond region between the two nuclei, while the (dominating) latter component implies an effective transverse contraction of the electron distribution, i.e., $\sigma_I^{eq.}(\text{transverse}) > 0$. The bonded system thus exhibits a net increase in the probability inhomogeneity, i.e., a higher gradient information compared to the separated-atoms limit (SAL). This is independently confirmed by a lowering of the system overall potential-energy displacement $\Delta W(R) = W(R) - W(\text{SAL})$ at the equilibrium bond-length R_e,

$$\Delta W(R_e) = 2\Delta E(R_e) = -2\Delta T(R_e) < 0 \tag{29}$$

Here,

$$\Delta W(R) = \Delta V(R) + [\Delta U_e(R) + \Delta U_n(R)] \equiv \Delta V(R) + \Delta U(R)$$

combines the (electron-nuclear) attraction (V) and repulsion ($U = U_e + U_n$) energies between electrons (U_e) and nuclei (U_n).

It is also of interest to examine variations of the resultant gradient information in specific geometrical displacements ΔR of this diatomic system. Its proportionality to the system kinetic-energy component again calls for using the molecular virial theorem, which allows one to partition the relative BO potential $\Delta E(R) = \Delta T(R) + \Delta W(R)$ into the SAL-related changes in the electronic kinetic energy

[$\Delta T(R)$] and its overall potential complement [$\Delta W(R)$]. In the BO approximation the virial theorem for diatomics reads:

$$2\Delta T(R)(+ \Delta W(R) + R[d\Delta E(R)/dR] = 0. \tag{30a}$$

It implies the following kinetic and potential energy components:

$$\Delta T(R) = -\Delta E(R) - R\,[d\Delta E(R)/dR] = -d[R\Delta E(R)]/dR \quad \text{and}$$
$$\Delta W(R) = 2\Delta E(R) + R\,[d\Delta E(R)/\partial R] = R^{-1}\,d[R^2\Delta E(R)]/dR. \tag{30b}$$

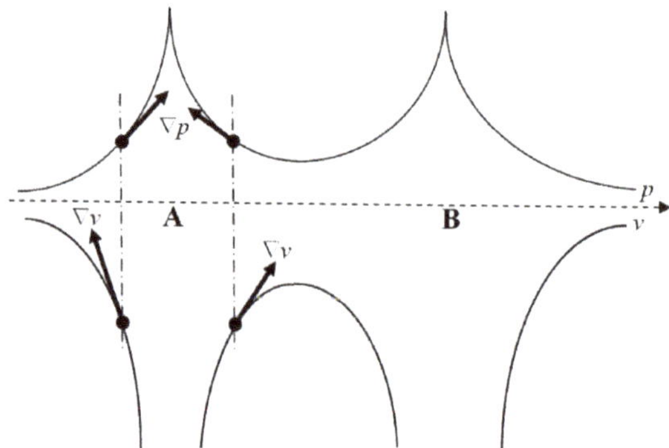

Figure 1. Schematic diagram of the axial (bond) profiles, in section containing the "z" direction of the coordinate system (along the bond axis), of the external potential (v) and electron probability (p) in a diatomic molecule A–B demonstrating a negative character of the scalar product $\nabla p(r)\cdot\nabla v(r)$. It confirms the negative equilibrium contribution $\sigma_I^{eq.}$(axial) of the resultant gradient information (Equations (A20) and (A21)) and positive source $\sigma_M^{eq.}$(axial) of the resultant gradient entropy (Equation (A24)) in the bond formation process, due to the equilibrium current of Equation (18), $j_{eq.}(r) \propto -\nabla p(r)$.

Figure 2 presents qualitative plots of the BO potential $\Delta E(R)$ and its kinetic-energy contribution $\Delta T(R)$ in diatomics. The latter also reflects the associated displacement plots for the resultant gradient information $\Delta I(R) = \sigma\,\Delta T(R)$. It follows from this qualitative diagram that, during a mutual approach by two constituent atoms, the kinetic-energy/gradient information is first diminished relative to SAL, due to the dominating longitudinal contribution related to Cartesian coordinate "z" (along the bond axis). However, at the equilibrium distance R_e the resultant information already rises above the SAL value, due to the dominating increase in transverse components of the kinetic-energy/information (corresponding to coordinates "x" and "y" perpendicular to the bond axis). Therefore, at the equilibrium separation R_e between atoms the bond-formation results in a net increase of the resultant gradient-information relative to SAL, due to—on average—more compact electron distribution in the field of both nuclei.

Consider next the (intrinsic) reaction coordinate $\boldsymbol{R}_c$, or the associated progress variable $P = |\boldsymbol{R}_c|$, of the arc length along this trajectory, for which the virial relations also assume the diatomic-like form. The virial theorem decomposition of the energy profile $E(P)$ along $\boldsymbol{R}_c$ in bimolecular reaction

$$A + B \rightarrow R^{\ddagger} \rightarrow C + D, \tag{31}$$

where the $R^{\ddagger}$ denotes the transition-state (TS) complex, then generates the associated profile of its kinetic-energy component $T(P)$, which also reflects the associated resultant gradient information $I(P)$. Such an application of the molecular virial theorem to endo- and exo-ergic reactions is presented in the upper panel of Figure 3, while the energy-neutral case of such a chemical process, on a "symmetric" potential energy surface (PES), refers to a lower panel in the figure.

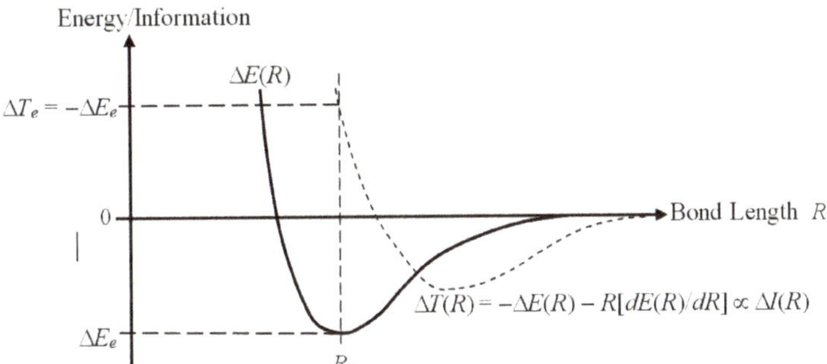

Figure 2. Variations of the electronic energy $\Delta E(R)$ (solid line) with the internuclear distance R in a diatomic molecule and of its kinetic energy component $\Delta T(R)$ (broken line) determined by the virial theorem partition.4. Reactivity Implications of Molecular Virial Theorem.

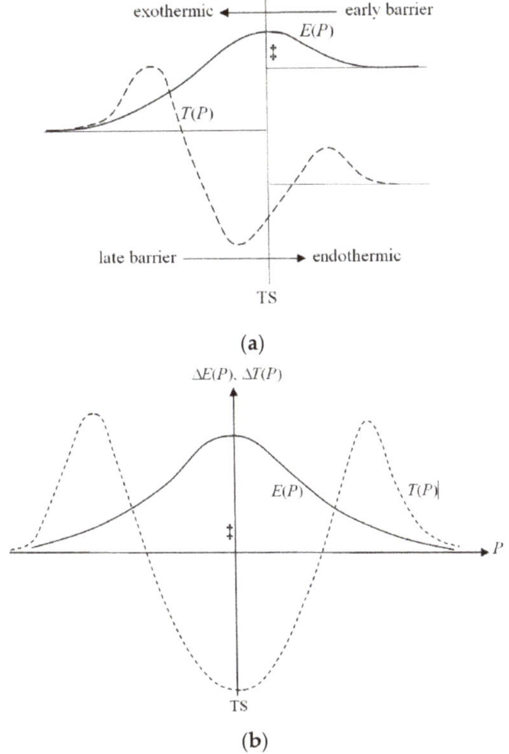

Figure 3. Variations of the electronic total (E) and kinetic (T) energies in exo-ergic ($\Delta E_r < 0$) or endo-ergic ($\Delta E_r > 0$) reactions (upper Panel (**a**)), and on the symmetrical BO potential energy surface (PES) ($\Delta E_r = 0$) (lower Panel (**b**)).

The (qualitative) Hammond postulate [80] of reactivity theory relates a general resemblance/proximity of the reaction TS complex $R^{\ddagger}$ to either its substrates $\alpha \in$ (A, B) or products $\beta \in$ (C, D) to the reaction energy $\Delta E_r = E(P_{prod.}) - E(P_{sub.})$: in *exo*-ergic ($\Delta E_r < 0$) processes, $R^{\ddagger} \approx \alpha$ and in endo-ergic ($\Delta E_r > 0$) reactions, $R^{\ddagger} \approx \beta$. Accordingly, for the vanishing reaction energy $\Delta E_r = 0$, the position of TS complex is expected to be located symmetrically between the reaction substrates and products. A reference to Figure 3 indeed shows that the activation barrier appears "early" in exo-ergic reaction, e.g., $H_2 + F \rightarrow H + HF$, with the reaction substrates being only slightly modified in TS, $R^{\ddagger} \approx$ [A–B]. Accordingly, in the endo-ergic bond-breaking–bond-forming process, e.g., $H + HF \rightarrow H_2 + F$, the barrier is "late" along the reaction coordinate P and the activated complex resembles more reaction products: $R^{\ddagger} \approx$ [C–D]. This qualitative statement has been subsequently given several more quantitative formulations and theoretical explanations using both the energetic and entropic arguments [93–100]

Previous virial-theorem analyses [1–3,5] have shown that this qualitative rule is fully indexed by the sign of the P-derivative of the average kinetic energy or of the resultant gradient information at TS complex. The energy profile along the reaction "progress" coordinate P, $\Delta E(P) = E(P) - E(P_{sub.})$ is again directly "translated" by the virial theorem into the associated displacement in its kinetic-energy contribution $\Delta T(P) = T(P) - T(P_{sub.})$, proportional to the corresponding change $\Delta I(P) = I(P) - I(P_{sub.})$ in the system resultant gradient information, $\Delta I(P) = \sigma \, \Delta T(P)$,

$$\Delta T(P) = -\Delta E(P) - P \, [d\Delta E(P)/dP] = -d[P\Delta E(P)]/dP. \tag{32}$$

A reference to qualitative plots in Figure 3 shows that the related $\Delta T(P)$ or $\Delta I(P)$ criteria distinguish these two directions by the sign of their geometrical derivative at TS complex:

$$\begin{aligned}
&\textit{endo}\text{-direction:} && (dI/dP)_{\ddagger} > 0 \text{ and } (dT/dP)_{\ddagger} > 0, \Delta E_r > 0; \\
&\textit{energy-neutral:} && (dI/dP)_{\ddagger} = 0 \text{ and } (dT/dP)_{\ddagger} = 0, \Delta E_r = 0; \\
&\textit{exo}\text{-direction:} && (dI/dP)_{\ddagger} < 0 \text{ and } (dT/dP)_{\ddagger} < 0, \Delta E_r < 0.
\end{aligned}$$

This observation demonstrates that the RC derivative of the resultant gradient information at TS complex, proportional to $dT/dP \, |_{\ddagger}$, can indeed serve as an alternative detector of the reaction energetic character: its positive/negative values respectively identify the endo/exo-ergic processes, exhibiting the late/early activation barriers, respectively, with the neutral case, $\Delta E_r = 0$ or $dT/dP \, |_{\ddagger} = 0$, exhibiting an "equidistant" position of TS between the reaction substrates and products on a symmetric PES, e.g., in the hydrogen exchange reaction $H + H_2 \rightarrow H_2 + H$.

The reaction energy ΔE_r determines the corresponding change in the resultant gradient information, $\Delta I_r = I(P_{prod.}) - I(P_{sub.}) = \sigma \, \Delta T_r$, proportional to $\Delta T_r = T(P_{prod.}) - T(P_{sub.}) = -\Delta E_r$. The virial theorem thus implies a net decrease of the resultant gradient information in endo-ergic processes, $\Delta I_r(endo) < 0$, its increase in exo-ergic reactions, $\Delta I_r(exo) > 0$, and a conservation of the resultant gradient information in the energy-neutral chemical processes: $\Delta I_r(neutral) = 0$. One also recalls that the classical part of this information displacement probes an average inhomogeneity of electronic density. Therefore, the endo-ergic processes, requiring a net supply of energy to R, give rise to more diffused electron distributions in the reaction products, compared to substrates. Accordingly, the exo-ergic transitions, which release the energy from R, generate a more compact electron distributions in products and no such change is predicted for the energy-neutral case.

4. Reactivity Criteria

The grand-ensemble basis of populational derivatives of the energy or information descritors in the externally open molecular systems [1–3,53,101,102] has been briefly summarized in Appendix C. The equilibrium energy function

$$\begin{aligned}
\mathcal{E}[\hat{D}_{eq.}] &= \mathcal{E}(\mu, T; v) = \textstyle\sum_i \sum_j P_j^i(\mu, T; v) E_j^i, \\
\hat{D}_{eq.} &= \textstyle\sum_i \sum_j |\psi_j^i\rangle P_j^i(\mu, T; v) \langle\psi_j^i| \equiv \hat{D}(\mu, T; v),
\end{aligned} \tag{33}$$

is determined by the optimum probabilities $\{P_j^i(\mu, T; v)\}$ of the ensemble stationary states $\{|\psi_j^i\rangle\}$, eigenstates of Hamiltonians $\{\hat{H}(N_i, v)\}$: $\hat{H}(N_i, v) |\psi_j^i\rangle = E_j^i \, |\psi_j^i\rangle$. These state probabilities correspond to the grand-potential minimum with respect to the ensemble density operator (see Equations (A29) and (A38)):

$$\min_{\hat{D}} \Omega[\hat{D}] = \Omega[\hat{D}_{eq.}]. \tag{34}$$

Thermodynamic energy of Equation (33) identifies the two Lagrange multipliers involved in this variational rule as corresponding partial derivatives with respect to the constraint values:

$$\mu = \left(\frac{\partial \mathcal{E}}{\partial \mathcal{N}}\right)_S \Big|_{\hat{D}_{eq.}} \quad \text{and} \quad T = \left(\frac{\partial \mathcal{E}}{\partial \mathcal{S}}\right)_{\mathcal{N}} \Big|_{\hat{D}_{eq.}}.$$

The minimum-energy principle of Equation (34) (see also Equation (A38)) can be alternatively interpreted as the associated extremum rule for the overall gradient information [1–3,19,45],

$$\sigma \min_{\hat{D}} \Omega[\hat{D}] = \sigma \Omega[\hat{D}_{eq.}] = \mathcal{I}[\hat{D}_{eq.}] + \frac{8m}{h^2}\{\mathcal{W}[\hat{D}_{eq.}] - \mu \mathcal{N}[\hat{D}_{eq.}] - T\mathcal{S}[\hat{D}_{eq.}]\}, \tag{35}$$

where the ensemble-average value of the system potential energy again combines the electron-nuclear attraction ($\mathcal{V}$) and the repulsion ($\mathcal{U}$) contributions: $\mathcal{W}[\hat{D}_{eq.}] = \mathcal{V}[\hat{D}_{eq.}] + \mathcal{U}[\hat{D}_{eq.}]$. This gradient-information/kinetic-energy principle is seen to contain the additional constraint of the fixed overall potential energy, $\langle W \rangle_{ens.} = \mathcal{W}$, multiplied by the Lagrange multiplier

$$\lambda_{\mathcal{W}} = -\sigma = \left(\frac{\partial \mathcal{I}}{\partial \mathcal{W}}\right)_{\mathcal{N},\mathcal{S}} \Big|_{\hat{D}_{eq.}} \equiv \kappa. \tag{36}$$

It also includes modified "intensities" associated with the remaining constraints: information potential

$$\lambda_{\mathcal{N}} = \sigma \mu = \left(\frac{\partial \mathcal{I}}{\partial \mathcal{N}}\right)_{\mathcal{W},\mathcal{S}} \Big|_{\hat{D}_{eq.}} \equiv \xi \quad \text{and} \tag{37}$$

information "temperature"

$$\lambda_{\mathcal{S}} = \sigma T = \left(\frac{\partial \mathcal{I}}{\partial \mathcal{S}}\right)_{\mathcal{W},\mathcal{N}} \Big|_{\hat{D}_{eq.}} \equiv \tau. \tag{38}$$

The conjugate thermodynamic principles, for the constrained extrema of the ensemble-average energy,

$$\delta(\mathcal{E}[\hat{D}] - \mu \mathcal{N}[\hat{D}] - T\mathcal{S}[\hat{D}])\hat{D}_{eq.} = 0, \tag{39}$$

and thermodynamic gradient information,

$$\delta(\mathcal{I}[\hat{D}] - \kappa \mathcal{W}[\hat{D}] - \xi \mathcal{N}[\hat{D}] - \tau \mathcal{S}[\hat{D}])\hat{D}_{eq.} = 0, \tag{40}$$

have the same state-probability solutions [1–3]. This manifests the physical equivalence of the energetic and entropic principles for determining the equilibrium states in thermodynamics [4].

The equilibrium value of resultant gradient information, given by the weighted expression in terms of the equilibrium probabilities in the grand-canonical mixed state,

$$\langle I \rangle_{ens.} \equiv \mathcal{I}[\hat{D}_{eq.}] = \mathrm{tr}(\hat{D}_{eq.}\hat{I}) = \sum_i \sum_j P_j^i(\mu, T; v)\langle \psi_j^i | \hat{I} | \psi_j^i \rangle \equiv \sum_i \sum_j P_j^i(\mu, T; v) I_j^i,$$
$$I_j^i = \sigma \langle \psi_j^i | \hat{T} | \psi_j^i \rangle \equiv \sigma T_j^i, \tag{41}$$

is related to the ensemble average kinetic energy $\mathcal{T}$:

$$\langle T \rangle_{ens.} \equiv \mathcal{T} = \mathrm{tr}(\hat{D}_{eq.}\hat{T}) = \sum_i \sum_j P_j^i(\mu, T; v)\langle \psi_j^i | \hat{T} | \psi_j^i \rangle = \sum_i \sum_j P_j^i(\mu, T; v) T_j^i, = \sigma^{-1}\mathcal{I}. \tag{42}$$

In this grand-ensemble approach the system chemical potential appears as the first (partial) populational derivative [53,73,74,101–105] of the system average energy. This interpretation also applies to the diagonal and mixed second derivatives of equilibrium electronic energy, which involve differentiation with respect to the electron-population variable $\mathcal{N}$. In this energy representation, the chemical hardness reflects $\mathcal{N}$-derivative of the chemical potential [53,73,74,106],

$$\eta = \left(\frac{\partial^2 \mathcal{E}}{\partial \mathcal{N}^2} \right)_{S} \bigg|_{\hat{D}_{eq.}} = \left(\frac{\partial u}{\partial \mathcal{N}} \right)_{S} \bigg|_{\hat{D}_{eq.}} > 0, \tag{43}$$

while the information hardness measures $\mathcal{N}$-derivative of the information potential:

$$\omega = \left(\frac{\partial^2 \mathcal{I}}{\partial \mathcal{N}^2} \right)_{W,S} \bigg|_{\hat{D}_{eq.}} = \left(\frac{\partial \xi}{\partial \mathcal{N}} \right)_{W,S} \bigg|_{\hat{D}_{eq.}} = \sigma \eta > 0. \tag{44}$$

By the Maxwell cross-differentiation relation, the mixed derivative of the energy,

$$f(\mathbf{r}) = \left(\frac{\partial^2 \mathcal{E}}{\partial \mathcal{N} \partial v(\mathbf{r})} \right)_{S} \bigg|_{\hat{D}_{eq.}} = \left(\frac{\partial u}{\partial v(\mathbf{r})} \right)_{S} \bigg|_{\hat{D}_{eq.}} = \left(\frac{\partial \rho(\mathbf{r})}{\partial \mathcal{N}} \right)_{S} \bigg|_{\hat{D}_{eq.}}, \tag{45}$$

measuring the global Fukui Function (FF) [53,73,74,107], can be alternatively interpreted as either the density response per unit populational displacement, or the global chemical-potential response per unit local change in the external potential. The associated mixed derivative of the resultant gradient information then reads:

$$\varphi(\mathbf{r}) = \left(\frac{\partial^2 \mathcal{I}}{\partial \mathcal{N} \partial v(\mathbf{r})} \right)_{W,S} \bigg|_{\hat{D}_{eq.}} = \left(\frac{\partial \xi}{\partial v(\mathbf{r})} \right)_{W,S} \bigg|_{\hat{D}_{eq.}} = \sigma f(\mathbf{r}). \tag{46}$$

The positive signs of the diagonal descriptors assure the external stability of a molecule with respect to external flows of electrons, between the molecular system and its electron reservoir. Indeed, they imply an increase (a decrease) of the global energetic and information "intensities" conjugate to $\mathcal{N} = N$, the chemical (μ), and information (ξ) potentials, in response to a perturbation created by an electron inflow (outflow) ΔN. This is in accordance with the familiar Le Châtelier and Le Châtelier–Braun principles of thermodynamics [4], that the secondary (spontaneous) responses in system intensities to an initial population displacement diminish effects of this primary perturbation.

Since reactivity phenomena involve electron flows between the mutually open substrates, only in such generalized, grand-ensemble framework can one precisely define the relevant CT criteria, determine the hypothetical "states" of subsystems, and eventually measure the effects of their mutual interaction. The open microscopic systems require the mixed-state description, in terms of the ensemble-average physical quantities, capable of reflecting the externally imposed thermodynamic conditions and defining the infinitesimal populational displacements invoked in reactivity theory. In this ensemble approach, the energetic and information principles are physically equivalent, giving rise to the same equilibrium probabilities. This basic equivalence is consistent with the alternative energetic and entropic principles invoked in equilibrium thermodynamics of macroscopic systems [4].

5. Donor-Acceptor Systems

In reactivity considerations one conventionally recognizes several hypothetical stages of chemical processes involving either the mutually closed (nonbonded, disentangled) or open (bonded, entangled) reactants $\alpha = \{A, B\}$ [1–3,73,74], e.g., substrates in a typical bimolecular reactive system R = A–B involving the acidic (A, electron acceptor) and basic (B, electron donor) subsystems. The nonbonded status of these fragments, when they conserve their initial (integer) overall numbers of electrons $\{N_\alpha = N_\alpha{}^0\}$ in the isolated (separated) reactants $\{\alpha^0\}$, is symbolized by the solid vertical line, e.g., in the intermediate, polarized reactive system $R^+ \equiv (A^+ | B^+)$ combining the internally polarized but mutually closed subsystems. It should be emphasized that only due to this mutual closure the substrate identity remains a meaningful concept. Their descriptors in the final, equilibrium-reactive system $R^* \equiv (A^* | B^*) \equiv R$, combining the mutually open (bonded) fragments, as symbolized by the vertical broken line separating the two subsystems, can be inferred only indirectly [1–3], by externally opening the two mutually closed subsystems of R^+ with respect to their separate (macroscopic) electron reservoirs $\{\mathcal{R}_\alpha\}$ in the composite polarized system

$$\mathcal{M}_R^+ = (\mathcal{R}_A | A^+ | B^+ | \mathcal{R}_B) \equiv [\mathcal{M}_A^+ | \mathcal{M}_B^+]. \tag{47}$$

The subsystem densities $\{\rho_\alpha = N_\alpha\, p_\alpha\}$, with p_α denoting the internal probability distribution in fragment α, are "frozen" in the promolecular reference $R^0 = (A^0 | B^0)$ consisting of the isolated-reactant distributions $\{\rho_\alpha^0 = N_\alpha^0\, p_\alpha^0\}$ shifted to their actual positions in the "molecular" system R. The polarized reactive system R^+ combines the relaxed subsystem densities, modified in presence of the reaction partner at finite separation between both subsystems: $\{\rho_\alpha^+ = N_\alpha^+ p_\alpha^+, N_\alpha^+ = \int \rho_\alpha^+ dr = N_\alpha^0\}$. In the global equilibrium state of R as a whole, these polarized subsystem densities are additionally modified by the effective inter-reactant CT: $\{\rho_\alpha^* = N_\alpha^* p_\alpha^*, N_\alpha^* = \int \rho_\alpha^* dr \neq N_\alpha^0\}$.

The overall electron density in the whole R^+ is given by the sum of reactant densities, polarized due to "molecular" external potential $v = v_A + v_B$ combining contributions due to the fixed nuclei in both substrates at their final mutual separation,

$$\rho_R^+ \equiv N_R\, p_R^+ = \rho_A^+ + \rho_B^+ \equiv N_A^+ p_A^+ + N_B^+ p_B^+, \quad N_\alpha^+ = \int \rho_\alpha^+ dr, \quad \sum_\alpha N_\alpha^+ = N_R. \tag{48}$$

Here, $\{p_\alpha^+ = \rho_\alpha^+/N_\alpha^+\}$ stand for the internal probability densities in such promoted fragments, and the global probability distribution reflects the "shape" factor of the overall electron density,

$$\begin{aligned} p_R^+ &= \rho_R^+/N_R = (N_A^+/N_R)\, p_A^+ + (N_B^+/N_R)\, p_B^+ \\ &\equiv P_A^+ p_A^+ + P_B^+ p_B^+, \quad \int p_R^+ dr = P_A^+ + P_B^+ = 1, \end{aligned} \tag{49}$$

where condensed reactant probabilities $\{P_\alpha^+ = N_\alpha^+/N_R = N_\alpha^0/N_R = P_\alpha^0\}$ denote fragment shares in N_R. At this polarization stage, both fragments exhibit internally equalized chemical potentials $\{\mu_\alpha^+ = \mu[N_\alpha^0, v]\}$, different from the separate-reactant levels $\{\mu_\alpha^0 = \mu[N_\alpha^0, v_\alpha]\}$.

The two molecular subsystems lose their identity in the bonded status, as the mutually open parts of the externally closed reactive system R^*, which allows for the inter-fragment (intra-R) flows of electrons. In such a global equilibrium each "part" effectively extends over the whole molecular system since the hypothetical boundary defining the fragment identity does not exists any more. Both subsystems then effectively exhaust the molecular electron distribution, their electron populations are equal to the global number of electrons,

$$\rho_A^* = \rho_B^* = \rho_R, \quad \{\mu_\alpha^* = \mu_R \equiv \mu[N_R, v], \quad N_\alpha^* = N_R, \quad p_\alpha^* = \rho_R/N_R \equiv p_R\}, \tag{50}$$

and subsystem chemical potentials in R are equalized at molecular level μ_R: $\{\mu_\alpha^* = \mu_R\}$.

One can contemplate, however, the external flows of electrons, between the mutually closed (nonbonded) but externally open reactants and their separate (macroscopic) reservoirs $\{\mathcal{R}_\alpha\}$. The mutual closure then implies the relevancy of subsystem identities established at the polarization stage of R^+, while the external openness in the composite subsystems $\{\mathcal{M}_\alpha^+ = (\alpha^+ | \mathcal{R}_\alpha^+)\}$ allows one to independently "regulate" the external chemical potentials of both parts, $\{\mu_\alpha^+ = \mu(\mathcal{R}_\alpha)\}$, and hence also their average densities $\{\rho_\alpha^+ = N_\alpha^+ p_\alpha^+\}$ and electron populations $N_\alpha^+ = \int \rho_\alpha^+ dr$. In particular, the substrate chemical potentials equalized at the molecular level in both subsystems, $\{\mu_\alpha^+ \equiv \mu[N_\alpha^*, v] = \mu_R \equiv \mu[N_R, v]\}$, which then also describes a common molecular reservoir $\{\mathcal{R}_\alpha(\mu_R) = \mathcal{R}(\mu_R)\}$ coupled to both reactants, $(\mathcal{R}(\mu_R) \mid A^* \mid B^*)$, formally define the global equilibrium state in the molecular part $R^* = (A^* \mid B^*)$ of the composite system which (indirectly) represents the global equilibrium in R as a whole:

$$\mathcal{M}_R^* = \mathcal{M}_R^+(\mu_R) \equiv [\mathcal{M}_A^*(\mu_R) \mid \mathcal{M}_B^*(\mu_R)]$$
$$= [\mathcal{R}(\mu_R) \mid A^+(\mu_R) \mid B^+(\mu_R) \mid \mathcal{R}(\mu_R)] \equiv [\mathcal{R}(\mu_R) \mid A^*(\mu_R) \mid B^*(\mu_R)] \equiv (\mathcal{R}^* \mid R^*). \tag{51}$$

This reactive system indeed implies an effective mutual openess of both reactants, realized through the direct external openness of both substrates to a common (molecular) reservoir $\mathcal{R}(\mu_R)$. It allows for an effective donor(B)$\rightarrow$acceptor(A) flow of electrons, between subsystems of the molecular fragment R^* of $\mathcal{M}_R^*$, while retaining the reactant identities assured by the direct mutual closeness of the polarized molecular part R^+ in $\mathcal{M}_R^+$.

The final reactant distributions $\{\rho_\alpha^* = \rho_\alpha^+(\mu_R)\}$ in R^* and the associated electron populations $\{N_\alpha^* = N_\alpha^+(\mu_R)\}$ are then modified by effects of the inter-reactant CT

$$N_{CT} = N_A^* - N_A^0 = N_B^0 - N_B^* > 0, \tag{52}$$

for the conserved overall (average) number of electrons in the globally isoelectronic processes in the reactive system as a whole:

$$N_A^* + N_B^* \equiv N_R^* = N(\mu_R) = N_R = N_A^+ + N_B^+ \equiv N_R^+ = N_A^0 + N_B^0 \equiv N_R^0. \tag{53}$$

Density changes due to these equilibrium redistributions of electrons are indexed by the corresponding in situ FF. In CSA [73,74], one introduces the reactant-resolved FF matrix of the substrate density responses to displacements in the fragment electron populations, for the fixed (molecular) external potential (geometry),

$$\mathbf{f}^+(\mathbf{r}) = \{f_{\alpha,\beta}(\mathbf{r}) = [\partial \rho_\beta^+(\mathbf{r})/\partial N_\alpha]_v\}, \tag{54}$$

which generate the FF indices of reactants in the B$\rightarrow$A CT:

$$f_\alpha^{CT}(\mathbf{r}) = \partial \rho_\alpha^+(\mathbf{r})/\partial N_{CT} = f_{\alpha,\alpha}(\mathbf{r}) - f_{\beta,\alpha}(\mathbf{r}); \quad (\alpha, \beta \neq \alpha) \in \{A, B\}. \tag{55}$$

One recalls, that these relative responses of reactants eventually combine into the corresponding global CT derivative, of the reactive system as a whole,

$$F_R^{CT} = \partial \rho_R^+/\partial N_{CT} = \textstyle\sum_{\alpha=A,B} \sum_{\beta=A,B} (\partial \rho_\beta^+/\partial N_\alpha) (\partial N_\alpha/\partial N_{CT})$$
$$= (f_{A,A} - f_{B,A}) - (f_{B,B} - f_{A,B}) \equiv f_A^{CT} - f_B^{CT}, \tag{56}$$

which represents the populational sensitivity of R^* with respect to the effective internal CT, between the externally open but mutually closed reactants.

To summarize, the fragment identity remains a meaningful concept only for the mutually closed (nonbonded) status of the acidic and basic reactants, e.g., in the polarized reactive system R^+ or in the R^+ part of $\mathcal{M}_R^+$. The global equilibrium in R as a whole, $R = R^*$, combining the effectively "bonded", externally open but mutually closed subsystems $\{\alpha^*\}$ in $\mathcal{M}_R^*$, accounts for the extra CT-induced polarization of reactants compared to R^+. Descriptors of this state, of the mutually "bonded" reactants, can be inferred only indirectly, by examining the chemical potential equalization in the equilibrium composite system $\mathcal{M}_R^*$. Similar external reservoirs are involved when one examines independent population displacements on reactants, e.g., in defining the fragment chemical potentials and their hardness tensor in the substrate fragment of $\mathcal{M}_R^+$. In this hypothetical chain of reaction "events", the polarized system R^+ thus appears as the launching stage for the subsequent CT and the accompanying induced polarization, after the hypothetical barrier for the flow of electrons between subsystems has been effectively lifted.

Thus, the equilibrium case $\mathcal{M}_R^*$ of $\mathcal{M}_R^+$ also represents the effectively open reactants in R^*. The equilibrium substrates $\{\alpha^*\}$ indeed display the final equilibrium densities $\{\rho_\alpha^* = \rho_\alpha(\mu_R)\}$ after the B$\to$A CT, giving rise to molecular electron distribution

$$\rho_A^* + \rho_B^* = \rho_A^+(\mu_R) + \rho_B^+(\mu_R) = \rho_R(\mu_R) \equiv \rho_R \tag{57}$$

and the associated populations $\{\int \rho_\alpha^+(\mu_R)\, d\boldsymbol{r} = N_\alpha^+(\mu_R) = N_\alpha^*\}$ corresponding to the chemical potential equalization in $R = R^*$ as a whole: $\mu_A^* = \mu_B^* = \mu_R$. One observes that the reactant chemical potentials have not been equalized at the preceding polarization stage in the molecular part $R_n^+ = (A^+ | B^+)$ of a general composite system

$$\mathcal{M}_R^+ = (\mathcal{M}_A^+ | \mathcal{M}_B^+) = (\mathcal{R}_A^+ | A^+ | \mathcal{R}^+ | R_B^+), \quad \{\mathcal{R}_\alpha^+ = \mathcal{R}_\alpha[\mu_\alpha^+]\}, \tag{58}$$

when $\mu_A^+[\rho_A^+] < \mu_B^+[\rho_B^+]$.

In the polarized reactive system, the fragment chemical potentials $\mu_R^+ = \{\mu_\alpha^+ = \mu_\alpha[N_\alpha^0, v]\}$ and the resulting elements of the hardness matrix $\eta_R^+ = \{\eta_{\alpha,\beta}\}$ represent populational derivatives of the system average electronic energy in reactant resolution, $E_v(\{N_\beta\})$. They are properly defined in $\mathcal{M}_R^+$, calculated for the fixed external potential v reflecting the "frozen" molecular geometry. These quantities represent the corresponding partials of the system ensemble-average energy with respect to ensemble-average populations $\{N_\alpha\}$ on subsystems in the mutually closed (externally open) composite subsystems $\{\mathcal{M}_\alpha^+ = (\alpha^+ | \mathcal{R}_\alpha^+)\}$ in $\mathcal{M}_R^+ = (\mathcal{M}_A^+ | \mathcal{M}_B^+)$:

$$\mu_\alpha \equiv \partial E_v(\{N_\gamma\})/\partial N_\alpha, \quad \eta_{\alpha,\beta} = \partial^2 E_v(\{N_\gamma\})/\partial N_\alpha \partial N_\beta = \partial \mu_\alpha/\partial N_\beta. \tag{59}$$

The global reactivity descriptors, of the $R = (A^* | B^*) \equiv R^*$ part of $\mathcal{M}_R^*$, similarly involve differentiations with respect to the average number of electrons of R in the combined system $\mathcal{M}_R^* = (R^* | \mathcal{R}^*) = (A^* | B^* | \mathcal{R}^*)$:

$$\mu_R = \partial E_v(N_R)/\partial N_R, \quad \eta_R = \partial^2 E_v(N_R)/\partial N_R^2 = \partial \mu_R/\partial N_R. \tag{60}$$

The optimum amount of the (fractional) CT is determined by the populational "force", measuring the difference between chemical potentials of the polarized acidic and basic reactants which defines the effective CT gradient,

$$\mu_{CT} = \partial E_v(N_{CT})/\partial N_{CT} = \mu_A^+ - \mu_B^+ = \sigma^{-1}[\xi_A^+ - \xi_B^+] \equiv \sigma^{-1}\xi_{CT} < 0, \tag{61}$$

and the in situ hardness (η_{CT}) or softness (S_{CT}) for this process,

$$\begin{aligned} \eta_{CT} = \partial \mu_{CT}/\partial N_{CT} &= (\eta_{A,A} - \eta_{A,B}) + (\eta_{B,B} - \eta_{B,A+}) \equiv \eta_A^R + \eta_B^R = S_{CT}^{-1} \\ &= \sigma^{-1}\partial \xi_{CT}/\partial N_{CT} = \sigma^{-1}\, \omega_{CT}, \end{aligned} \tag{62}$$

representing the effective CT Hessian and its inverse, respectively, proportional to the CT information hardness Hessian ω_{CT}. The optimum amount of the inter-reactant CT,

$$N_{CT} = -\mu_{CT}/\eta_{CT} = -\xi_{CT}/\omega_{CT} > 0, \tag{63}$$

then generates the associated second-order stabilization energy due to CT,

$$E_{CT} = \mu_{CT}\, N_{CT}/2 = -\mu_{CT}^2/(2\eta_{CT}) = \sigma^{-1}\,[-\xi_{CT}^2/(2\omega_{CT})] \equiv \sigma^{-1}\, I_{CT} < 0, \tag{64}$$

proportional to the associated change I_{CT} in the resultant gradient information.

6. HSAB Principle Revisited

The physical equivalence of reactivity concepts formulated in the energy and resultant gradient-information representations has direct implications for OCT, in which one treats a molecule as an information network propagating signals of the AO origins of electrons in the bond system determined by the configuration occupied MO. It has been argued that elements of the CBO matrix $\gamma = \{\gamma_{k,l}\}$ of Equation (26), weighting factors in Equation (24), determine amplitudes of conditional probabilities defining molecular (direct) communications between AO. Entropic descriptors of such information channel generate the entropic bond orders and measures of their covalent/ionic components, which ultimately facilitate an IT understanding of molecular electronic structure in chemical terms [7,19–21,32–42]. The communication "noise" (orbital indeterminicity) in this network, measured by the channel conditional entropy, is due to the electron delocalization in the bond system of a molecule. It represents the system overall bond "covalency", while the channel information "capacity" (orbital determinicity), reflected by the mutual information of the molecular communication network, measures its resultant bond "iconicity". The more scattering (indeterminate) is the molecular information system, the higher its covalent character; a more deterministic (less noisy) channel thus represents a more ionic molecular system. These two bond attributes thus compete with one another.

In chemistry, the bond covalency, a common possession of electrons by interacting atoms, is indeed synonymous with an electron delocalization generating a communication noise. A classical example of a purely covalent interaction constitute bonds connecting identical atoms, e.g., hydrogens in H_2 or carbons in ethylene, when the interacting AO in the familiar MO diagrams of chemistry exhibit the same AO energies. The bond ionicity accompanies large differences in electronegativities, generating a substantial CT between the interacting atoms. Such bonds correspond to a wide separation of AO energies in MO diagrams. The ionic bond component diminishes noise and introduces more determinicity into AO communications, thus representing the bond mechanism competitive with the bond covalency.

One of the celebrated (qualitative) rules of chemistry deals with stability preferences in molecular coordinations. The HSAB principle [84,85,106] predicts that chemically hard (H) acids prefer to coordinate hard bases in the [HH] complexes A–B, and soft (S) acids prefer to coordinate soft bases in the [SS] complexes, whereas the "mixed" [HS] or [SH] coordinations, of hard acids with soft bases or of soft acids with hard bases, are relatively unstable. Little is known about the communication implications of this principle. In such a communication perspective on reactive systems the H and S reactants correspond to internally ionic (deterministic) and covalent (noisy) substrate channels, respectively. The former involves relatively localized communications between AO, while the latter corresponds to delocalized probability scatterings between the basis states.

In the reactivity context, the following additional questions arise:

What is an overall character of communications responsible for the mutual interaction between reactants?

How does the HSAB rule influence the inter-reactant propagations of information, i.e., how do the [HH] and [SS] preferences shape the inter-reactant communications in these complexes?

Do the S substrates in [SS] complex predominantly interact "covalently", and H substrates in the [HH] complex "ionically"?

In the frontier electron (FE) [81–83] approach to molecular interactions and CT phenomena, the orbital energy of the substrate highest occupied MO (HOMO) determines its donor (basic) level of the chemical potential, while the lowest unoccupied MO (LUMO) energy establishes its acceptor (acidic) capacity (see Figure 4). The HOMO–LUMO energy gaps in subsystems then reflect the substrate molecular hardnesses. The interaction between the reactant frontier MO of comparable orbital energies is predominantly covalent (chemically "soft") in character, while that between the subsystem MO of distinctly different energies becomes mostly ionic (chemically "hard"). Figure 4 summarizes the alternative relative positions of the donor (HOMO) levels of the basic reactant, relative to the acceptor (LUMO) levels of its acidic partner, for all admissible hardness combinations in the reactive system

R = A–B. In view of the proportionality relations between the energetic and information reactivity criteria, these relative MO energy levels also reflect the corresponding information potential and hardness descriptors of subsystems, including the in situ derivatives driving the information transfer between reactants.

A magnitude of the ionic, CT-stabilization energy in A–B systems is then determined by the corresponding in situ populational derivatives in R,

$$\Delta\varepsilon_{ion.} = |E_{CT}| = \mu_{CT}^2/(2\eta_{CT}) > 0, \tag{65}$$

where μ_{CT} and η_{CT} stand for the effective chemical potential and hardness descriptors of R involving the FE of reactants. Since the donor/acceptor properties of reactants are already implied by their known relative acidic or basic character, one applies the biased estimate of the CT chemical potential. In FE approximation, the chemical potential difference μ_{CT} for the effective internal B→A CT then reads (see Figure 4):

$$\mu_{CT}(B{\to}A) = \mu_A^{(-)} - \mu_B^{(+)} = \varepsilon_A(LUMO) - \varepsilon_B(HOMO) \approx I_B - A_A > 0. \tag{66}$$

It determines the associated first-order energy change for the B→A transfer of N_{CT} electrons:

$$\Delta E_{B{\to}A}(N_{CT}) = \mu_{CT}(B{\to}A)\,N_{CT} > 0. \tag{67}$$

The CT chemical potential of Equation (66) thus combines the electron-removal potential of the basic reactant, i.e., its negative ionization potential $I_B = E(B^{+1}) - E(B^0) > 0$,

$$\mu_B^{(+)} = \varepsilon_B(HOMO) \approx -I_B, \tag{68}$$

and the electron-insertion potential of the acidic substrate, i.e., its negative electron affinity

$$A_A = E(A^0) - E(A^{-1}) > 0,$$
$$\mu_A^{(-)} = \varepsilon_A(LUMO) \approx -A_A. \tag{69}$$

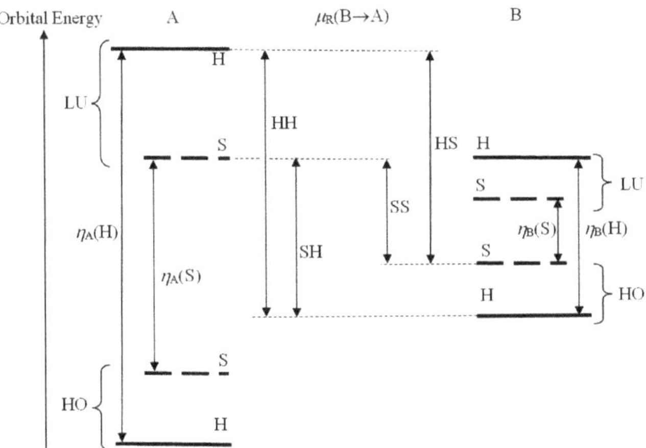

Figure 4. Schematic diagram of the in situ chemical potentials $\mu_{CT}(B{\to}A)$, determining the effective internal charge transfer (CT) from basic (B) reactant to its acidic (A) partner in A–B complexes, for their alternative hard (H) and soft (S) combinations. The subsystem hardnesses reflect the HOMO-LUMO gaps in their orbital energies.

The energy of the CT disproportionation process,

$$[A\text{---}B] + [A\text{---}B] \rightarrow [A^{-1}\text{---}B^{+1}] + [A^{+1}\text{---}B^{-1}], \tag{70}$$

then generates the (unbiased) finite-difference measure of the effective hardness descriptor for this implicit CT:

$$\eta_{CT} = (I_A - A_A) + (I_B - A_B)$$
$$\approx [\varepsilon_A(\text{LUMO}) - \varepsilon_A(\text{HOMO})] + [\varepsilon_B(\text{LUMO}) - \varepsilon_B(\text{HOMO})] \tag{71}$$
$$= \eta_A + \eta_B > 0.$$

These in situ derivatives ultimately determine a magnitude of the CT stabilization energy of Equation (65), which reflects the ionic part of the overall interaction energy,

$$\Delta\varepsilon_{ion.} = \mu_{CT}^2/(2\eta_{CT}) = [\varepsilon_A(\text{LUMO}) - \varepsilon_B(\text{HOMO})]^2/[2(\eta_A + \eta_B)]. \tag{72}$$

In the FE framework of Figure 4, the CT-interaction energy is thus proportional to the squared gap between the LUMO orbital energy of the acidic reactant and the HOMO level of the basic substrate. This ionic interaction is thus predicted to be strongest in [HH] pairs of subsystems and weakest in [SS] arrangements, with the mixed [HS] and [SH] combinations representing the intermediate magnitudes of this ionic-stabilization effect.

It should be realized, however, that ionic and covalent energy contributions complement each other in the resultant bond energy [85]. Therefore, the [SS] complex, for which the energy gap $\varepsilon_A(\text{LUMO})-\varepsilon_B(\text{HOMO})$ between the interacting orbitals reaches the minimum value, implies the strongest covalent stabilization in the reactive complex. Indeed, the lowest (bonding) energy level ε_b of this FE interaction, corresponding to the bonding combination of the (positively overlapping) frontier MO of subsystems,

$$\varphi_b = N_b\,[\varphi_B(\text{HOMO}) + \lambda\varphi_A(\text{LUMO})], \quad S = \langle \varphi_A(\text{LUMO}) \mid \varphi_B(\text{HOMO}) \rangle > 0, \tag{73}$$

then exhibits the maximum bonding energy due to the covalent interaction:

$$\Delta\varepsilon_{cov.} = \varepsilon_B(\text{HOMO}) - \varepsilon_b > 0. \tag{74}$$

It follows from the familiar secular equations of the Ritz method that this covalent energy can be approximated by the limiting MO expression

$$\Delta\varepsilon_{cov.} \cong (\beta - \varepsilon_b\,S)^2/[\varepsilon_A(\text{LUMO}) - \varepsilon_B(\text{HOMO})], \tag{75}$$

where the matrix element of the system electronic Hamiltonian, which couples the two states,

$$\beta = \langle \varphi_A(\text{LUMO})|\hat{H}|\varphi_B(\text{HOMO})\rangle \tag{76}$$

is expected to be proportional to the overlap integral S between the frontier MO involved. It indeed follows from Equation (75) that the maximum covalent component of the inter-reactant chemical bond is expected in interactions between soft, strongly overlapping reactants, since then the numerator assumes the highest value while the denominator reaches its minimum. For the same reason one predicts the smallest covalent stabilization in interactions between the hard, weakly overlapping substrates, with the mixed hardness combinations giving rise to intermediate bond covalencies.

To summarize, the [HH] complex exhibits the maximum ionic stabilization, the [SS] complex the maximum covalent interaction, while the mixed combinations of reactant hardnesses in [HS] and [SH] coordinations exhibit a mixture of moderate covalent and ionic bond components between the acidic and basic subsystems. Therefore, communications representing the inter-reactant bonds between the chemically soft (covalent) reactants are also expected to be predominantly "soft" (delocalized, indeterministic) in character, while those between the chemically hard (ionic) subsystems are predicted to be dominated by the "hard" (localized, deterministic) propagations in the communication system for R as a whole.

The electron communications between reactants $\{\alpha = A, B\}$ in the acceptor-donor reactive system R = A–B are determined by the corresponding matrix of conditional probabilities (or of their amplitudes) in AO resolution, which can be partitioned into diagonal blocks of the intra-reactant (internal) communications within individual substrates and off-diagonal blocks of the inter-reactant (external) communications between different subsystems:

$$[R \rightarrow R] = \{[\alpha \rightarrow \beta]\} = \{[\alpha \rightarrow \alpha]\delta_{\alpha,\beta}\} + \{[\alpha \rightarrow \beta](1 - \delta_{\alpha,\beta})\} = \{intra\} + \{inter\}. \tag{77}$$

The [SS] complexes combining the "soft" (noisy, delocalized) internal blocks of such probability propagations imply similar covalent character of the external blocks of electron AO communications between reactanta, i.e., strongly indeterministic scatterings between subsystems: $\{intra\text{-}S\} \Rightarrow \{inter\text{-}S\}$. The "hard" (ionic) internal channels are similarly associated with the ionic (localized) external communications: $\{intra\text{-}H\} \Rightarrow \{inter\text{-}H\}$. This observation adds a new communication angle to the classical HSAB principle of chemistry.

7. Regional HSAB versus Complementary Coordinations

One of the still problematic issues in reactivity theory is the most favorable mutual arrangement of the acidic and basic parts of molecular reactants in donor-acceptor systems. It appears that the global HSAB preference is no longer valid regionally, in interactions between fragments of reactants, where the complementarity principle [108,109] establishes the preferred arrangement between the acidic and basic parts of both substrates. These more subtle reactivity preferences result from the induced electron flows in reactive systems, reflecting responses to the primary displacements in molecular complexes. Such flow patterns can be diagnosed, estimated, and compared using either the energetical or information reactivity criteria defined above.

Consider the reactive A–B complex consisting of the basic reactant B = $(a_B | \ldots | b_B) \equiv (a_B | b_B)$ and the acidic substrate A = $(a_A | \ldots | b_A) \equiv (a_A | b_A)$, where a_X and b_X denote the acidic and basic parts of X, respectively. The acidic (electron acceptor) part is relatively hard, i.e., less responsive to external perturbations, exhibiting lower values of the fragment FF descriptor, while the basic (electron donor) fragment is relatively soft and more polarizable, as reflected by its higher density or population response descriptors. The acidic part a_X exerts an electron-accepting (stabilizing) influence on the neighboring part of another reactant Y, while the basic fragment b_X produces an electron-donor (destabilizing) effect on a fragment of Y in its vicinity.

There are two ways in which both reactants can mutually coordinate in the reactive complexes. In the complementary (c) arrangement of Figure 5,

$$R_c \equiv \begin{bmatrix} a_A - b_B \\ b_A - a_B \end{bmatrix}, \tag{78}$$

the reactants orient themselves in such a way that geometrically accessible a-fragment of one reactant faces the geometrically accessible b-fragment of the other substrate. This pattern follows from the maximum complementarity (MC) rule of chemical reactivity [108,109], which reflects an electrostatic preference: an electron-rich (repulsive, basic) fragment of one reactant prefers to face an electron-deficient (attractive, acidic) part of the reaction partner. In the alternative regional HSAB-type structure of Figure 6,

$$R_{HSAB} \equiv \begin{bmatrix} a_A - a_B \\ b_A - b_B \end{bmatrix}, \tag{79}$$

the acidic (basic) fragment of one reactant faces the like-fragment of the other substrate.

The complementary complex, in which the "excessive" electrons of b_X are placed in the attractive field generated by the electron "deficient" a_Y, is expected to be favored electrostatically since the other arrangement produces regional repulsions between two acidic and two basic sites of reactants. Additional rationale for this complementary preference over the regional HSAB alignment comes from examining charge flows created by the dominating shifts in the site chemical potential due to the presence of the ("frozen") coordinated site of the nearby part of the reaction partner. At finite separations between the two subsystems, these displacements trigger the polarizational flows $\{P_X\}$ shown in Figures 5 and 6, which restore the internal equilibria in subsystems, initially displaced by the presence of the other reactant.

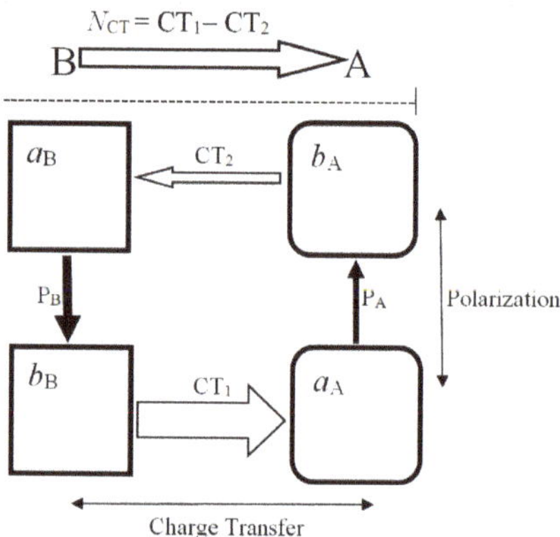

Figure 5. Polarizational $\{P_\alpha = (a_\alpha \rightarrow b_\alpha)\}$ and charge-transfer $\{CT_\alpha = (b_\alpha \rightarrow a_\beta)\}$ electron flows, $(\alpha, \beta \neq \alpha)$ $\in \{A, B\}$, involving the acidic $A = (a_A \mid b_A)$ and basic $B = (a_B \mid b_B)$ reactants in the complementary arrangement R_c of their acidic (a) and basic (b) fragments, with the chemically "hard" (acidic) fragment of one substrate facing the chemically "soft" (basic) part of its reaction partner. The polarizational flows $\{P_\alpha\}$ (black arrows) in the mutually closed substrates, relative to the substrate "promolecular" references, preserve the overall numbers of electrons of isolated reactants $\{\alpha^0\}$, while the two partial $\{CT_i\}$ fluxes (white arrows), from the basic fragment of one reactant to the acidic part of the other reactant, generate a substantial resultant B→A transfer of $N_{CT} = CT_1 - CT_2$ electrons between the mutually open reactants. These hypothetical electron flows in such a "complementary complex" are seen to produce an effective concerted ("circular") flux of electrons between the four fragments invoked in this regional "functional" partition, which precludes an exaggerated depletion or concentration of electrons on any fragment of reactive system.

In R_c, the harder (acidic) site a_Y initially lowers the chemical potential of the softer (basic) site b_X, while b_Y rises the chemical potential level of a_X. These shifts trigger the internal polariaztional flows $\{a_X \rightarrow b_X\}$ in both reactants, which enhance the acceptor capacity of a_X and donor ability of b_X, thus creating more favourable conditions for the subsequent inter-reactant CT of Figure 5. A similar analysis of R_{HSAB} (Figure 6) predicts the $b_X \rightarrow a_X$ polarizational flows, which lower the acceptor capacity of a_X and donor ability of b_X, i.e., produce an excess electron accumulation on a_X and stronger electron depletion on b_X, thus creating less favourable conditions for the subsequent inter-reactant CT.

The complementary preference also follows from the electronic stability considerations, in spirit of the familiar Le Châtelier–Braun principle in ordinary thermodynamics [4]. In Figures 5 and 6, the CT responses follow the preceding polarizations of reactants, the equilibrium responses to displacements $\{\Delta v_X\}$ in the external potential on subsystems. In stability considerations one first assumes the primary (inter-reactant) CT displacements $\{\Delta CT_1, \Delta CT_2\}$ of Figures 5 and 6, in the internally "frozen" but externally open reactants, and then examines the induced secondary (intra-reactant) relaxational responses $\{I_X\}$ to these populational shifts in the CT-perturbed substrates.

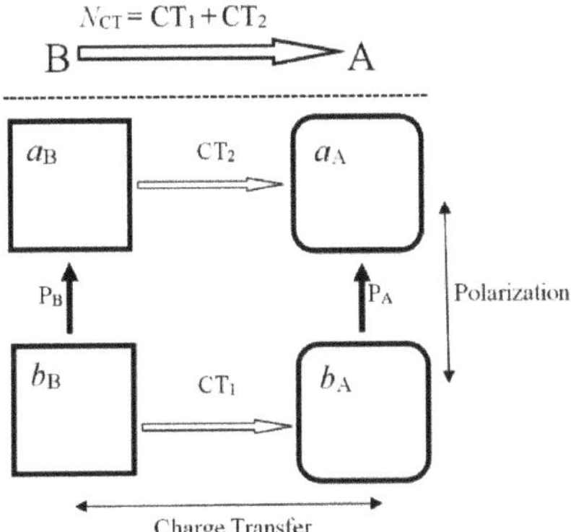

Figure 6. Polarizational $\{P_\alpha = (b_\alpha \rightarrow a_\alpha)\}$ and charge-transfer, $CT_1 = (b_B \rightarrow b_A)$ and $CT_2 = (a_B \rightarrow a_A)$, electron flows, involving the acidic $A = (a_A \mid b_A)$ and basic $B = (a_B \mid b_B)$ reactants in the regional HSAB complex R_{HSAB}, in which the chemically hard (acidic) and soft (basic) fragments of one reactant coordinate to the like fragment of the other substrate. The two partial $\{CT_i\}$ fluxes (white arrows) now generate a moderate overall $B \rightarrow A$ transfer of $N_{CT} = CT_1 + CT_2$ electrons between the mutually open reactants. These hypothetical electron flows in the regional HSAB complex are seen to produce a disconcerted pattern of fluxes producing an exaggerated outflow of electrons from b_B and and their accentuated inflow to a_A.

Let us first examine the CT-displaced complementary structure R_c,

$$R_c^{CT} \equiv \begin{bmatrix} a_A & \overset{\Delta CT_1}{\leftarrow} & b_B \\ \downarrow I_A & & \uparrow I_B \\ b_A & \overset{\Delta CT_2}{\rightarrow} & a_B \end{bmatrix}. \tag{80}$$

It corresponds to the primary CT perturbations of Figure 5:

$$[\Delta(CT_1) = \Delta N(a_A) = -\Delta N(b_B)] > [\Delta(CT_2) = \Delta N(a_B) = -\Delta N(b_A)]. \tag{81}$$

In accordance with the Le Châtelier principle, an inflow/outflow of electrons to/from a given site x increases/decreases the site chemical potential, as indeed reflected by the positive value of the site diagonal hardness descriptor

$$\eta_{x,x} = \partial \mu_x / \partial N_x \equiv \eta_x > 0. \tag{82}$$

The initial perturbations of the partial CT flows $\{\Delta CT_k\}$ thus create the following shifts in the site chemical potentials, compared to the initially equalized levels in isolated reactants $A^0 = (a_A^0 \mid b_A^0)$ and $B^0 = (a_B^0 \mid b_B^0)$,

$$[\Delta \mu_{a_A}(CT_1) > 0] > [\Delta \mu_{b_A}(CT_2) < 0], \quad [\Delta \mu_{a_B}(CT_2) > 0] > [\Delta \mu_{b_B}(CT_1) < 0]. \tag{83}$$

These CT-induced shifts in the fragment electronegativities thus trigger the following secondary (induced) relaxational flows $\{I_X\}$ in R_c^{CT},

$$a_A \overset{I_A}{\rightarrow} b_A \quad \text{and} \quad a_B \overset{I_B}{\rightarrow} b_B, \tag{84}$$

which diminish effects of the CT perturbations by reducing the extra charge accumulations/depletions created by the primary populational displacements.

Consider next the CT-displaced HSAB complex R_{HSAB},

$$R_{HSAB}^{CT} \equiv \begin{bmatrix} a_A & \overset{\Delta CT_2}{\leftarrow} & a_B \\ \downarrow I_A & & \uparrow I_B \\ b_A & \overset{\Delta CT_1}{\leftarrow} & b_B \end{bmatrix}. \tag{85}$$

The primary CT shifts in the site electron populations,

$$[\Delta(CT_1) = \Delta N(b_A) = -\Delta N(b_B)] < [\Delta(CT_2) = \Delta N(a_A) = -\Delta N(a_B)], \tag{86}$$

where inequality reflects expected magnitudes of the associated in situ chemical potentials,

$$[|\mu(CT_1)| = \mu(b_B) - \mu(b_A)] < [|\mu(CT_2)| = \mu(a_B) - \mu(a_A)], \tag{87}$$

now induce the following internal *relaxations* in reactants:

$$a_A \overset{I_A}{\to} b_A \quad \text{and} \quad b_B \overset{I_B}{\to} a_B. \tag{88}$$

These charge responses further exaggerate charge depletions/accumulations created by the primary CT perturbations, thus giving rise to less stable reactive complex.

8. Conclusions

In this work we have emphasized the information contributions due to the modulus (probability) and phase (current) components of general (complex) electronic states in molecules, accounted for in generalized entropy/information concepts of the QIT description. The relevant continuity relations for these elementary physical distributions are summarized in Appendix A. In this overview, the physical equivalence of variational principles formulated in terms of the average electronic energy and the resultant-information/kinetic energy has been stressed. The quantum dynamics of the resultant gradient information measure is examined in Appendix B. The proportionality of the state average gradient information to electronic kinetic energy allows one to use the molecular virial theorem in the information exploration of general (energetical) rules of structural chemistry and chemical reactivity. The phase aspect of molecular states was shown to be vital for distinguishing the hypothetical bonded (entangled) and nonbonded (disentangled) states of molecular subsystems in reactive systems, for the same set of the fragment electron densities. The resultant-information analysis of reactivity phenomena complements earlier classical IT approaches to chemical reactions [110–112] as well as the catastrophe-theory or quantum-topology descriptions [113–115]. One also recalls a close connection between the ELF criterion of electron localization and the nonadditive component of the molecular Fisher information [46,56].

The grand-ensemble description of thermodynamic equilibria in externally open molecular systems is outlined in Appendix C. It has been argued that thermodynamic variational principles for the ensemble-average electronic energy of molecular systems and their resultant gradient information are physically equivalent. The populational derivatives of the resultant gradient information, related to the system average kinetic energy, have been suggested as alternative reactivity criteria, proportional to their energetical analogs and fully equivalent in describing the CT phenomena. Indeed, they were shown to correctly predict both the direction and magnitude of the electron flows in reactive systems. The virial theorem has been used in an information exploration of the bond-formation process and in probing the qualitative Hammond postulate of reactivity theory. The information production in chemical reactions has been addressed, and the ionic/covalent interactions between frontier-electrons of the acidic and basic reactants have been examined. The HSAB preferences in molecular coordinations have been explained and the communication perspective on interaction between reactants has been addressed. It has been argued that the internally soft and hard reactants prefer to externally communicate in the like manner, consistent with their internal communications. The HSAB preference is thus reflected by the predicted character of the inter-reactant bonds/communications: Covalent/noisy in [SS] and ionic/localized in [HH] complexes. The regional complementary and HSAB coordinations, between acidic and basic fragments of the donor and acceptor reactants, have been compared, and the complementary preference has been explained using the electrostatic, polarizational, and relaxational (stability) arguments.

Conflicts of Interest: The author declares no conflict of interest.

Nomenclature

The following notation is adopted: A denotes a scalar, A is the row or column vector, $\mathbf{A}$ represents a square or rectangular matrix, and the dashed symbol $\hat{A}$ stands for the quantum-mechanical operator of the physical property A. The logarithm of Shannon's information measure is taken to an arbitrary but fixed base: $\log = \log_2$ corresponds to the information content measured in bits (binary digits), while $\log = \ln$ expresses the amount of information in nats (natural units): 1 nat = 1.44 bits.

Appendix A. Continuities of Probability and Phase Distributions

Consider a general quantum state in mono-electronic system represented by the wavefunction of Equation (1). The total time derivative of the particle probability density $p(t) = p[r(t), t]$ reads

$$dp(t)/dt = \partial p(t)/\partial t + (dr/dt) \cdot \partial p(t)/\partial r = \partial p(t)/\partial t + V(t) \cdot \nabla p(t) \equiv \sigma_p(t). \tag{A1}$$

It determines the local probability "source" $\sigma_p(r, t) \equiv \sigma_p(t)$, which measures the time rate of change in an infinitesimal volume element around r in the probability fluid, moving with the local velocity $dr/dt = V(t)$, while the partial derivative $\partial p(t)/\partial t$ refers to the volume element around the fixed point in space:

$$\begin{aligned}
\partial p(t)/\partial t &= \sigma_p(t) - V(t) \cdot \nabla p(t) \\
&= \sigma_p(t) - (\hbar/2mi)[\psi(t)^* \Delta\psi(t) - \psi(t) \Delta\psi(t)^*] = \sigma_p(t) - \nabla \cdot j(t) \\
&= \sigma_p(t) - [V(t) \cdot \nabla p(t) + p(t) \nabla \cdot V(t)] \\
&= \sigma_p(t) - (\hbar/m) [\nabla\phi(t) \cdot \nabla p(t) + p(t) \nabla^2\phi(t)].
\end{aligned} \tag{A2}$$

One also recalls that probability dynamics from SE expresses the sourceless continuity relation for the particle probability "fluid", $\sigma_p(t) = 0$, or

$$\partial p(t)/\partial t + \nabla \cdot j(t) = \partial p(t)/\partial t + \nabla p \cdot V(t) + p \nabla \cdot V(t) = 0. \tag{A3}$$

This relation thus implies the vanishing divergence of effective velocity field determined by the phase-Laplacian:

$$\nabla \cdot V(t) = (\hbar/m) \nabla^2\phi(t) = 0. \tag{A4}$$

The partial derivative of Equation (A2) expressing the probability dynamics thus reads:

$$\partial p(t)/\partial t = - V(t) \cdot \nabla p(t) = - (\hbar/m) \nabla\phi(t) \cdot \nabla p(t) \tag{A5}$$

or

$$dp(t)/dt = \partial p(t)/\partial t + \nabla \cdot j(t) = \partial p(t)/\partial t + V(t) \cdot \nabla p(t) = 0. \tag{A6}$$

The probability continuity also determines the dynamics of the state modulus component:

$$\partial R(t)/\partial t = - (\hbar/m) \nabla\phi(t) \cdot \nabla R(t). \tag{A7}$$

For example, for a general local-phase expression [86,87]

$$\phi(r) = k \cdot f(r) + C, \tag{A8}$$

where the constant C remains unspecified in QM, one determines the probability-velocity field at point $r = \{x_\alpha\}$: $V(r) = (\hbar/m) k \nabla \cdot f(r)$. Its vanishing divergence then implies a local condition

$$k \cdot \nabla[\nabla \cdot f(r)] \equiv k \cdot \nabla^2 f(r) = \sum_\alpha \sum_\beta k_\alpha [\partial^2 f_\beta(r)/\partial x_\alpha \, \partial x_\beta] = 0. \tag{A9}$$

The effective velocity $V(t)$ also determines the phase current, the flux concept associated with the state phase: $J(t) = \phi(t)\,V(t)$. The scalar field $\phi(t)$ and its conjugate current density $J(t)$ then generate a nonvanishing phase source in the associated continuity equation:

$$\sigma_\phi(t) \equiv d\phi(t)/dt = \partial\phi(t)/\partial t + \nabla\cdot J(t) = \partial\phi(t)/\partial t + V(t)\cdot\nabla\phi(t) \neq 0 \quad \text{or}$$
$$\partial\phi(t)/\partial t - \sigma_\phi(t) = -\nabla\cdot J(t) = -(\hbar/m)\,[\nabla\phi(t)]^2. \tag{A10}$$

The phase dynamics from SE [7],

$$\partial\phi/\partial t = [\hbar/(2m)]\,[R^{-1}\Delta R - (\nabla\phi)^2] - v/\hbar, \tag{A11}$$

finally identifies the phase source:

$$\sigma_\phi = [\hbar/(2m)]\,[R^{-1}\nabla^2 R + (\nabla\phi)^2] - v/\hbar. \tag{A12}$$

As an illustration consider the stationary wavefunction corresponding to energy E_s,

$$\psi_s(t) = R_s(r)\,\exp[i\phi_s(t)], \quad \phi_s(t) = -(E_s/\hbar)\,t = -\omega_s\,t, \tag{A13}$$

representing the eigenstate of electronic Hamiltonian of Equation (6):

$$\hat{H}(r)R_s(r) - -(h^2/2m)\nabla^2 R_s(r) + v(r)R_s(r) = E_s R_s(r). \tag{A14}$$

It exhibits the stationary probability distribution, $p_s(t) \equiv R_s(r)^2 = p_s(r)$, the purely time-dependent phase $\phi_s(t)$, $V_s(t) = 0$ and hence also $j_s(t) = J_s(t) = 0$. The phase-dynamics Equations (A11) and (A12) then recover the preceding stationary SE and identify the constant phase source (see Equation (A13)):

$$\sigma_\phi[\psi_s] = [\hbar/(2m)]\,(R_s^{-1}\nabla^2 R_s) - v/\hbar = -\omega_s = -(E_s/\hbar) = const. \tag{A15}$$

Appendix B. Information Dynamics

Let us now examine a temporal evolution of the overall integral measure of the gradient-information content in the specified (pure) quantum state $|\psi(t)\rangle$ of such a mono-electronic system (Equation (11)). In the Schrödinger dynamical picture, the time change of the resultant gradient information, the operator of which does not depend on time explicitly,

$$\hat{I}(r) = -4\nabla^2 = (8m/h^2)\hat{T}(r) \equiv \sigma\hat{T}(r), \tag{A16}$$

results solely from the time dependence of the system state vector itself. The time derivative of the average (Fisher-type) gradient information is then generated by the expectation value of the commutator

$$[\hat{H},\hat{I}] = [v,\hat{I}] = 4[\nabla^2, v] = 4\{[\nabla, v]\cdot\nabla + \nabla\cdot[\nabla, v]\}, \quad [\nabla, v] = \nabla v, \tag{A17}$$

$$dI(t)/dt = (i/h)\langle\psi(t)|[\hat{H},\hat{I}]|\psi(t)\rangle, \tag{A18}$$

and the integration by parts implies:

$$\langle\psi(t)|\nabla\psi(t)\rangle = -\langle\nabla\psi(t)|\psi(t)\rangle \equiv \langle\nabla^\dagger\psi(t)|\psi(t)\rangle \quad \text{or} \quad \nabla^\dagger = -\nabla. \tag{A19}$$

Hence, the total time derivative of the overall gradient information, i.e., the integral production (source) of this information descriptor, reads:

$$\begin{aligned}
\sigma_I(t) \equiv dI(t)/dt &= (4i/\hbar)\,\{\langle\psi(t)|\nabla v\cdot|\nabla\psi(t)\rangle - \langle\nabla\psi(t)|\cdot\nabla v|\psi(t)\rangle\} \\
&= -(8/\hbar)\,\mathrm{Im}\,\langle\psi(t)|\nabla v\cdot|\nabla\psi(t)\rangle \\
&= -(8/\hbar)\,\mathrm{Im}[\int\psi(t)^*\,\nabla v\cdot\nabla\psi(t)\,dr] \\
&= -(8/\hbar)\int p(t)\,\nabla\phi(t)\cdot\nabla v\,dr = -\sigma\int j(t)\cdot\nabla v\,dr.
\end{aligned} \tag{A20}$$

This derivative is seen to be determined by the current content of the molecular electronic state. Therefore, it identically vanishes for the zero current density everywhere, when the local component of the state phase identically vanishes, thus confirming its nonclassical origin.

This conclusion can be also demonstrated directly (see Equation (11)):

$$\sigma_I(t) \equiv dI(t)/dt = dI[p]/dt + dI[\phi]/dt = dI[\phi]/dt, \tag{A21}$$

since by the sourceless probability continuity of Equation (A6),

$$\sigma_p(r) = dp(r)/dt = 0, \tag{A22}$$

and hence

$$dI[p]/dt = \int [dp(r)/dt]\,[\delta I[p]/\delta p(r)]\,dr = 0. \tag{A23}$$

Therefore, the integral source of resultant gradient information in fact reflects the total time derivative of its nonclassical contribution $I[\phi]$. Hence the associated derivative of the overall gradient entropy of Equation (16):

$$\sigma_M(t) \equiv dM(t)/dt = dI[p]/dt - dI[\phi]/dt = -dI[\phi]/dt = -\sigma_I(t). \tag{A24}$$

This result is in accordance with the intuitive expectation that an increase in the state overall structural determinicity (order) information, $\sigma_I(t) > 0$, implies the associated decrease in its structural indeterminicity (disorder) information (entropy): $\sigma_M(t) < 0$.

Appendix C. Ensemble Representation of Thermodynamic Conditions

Only the ensemble average overall number of electrons $\langle N \rangle_{ens.} \equiv \mathcal{N}$ of the (open) molecular part M(v), identified by the external potential v of the specified (microscopic) system, in the composite (macroscopic) system $\mathcal{M} = [M(v) \mid \mathcal{R}]$ including the external electron-reservoir $\mathcal{R}$,

$$\mathcal{N} = \text{tr}(\hat{D}\hat{N}) = \sum_i N_i (\sum_j P_j^i) \equiv \sum_i N_i P^i, \quad \sum_i P^i = 1, \tag{A25}$$

exhibits a continuous (fractional) spectrum of values, thus justifying the very concept of the populational ($\mathcal{N}$) derivative itself. Here, $\hat{N} = \sum_i N_i (\sum_j |\psi_j^i\rangle \langle\psi_j^i|)$ stands for the particle-number operator in Fock's space and the density operator $\hat{D} = \sum_i \sum_j |\psi_j^i\rangle P_j^i \langle\psi_j^i|$ identifies the statistical mixture of the system (pure) stationary states $\{|\psi_j^i\rangle \equiv |\psi_j(N_i)\rangle\}$ defined for different (integer) numbers of electrons $\{N_i\}$, which appear with the (external) probabilities $\{P_j^i\}$ in the ensemble. Such $\mathcal{N}$-derivatives are involved in definitions of familiar CT criteria of chemical reactivity, e.g., the chemical potential (negative electronegativity) [53,101–105] or the chemical hardness/softness [106], and Fukui function (FF) [107] descriptors (see also References [53,73,74]).

Such $\mathcal{N}$-derivatives are thus definable only in the mixed electronic states, e.g., those corresponding to thermodynamic equilibria in externally open molecules. In the grand ensemble, this state is determined by the density operator specified by the relevant (externally imposed) intensive parameters, the chemical potential of the electron reservoir, $\mu = \mu_{\mathcal{R}}$, and the absolute temperature T of a heat bath $\mathcal{B}$, $T = T_B$: $\hat{D}_{eq.} \equiv \hat{D}(\mu, T; v)$. The optimum state probabilities $\{P_j^i(\mu, T; v)\}$ correspond to the minimum of the associated thermodynamic potential, the Legendre transform

$$\Omega = \mathcal{E} - (\partial\mathcal{E}/\partial\mathcal{N})\mathcal{N} - (\partial\mathcal{E}/\partial\mathcal{S})\mathcal{S} = \mathcal{E} - \mu\mathcal{N} - T\mathcal{S} \tag{A26}$$

of the ensemble-average energy

$$\mathcal{E}[\hat{D}] \equiv (\mathcal{N}, \mathcal{S}; v) = \text{tr}(\hat{D}\hat{H}) = \sum_i \sum_j P_j^i E_j^i, \tag{A27}$$

called the grand potential. The latter corresponds to replacing the "extensive" (ensemble-average) state parameters, the particle number $\mathcal{N}$ and thermodynamic entropy

$$S[\hat{D}] = \text{tr}(\hat{D}\hat{S}) = \sum_i \sum_j P_j^i S_j^i(P_j^i)\}, \quad S_j^i(P_j^i) = -k_B \ln P_j^i, \tag{A28}$$

where k_B denotes the Boltzmann constant, by their respective "intensive" conjugates defining the applied thermodynamic conditions: $\mu = \mu_{\mathcal{R}}$ and $T = T_B$. The grand potential of Equation (A26) includes these externally imposed intensities as Lagrange multipliers enforcing constraints of the specified values of system's average number of electrons, $\langle N \rangle_{ens.} = \mathcal{N}$, and its thermodynamic-entropy, $\langle S \rangle_{ens.} = \mathcal{S}$, at the grand-potential minimum:

$$\min_{\hat{D}} \Omega\hat{D} = \Omega[D(\mu, T; v)] \equiv \Omega(\mu, T; v). \tag{A29}$$

The externally imposed parameters (μ, T) then determine the optimum probabilities of the ensemble stationary states, eigenstates of the Hamiltonians $\{\hat{H}(N_i, v)\}$, in the (mixed) equilibrium state of the grand ensemble,

$$P_j^i(\mu, T; v) = \Xi^{-1} \exp[\beta(\mu N_i - E_j^i)], \tag{A30}$$

and the associated density operator

$$\hat{D}(\mu, T; v) = \sum_i \sum_j |\psi_j^i\rangle P_j^i(\mu, T; v)\langle\psi_j^i| \equiv \hat{D}_{eq.}. \tag{A31}$$

Here, Ξ stands for the grand-ensemble partition function and $\beta = (k_B T)^{-1}$. The equilibrium probabilities of Equation (A30) represent eigenvalues of the grand-canonical statistical operator acting in Fock's space:

$$\hat{d}(\mu, T; v) = \exp\{\beta[\mu\hat{N} - \hat{H}(v)]\}/\mathrm{tr}\{\beta[\mu\hat{N} - \hat{H}(v)]\}. \tag{A32}$$

In the $T{\to}0$ limit [53,101,102] only two ground states ($j = 0$), $|\psi_0^i\rangle$ and $|\psi_0^{i+1}\rangle$, corresponding to the neighboring integers "bracketing" the given (fractional) $\langle N\rangle_{ens.} = \mathcal{N}$,

$$N_i \leq \mathcal{N} \leq N_i + 1, \tag{A33}$$

appear in the equilibrium mixed state. Their ensemble probabilities for the specified

$$\langle N\rangle_{ens.} = i\, P_i(T{\to}0) + (i + 1)[1 - P_i(T{\to}0)] = \mathcal{N} \tag{A34}$$

read:

$$P_i(T{\to}0) = 1 + i - \mathcal{N} \equiv 1 - \omega \quad \text{and} \quad P_{i+1}(T{\to}0) = \mathcal{N} - i \equiv \omega. \tag{A35}$$

The continuous energy function $\mathcal{E}(\mathcal{N}, \mathcal{S})$ then consists of the straight-line segments between the neighboring integer values of $\mathcal{N}$. This implies constant values of the chemical potential in all such admissible (partial) ranges of the average number of electrons, and the μ-discontinuity at the integer values of the average electron-number $\{\mathcal{N} = N_i\}$.

This zero-temperature mixture of the molecular ground states $\{|\psi_0^i\rangle = \psi[N_i, v]\}$, defined for the integer number of electrons $N_i = \langle\psi_i|\hat{N}|\psi_i\rangle$ and corresponding to energies

$$E_0^i = \langle\psi_0^i|\hat{H}(N_i, v)|\psi_0^i\rangle = E[N_i, v], \tag{A36}$$

which appear in the ensemble with the equilibrium thermodynamic probabilities $P_i(\mu, T{\to}0; v)$, represents an externally open molecule $\langle M(\mu, T{\to}0; v)\rangle_{ens.}$ in these thermodynamic conditions.

In this thermodynamic scenario the *pure*-state probabilities thus result from the variational principle for the thermodynamic potential

$$\Omega[\hat{D}] = \mathcal{E}[\hat{D}] - \mu\mathcal{N}[\hat{D}] - T\mathcal{S}[\hat{D}] = \mathrm{tr}(\hat{D}\hat{\Omega}),$$
$$\hat{\Omega}(\mu, T; v) = \hat{H}(v) - \mu\hat{N} - T\hat{\mathcal{S}}, \quad \hat{\mathcal{S}} = -k_B \ln \hat{d}, \tag{A37}$$

$$\min_{\hat{D}}\Omega[\hat{D}] = \Omega[\hat{D}(\mu, T; v)] = \mathcal{E}[\hat{D}_{eq.}] - \mu\mathcal{N}[\hat{D}_{eq.}] - T\mathcal{S}[\hat{D}_{eq.}]. \tag{A38}$$

The relevant ensemble averages of the system energy and electron number read:

$$\langle E(\mu, T)\rangle_{ens.} = \mathcal{E}[\hat{D}_{eq.}] = \Sigma_i \Sigma_j p_j^i(\mu, T; v)E_j^i,$$
$$\langle N(\mu, T)\rangle_{ens.} = \mathcal{N}[\hat{D}_{eq.}] = \Sigma_i[\Sigma_j p_j^i(\mu, T; v)]N_i = \Sigma_i p_i(\mu, T; v)N_i \tag{A39}$$

and von Neumann's [116] ensemble entropy

$$\langle S(\mu, T)\rangle_{ens.} = \mathcal{S}[\hat{D}_{eq.}] = \mathrm{tr}[\hat{D}_{eq.}\hat{\mathcal{S}}(\mu, T; v)]$$
$$= -k_B\mathrm{tr}(\hat{D}_{eq.}\ln\hat{D}_{eq.}) - k_B \Sigma_i \Sigma_j P_j^i(\mu, T; v) \Sigma\Sigma_i \Sigma_j P_j^i(\mu, T; v)\mathcal{S}_j^i(\mu, T; v),$$
$$\hat{\mathcal{S}}(\mu, T; v) = \Sigma_i \Sigma_i |\psi_j^i\rangle \mathcal{S}_j^i(\mu, T; v)\langle\psi_j^i|,$$
$$\mathcal{S}_j^i(\mu, T; v) = \langle\psi_j^i|\hat{\mathcal{S}}(\mu, T; v)|\psi_j^i\rangle = -k_B \ln P_j^i(\mu, T; v). \tag{A40}$$

The latter identically vanishes in the pure quantum state ψ_j^i, when $P_j^i = 1$ for the vanishing remaining state probabilities.

References

1. Nalewajski, R.F. On Entropy/Information Description of Reactivity Phenomena. In *Advances in Mathematics Research*; Baswell, A.R., Ed.; Nova Science Publishers: New York, NY, USA, 2019; Volume 26, in press.
2. Nalewajski, R.F. Information description of chemical reactivity. *Curr. Phys. Chem.* **2019**, in press.

3. Nalewajski, R.F. Role of electronic kinetic energy (resultant gradient information) in chemical reactivity. *J. Mol. Model.* **2019**. submitted.

4. Callen, H.B. *Thermodynamics: An Introduction to the Physical Theories of Equilibrium Thermostatics and Irreversible Thermodynamics*; Wiley: New York, NY, USA, 1962.

5. Nalewajski, R.F. Virial theorem implications for the minimum energy reaction paths. *Chem. Phys.* **1980**, *50*, 127–136. [CrossRef]

6. Prigogine, I. *From Being to Becoming: Time and Complexity in the Physical Sciences*; Freeman WH & Co.: San Francisco, CA, USA, 1980.

7. Nalewajski, R.F. *Quantum Information Theory of Molecular States*; Nova Science Publishers: New York, NY, USA, 2016.

8. Nalewajski, R.F. Complex entropy and resultant information measures. *J. Math. Chem.* **2016**, *54*, 1777–1782. [CrossRef]

9. Nalewajski, R.F. On phase/current components of entropy/information descriptors of molecular states. *Mol. Phys.* **2014**, *112*, 2587–2601. [CrossRef]

10. Nalewajski, R.F. Quantum information measures and their use in chemistry. *Curr. Phys. Chem.* **2017**, *7*, 94–117. [CrossRef]

11. Fisher, R.A. Theory of statistical estimation. *Proc. Camb. Phil. Soc.* **1925**, *22*, 700–725. [CrossRef]

12. Frieden, B.R. *Physics from the Fisher Information—A Unification*; Cambridge University Press: Cambridge, UK, 2004.

13. Shannon, C.E. The mathematical theory of communication. *Bell Syst. Tech. J.* **1948**, *27*, 379–493, 623–656. [CrossRef]

14. Shannon, C.E.; Weaver, W. *The Mathematical Theory of Communication*; University of Illinois, Urbana: Champaign, IL, USA, 1949.

15. Kullback, S.; Leibler, R.A. On information and sufficiency. *Ann. Math. Stat.* **1951**, *22*, 79–86. [CrossRef]

16. Kullback, S. *Information Theory and Statistics*; Wiley: New York, NY, USA, 1959.

17. Abramson, N. *Information Theory and Coding*; McGraw-Hill: New York, NY, USA, 1963.

18. Pfeifer, P.E. *Concepts of Probability Theory*; Dover: New York, NY, USA, 1978.

19. Nalewajski, R.F. *Information Theory of Molecular Systems*; Elsevier: Amsterdam, The Netherlands, 2006.

20. Nalewajski, R.F. *Information Origins of the Chemical Bond.*; Nova Science Publishers: New York, NY, USA, 2010.

21. Nalewajski, R.F. *Perspectives in Electronic Structure Theory*; Springer: Heidelberg, Germany, 2012.

22. Nalewajski, R.F.; Parr, R.G. Information theory, atoms-in-molecules and molecular similarity. *Proc. Natl. Acad. Sci. USA* **2000**, *97*, 8879–8882. [CrossRef]

23. Nalewajski, R.F. Information principles in the theory of electronic structure. *Chem. Phys. Lett.* **2003**, *272*, 28–34. [CrossRef]

24. Nalewajski, R.F. Information principles in the Loge Theory. *Chem. Phys. Lett.* **2003**, *375*, 196–203. [CrossRef]

25. Nalewajski, R.F. Electronic structure and chemical reactivity: density functional and information theoretic perspectives. *Adv. Quantum Chem.* **2003**, *43*, 119–184.

26. Nalewajski, R.F.; Parr, R.G. Information-theoretic thermodynamics of molecules and their Hirshfeld fragments. *J. Phys. Chem. A* **2001**, *105*, 7391–7400. [CrossRef]

27. Nalewajski, R.F. Hirschfeld analysis of molecular densities: subsystem probabilities and charge sensitivities. *Phys. Chem. Chem. Phys.* **2002**, *4*, 1710–1721. [CrossRef]

28. Parr, R.G.; Ayers, P.W.; Nalewajski, R.F. What is an atom in a molecule? *J. Phys. Chem. A* **2005**, *109*, 3957–3959. [CrossRef] [PubMed]

29. Nalewajski, R.F.; Broniatowska, E. Atoms-in-Molecules from the stockholder partition of molecular two-electron distribution. *Theor. Chem. Acc.* **2007**, *117*, 7–27. [CrossRef]

30. Heidar-Zadeh, F.; Ayers, P.W.; Verstraelen, T.; Vinogradov, I.; Vöhringer-Martinez, E.; Bultinck, P. Information-theoretic approaches to Atoms-in-Molecules: Hirshfeld family of partitioning schemes. *J. Phys. Chem. A* **2018**, *122*, 4219–4245. [CrossRef]

31. Hirshfeld, F.L. Bonded-atom fragments for describing molecular charge densities. *Theor. Chim. Acta* **1977**, *44*, 129–138. [CrossRef]

32. Nalewajski, R.F. Entropic measures of bond multiplicity from the information theory. *J. Phys. Chem. A* **2000**, *104*, 11940–11951. [CrossRef]

33. Nalewajski, R.F. Entropy descriptors of the chemical bond in Information Theory: I. Basic concepts and relations. II. Application to simple orbital models. *Mol. Phys.* **2004**, *102*, 531–546, 547–566. [CrossRef]
34. Nalewajski, R.F. Entropic and difference bond multiplicities from the two-electron probabilities in orbital resolution. *Chem. Phys. Lett.* **2004**, *386*, 265–271. [CrossRef]
35. Nalewajski, R.F. Reduced communication channels of molecular fragments and their entropy/information bond indices. *Theor. Chem. Acc.* **2005**, *114*, 4–18. [CrossRef]
36. Nalewajski, R.F. Partial communication channels of molecular fragments and their entropy/information indices. *Mol. Phys.* **2005**, *103*, 451–470. [CrossRef]
37. Nalewajski, R.F. Entropy/information descriptors of the chemical bond revisited. *J. Math. Chem.* **2011**, *49*, 2308–2329. [CrossRef]
38. Nalewajski, R.F. Quantum information descriptors and communications in molecules. *J. Math. Chem.* **2014**, *52*, 1292–1323. [CrossRef]
39. Nalewajski, R.F. Multiple, localized and delocalized/conjugated bonds in the orbital-communication theory of molecular systems. *Adv. Quantum Chem.* **2009**, *56*, 217–250.
40. Nalewajski, R.F.; Szczepanik, D.; Mrozek, J. Bond differentiation and orbital decoupling in the orbital communication theory of the chemical bond. *Adv. Quant. Chem.* **2011**, *61*, 1–48.
41. Nalewajski, R.F.; Szczepanik, D.; Mrozek, J. Basis set dependence of molecular information channels and their entropic bond descriptors. *J. Math. Chem.* **2012**, *50*, 1437–1457. [CrossRef]
42. Nalewajski, R.F. Electron Communications and Chemical Bonds. In *Frontiers of Quantum Chemistry*; Wójcik, M., Nakatsuji, H., Kirtman, B., Ozaki, Y., Eds.; Springer: Singapore, 2017; pp. 315–351.
43. Nalewajski, R.F.; Świtka, E.; Michalak, A. Information distance analysis of molecular electron densities. *Int. J. Quantum Chem.* **2002**, *87*, 198–213. [CrossRef]
44. Nalewajski, R.F.; Broniatowska, E. Entropy displacement analysis of electron distributions in molecules and their Hirshfeld atoms. *J. Phys. Chem. A* **2003**, *107*, 6270–6280. [CrossRef]
45. Nalewajski, R.F. Use of Fisher information in quantum chemistry. *Int. J. Quantum Chem.* **2008**, *108*, 2230–2252. [CrossRef]
46. Nalewajski, R.F.; Köster, A.M.; Escalante, S. Electron localization function as information measure. *J. Phys. Chem. A* **2005**, *109*, 10038–10043. [CrossRef] [PubMed]
47. Becke, A.D.; Edgecombe, K.E. A simple measure of electron localization in atomic and molecular systems. *J. Chem. Phys.* **1990**, *92*, 5397–5403. [CrossRef]
48. Silvi, B.; Savin, A. Classification of chemical bonds based on topological analysis of electron localization functions. *Nature* **1994**, *371*, 683–686. [CrossRef]
49. Savin, A.; Nesper, R.; Wengert, S.; Fässler, T.F. ELF: The electron localization function. *Angew. Chem. Int. Ed. Engl.* **1997**, *36*, 1808–1832. [CrossRef]
50. Hohenberg, P.; Kohn, W. Inhomogeneous electron gas. *Phys. Rev.* **1964**, *136B*, 864–971. [CrossRef]
51. Kohn, W.; Sham, L.J. Self-consistent equations including exchange and correlation effects. *Phys. Rev.* **1965**, *140A*, 133–1138. [CrossRef]
52. Levy, M. Universal variational functionals of electron densities, first-order density matrices, and natural spin-orbitals and solution of the v-representability problem. *Proc. Natl. Acad. Sci. USA* **1979**, *76*, 6062–6065. [CrossRef]
53. Parr, R.G.; Yang, W. *Density-Functional Theory of Atoms and Molecules*; Oxford University Press: New York, NY, USA, 1989.
54. Dreizler, R.M.; Gross, E.K.U. *Density Functional Theory: An Approach to the Quantum Many-Body Problem*; Springer: Berlin, Germany, 1990.
55. Nalewajski, R.F. (Ed.) *Density Functional Theory I-IV. Topics in Current Chemistry*; Springer: Berlin/Heidelberg, Germany, 1996; Volume 180–183.
56. Nalewajski, R.F.; de Silva, P.; Mrozek, J. Use of nonadditive Fisher information in probing the chemical bonds. *J. Mol. Struct.* **2010**, *954*, 57–74. [CrossRef]
57. Nalewajski, R.F. Through-space and through-bridge components of chemical bonds. *J. Math. Chem.* **2011**, *49*, 371–392. [CrossRef]
58. Nalewajski, R.F. Chemical bonds from through-bridge orbital communications in prototype molecular systems. *J. Math. Chem.* **2011**, *49*, 546–561. [CrossRef]

59. Nalewajski, R.F. On interference of orbital communications in molecular systems. *J. Math. Chem.* **2011**, *49*, 806–815. [CrossRef]

60. Nalewajski, R.F.; Gurdek, P. On the implicit bond-dependency origins of bridge interactions. *J. Math. Chem.* **2011**, *49*, 1226–1237. [CrossRef]

61. Nalewajski, R.F. Direct (through-space) and indirect (through-bridge) components of molecular bond multiplicities. *Int. J. Quantum Chem.* **2012**, *112*, 2355–2370. [CrossRef]

62. Nalewajski, R.F.; Gurdek, P. Bond-order and entropic probes of the chemical bonds. *Struct. Chem.* **2012**, *23*, 1383–1398. [CrossRef]

63. Nalewajski, R.F. Exploring molecular equilibria using quantum information measures. *Ann. Phys.* **2013**, *525*, 256–268. [CrossRef]

64. Nalewajski, R.F. On phase equilibria in molecules. *J. Math. Chem.* **2014**, *52*, 588–612. [CrossRef]

65. Nalewajski, R.F. Quantum information approach to electronic equilibria: Molecular fragments and elements of non-equilibrium thermodynamic description. *J. Math. Chem.* **2014**, *52*, 1921–1948. [CrossRef]

66. Nalewajski, R.F. Phase/current information descriptors and equilibrium states in molecules. *Int. J. Quantum Chem.* **2015**, *115*, 1274–1288. [CrossRef]

67. Nalewajski, R.F. Quantum Information Measures and Molecular Phase Equilibria. In *Advances in Mathematics Research*; Baswell, A.R., Ed.; Nova Science Publishers: New York, NY, USA, 2015; Volume 19, pp. 53–86.

68. Nalewajski, R.F. Phase Description of Reactive Systems. In *Conceptual Density Functional Theory*; Islam, N., Kaya, S., Eds.; Apple Academic Press: Waretown, NJ, USA, 2018; pp. 217–249.

69. Nalewajski, R.F. Entropy Continuity, Electron Diffusion and Fragment Entanglement in Equilibrium States. In *Advances in Mathematics Research*; Baswell, A.R., Ed.; Nova Science Publishers: New York, NY, USA, 2017; Volume 22, pp. 1–42.

70. Nalewajski, R.F. On entangled states of molecular fragments. *Trends Phys. Chem.* **2016**, *16*, 71–85.

71. Nalewajski, R.F. Chemical reactivity description in density-functional and information theories. *Acta Phys.-Chim. Sin.* **2017**, *33*, 2491–2509.

72. Nalewajski, R.F. Information equilibria, subsystem entanglement and dynamics of overall entropic descriptors of molecular electronic structure. *J. Mol. Model.* **2018**, *24*, 212–227. [CrossRef]

73. Nalewajski, R.F.; Korchowiec, J.; Michalak, A. *Reactivity Criteria in Charge Sensitivity Analysis. Topics in Current Chemistry: Density Functional Theory IV*; Nalewajski, R.F., Ed.; Springer: Berlin/Heidelberg, Germany, 1996; Volume 183, pp. 25–141.

74. Nalewajski, R.F.; Korchowiec, J. *Charge Sensitivity Approach to Electronic Structure and Chemical Reactivity*; World Scientific: Singapore, 1997.

75. Geerlings, P.; De Proft, F.; Langenaeker, W. Conceptual density functional theory. *Chem. Rev.* **2003**, *103*, 1793–1873. [CrossRef] [PubMed]

76. Nalewajski, R.F. Sensitivity analysis of charge transfer systems: In situ quantities, intersecting state model and ist implications. *Int. J. Quantum Chem.* **1994**, *49*, 675–703. [CrossRef]

77. Nalewajski, R.F. Charge sensitivity analysis as diagnostic tool for predicting trends in chemical reactivity. In Proceedings of the NATO ASI on Density Functional Theory (Il Ciocco, 1993); Dreizler, R.M., Gross, E.K.U., Eds.; Plenum: New York, NY, USA, 1995; pp. 339–389.

78. Chattaraj, P.K. (Ed.) *Chemical Reactivity Theory: A Density Functional View*; CRC Press: Boca Raton, FL, USA, 2009.

79. Gatti, C.; Macchi, P. *Modern Charge-Density Analysis*; Springer: Berlin, Germany, 2012.

80. Hammond, G.S. A correlation of reaction rates. *J. Am. Chem. Soc.* **1955**, *77*, 334–338. [CrossRef]

81. *Fukui K Theory of Orientation and Stereoselection*; Springer-Verlag: Berlin, Germany, 1975.

82. Fukui, K. Role of frontier orbitals in chemical reactions. *Science* **1987**, *218*, 747–754. [CrossRef] [PubMed]

83. Fujimoto, H.; Fukui, K. Intermolecular Interactions and Chemical Reactivity. In *Chemical Reactivity and Reaction Paths*; Klopman, G., Ed.; Wiley-Interscience: New York, NY, USA, 1974; pp. 23–54.

84. Pearson, R.G. *Hard and Soft Acids and Bases*; Dowden, Hutchinson and Ross: Stroudsburg, PA, USA, 1973.

85. Nalewajski, R.F. Electrostatic effects in interactions between hard (soft) acids and bases. *J. Am. Chem. Soc.* **1984**, *106*, 944–945. [CrossRef]

86. von Weizsäcker, C.F. Zur theorie der kernmassen. *Z. Phys. Hadron. Nucl.* **1935**, *96*, 431–458.

87. Harriman, J.E. Orthonormal orbitals fort the representation of an arbitrary density. *Phys. Rev.* **1980**, *A24*, 680–682.

88. Zumbach, G.; Maschke, K. New approach to the calculation of density functionals. *Phys. Rev.* **1983**, *A28*, 544–554, Erratum in **1984**, *A29*, 1585–1587. [CrossRef]

89. Ruedenberg, K. The physical nature of the chemical bond. *Rev. Mod. Phys.* **1962**, *34*, 326–376. [CrossRef]

90. Feinberg, M.J.; Ruedenberg, K. Paradoxical role of the kinetic-energy operator in the formation of the covalent bond. *J. Chem. Phys.* **1971**, *54*, 1495–1512. [CrossRef]

91. Feinberg, M.J.; Ruedenberg, K. Heteropolar one-electron bond. *J. Chem. Phys.* **1971**, *55*, 5805–5818. [CrossRef]

92. Bacskay, G.B.; Nordholm, S.; Ruedenberg, K. The virial theorem and covalent bonding. *J. Phys. Chem.* **2018**, *A122*, 7880–7893. [CrossRef]

93. Marcus, R.A. Theoretical relations among rate constants, barriers, and Broensted slopes of chemical reactions. *J. Phys. Chem.* **1968**, *72*, 891–899. [CrossRef]

94. Agmon, N.; Levine, R.D. Energy, entropy and the reaction coordinate: Thermodynamic-like relations in chemical kinetics. *Chem. Phys. Lett.* **1977**, *52*, 197–201. [CrossRef]

95. Agmon, N.; Levine, R.D. Empirical triatomic potential energy surfaces defined over orthogonal bond-order coordinates. *J. Chem. Phys.* **1979**, *71*, 3034–3041. [CrossRef]

96. Miller, A.R. A theoretical relation for the position of the energy barrier between initial and final states of chemical reactions. *J. Am. Chem. Soc.* **1978**, *100*, 1984–1992. [CrossRef]

97. Ciosłowski, J. Quantifying the Hammond Postulate: Intramolecular proton transfer in substituted hydrogen catecholate anions. *J. Am. Chem. Soc.* **1991**, *113*, 6756–6761. [CrossRef]

98. Nalewajski, R.F.; Formosinho, S.J.; Varandas, A.J.C.; Mrozek, J. Quantum mechanical valence study of a bond breaking—Bond forming process in triatomic systems. *Int. J. Quantum Chem.* **1994**, *52*, 1153–1176. [CrossRef]

99. Nalewajski, R.F.; Broniatowska, E. Information distance approach to Hammond postulate. *Chem. Phys. Lett.* **2003**, *376*, 33–39. [CrossRef]

100. Dunning, T.H., Jr. Theoretical studies of the energetics of the abstraction and exchange reactions in H + HX, with X = F−I. *J. Phys. Chem.* **1984**, *88*, 2469–2477. [CrossRef]

101. Gyftopoulos, E.P.; Hatsopoulos, G.N. Quantum-thermodynamic definition of electronegativity. *Proc. Natl. Acad. Sci. USA* **1965**, *60*, 786–793. [CrossRef]

102. Perdew, J.P.; Parr, R.G.; Levy, M.; Balduz, J.L. Density functional theory for fractional particle number: Derivative discontinuities of the energy. *Phys. Rev. Lett.* **1982**, *49*, 1691–1694. [CrossRef]

103. Mulliken, R.S. A new electronegativity scale: Together with data on valence states and on ionization potentials and electron affinities. *J. Chem. Phys.* **1934**, *2*, 782–793. [CrossRef]

104. Iczkowski, R.P.; Margrave, J.L. Electronegativity. *J. Am. Chem. Soc.* **1961**, *83*, 3547–3551. [CrossRef]

105. Parr, R.G.; Donnelly, R.A.; Levy, M.; Palke, W.E. Electronegativity: The density functional viewpoint. *J. Chem. Phys.* **1978**, *69*, 4431–4439. [CrossRef]

106. Parr, R.G.; Pearson, R.G. Absolute hardness: Companion parameter to absolute electronegativity. *J. Am. Chem. Soc.* **1983**, *105*, 7512–7516. [CrossRef]

107. Parr, R.G.; Yang, W. Density functional approach to the frontier-electron theory of chemical reactivity. *J. Am. Chem. Soc.* **1984**, *106*, 4049–4050. [CrossRef]

108. Nalewajski, R.F. Manifestations of the maximum complementarity principle for matching atomic softnesses in model chemisorption systems. *Top. Catal.* **2000**, *11*, 469–485. [CrossRef]

109. Chandra, A.K.; Michalak, A.; Nguyen, M.T.; Nalewajski, R.F. On regional matching of atomic softnesses in chemical reactions: Two-reactant charge sensitivity study. *J. Phys. Chem.* **1998**, *A102*, 10182–10188. [CrossRef]

110. Hô, M.; Schmider, H.L.; Weaver, D.F.; Smith, V.H., Jr.; Sagar, R.P.; Esquivel, R.O. Shannon entropy of chemical changes: S_N2 displacement reactions. *Int. J. Quantum Chem.* **2000**, *77*, 376–382. [CrossRef]

111. López-Rosa, S.; Esquivel, R.O.; Angulo, J.C.; Antolin, J.; Dehesa, J.S.; Flores-Gallegos, N. Fisher information study in position and momentum spaces for elementary chemical reactions. *J. Chem. Theory Comput.* **2010**, *6*, 145–154. [CrossRef] [PubMed]

112. Esquivel, R.O.; Liu, S.B.; Angulo, J.C.; Dehesa, J.S.; Antolin, J.; Molina-Espiritu, M. Fisher information and steric effect: Study of the internal rotation barrier in ethane. *J. Phys. Chem.* **2011**, *A115*, 4406–4415. [CrossRef]

113. Krokidis, X.; Noury, S.; Silvi, B. Characterization of elementary chemical processes by catastrophe theory. *J. Phys. Chem.* **1997**, *A101*, 7277–7282. [CrossRef]

114. Ndassa, I.M.; Silvi, B.; Volatron, F. Understanding reaction mechanisms in organic chemistry from catastrophe theory: Ozone addition on benzene. *J. Phys. Chem.* **2010**, *A114*, 12900–12906. [CrossRef] [PubMed]

115. Domingo, L.R. A new C-C bond formation model based on the quantum chemical topology of electron density. *Rsc Adv.* **2014**, *4*, 32415–32428. [CrossRef]
116. Von Neumann, J. *Mathematical Foundations of Quantum Mechanics*; Princeton University Press: Princeton, NJ, USA, 1955.

MDPI

St. Alban-Anlage 66

4052 Basel

Switzerland

Tel. +41 61 683 77 34

Fax +41 61 302 89 18

www.mdpi.com

Applied Sciences Editorial Office

E-mail: applsci@mdpi.com

www.mdpi.com/journal/applsci